Nonlinear Dynamics of Structures

Lecture Notes on Numerical Methods in Engineering and Sciences

Aims and Scope of the Series

This series publishes text books on topics of general interest in the field of computational engineering sciences.

The books will focus on subjects in which numerical methods play a fundamental role for solving problems in engineering and applied sciences. Advances in finite element, finite volume, finite differences, discrete and particle methods and their applications are examples of the topics covered by the series.

The main intended audience is the first year graduate student. Some books define the current state of a field to a highly specialised readership; others are accessible to final year undergraduates, but essentially the emphasis is on accessibility and clarity.

The books will be also useful for practising engineers and scientists interested in state of the art information on the theory and application of numerical methods.

Series Editor
Eugenio Oñate
International Center for Numerical Methods in Engineering (CIMNE)
School of Civil Engineering, Technical University of Catalonia (UPC), Barcelona, Spain

Editorial Board
Francisco Chinesta, Ecole Nationale Supérieure d'Arts et Métiers, Paris, France
Charbel Farhat, Stanford University, Stanford, USA
Carlos Felippa, University of Colorado at Boulder, Colorado, USA
Antonio Huerta, Technical University of Catalonia (UPC), Barcelona, Spain
Thomas J.R. Hughes, The University of Texas at Austin, Austin, USA
Sergio R. Idelsohn, CIMNE-ICREA, Barcelona, Spain
Pierre Ladeveze, ENS de Cachan-LMT-Cachan, France
Wing Kam Liu, Northwestern University, Evanston, USA
Xavier Oliver, Technical University of Catalonia (UPC), Barcelona, Spain
Manolis Papadrakakis, National Technical University of Athens, Greece
Jacques Périaux, CIMNE-UPC Barcelona, Spain & Univ. of Jyväskylä, Finland
Bernhard Schrefler, Università degli Studi di Padova, Padova, Italy
Genki Yagawa, Tokyo University, Tokyo, Japan
Mingwu Yuan, Peking University, China

Titles:

1. E. Oñate, Structural Analysis with the Finite Element Method. Linear Statics. Volume 1. Basis and Solids, 2009
2. K. Wiśniewski, Finite Rotation Shells. Basic Equations and Finite Elements for Reissner Kinematics, 2010
3. E. Oñate, Structural Analysis with the Finite Element Method. Linear Statics. Volume 2. Beams, Plates and Shells, 2013
4. E.W.V. Chaves. Notes on Continuum Mechanics. 2013
5. S. Oller. Numerical Simulation of Mechanical Behavior of Composite Materials, 2014
6. S. Oller. Nonlinear Dynamics of Structures, 2014

Nonlinear Dynamics of Structures

Sergio Oller
International Center for Numerical Methods in Engineering (CIMNE)
School of Civil Engineering
Universitat Politècnica de Catalunya (UPC)
Barcelona, Spain

ISBN 978-3-319-05193-2 (HB)
ISBN 978-3-319-05194-9 (eBook)

Depósito legal: B-4031-2014

A C.I.P. Catalogue record for this book is available from the Library of Congress

Lecture Notes Series Manager: **Mª Jesús Samper,** CIMNE, Barcelona, Spain

Cover page: **Pallí Disseny i Comunicació,** www.pallidisseny.com

Printed by: **Artes Gráficas Torres S.L.**
Huelva 9, 08940 Cornellà de Llobregat (Barcelona), España
www.agraficastorres.es

Printed on elemental chlorine-free paper

Nonlinear Dynamics of Structures
Sergio Oller

First edition, 2014

© International Center for Numerical Methods in Engineering (CIMNE), 2014
Gran Capitán s/n, 08034 Barcelona, Spain
www.cimne.com

This work is dedicated to my wife and son, and also to all my loved ones

Preface

This book has been written to present the conceptual basis of "Nonlinear Dynamics" of structural systems. Although there are many papers on this subject, I have decided to write this book for educational purposes addressed to students with an academic level equivalent to a master's degree.

The book is divided into three main parts: the first one sets up the basis on which the nonlinear dynamics applied to discrete structures is based on; the second one shows the effect of time-independent constitutive model behavior within the nonlinear dynamic response; and finally, the third part analyzes the effect of time-dependent constitutive models in a nonlinear dynamic behavior.

This work has been possible thanks to the institutional support of CIMNE (International Center for Numerical Methods in Engineering), which has financially supported this book since its first edition in Spanish in 2002, and later in its English edition. Many people have participated in the latter, and I would particularly like to thank Ms. Hamdy Briceño, Prof. Miguel Cerrolaza and Cristina Pérez Arias for their careful translation and revision of this text. I would also like to thank all my students who have contributed to the correction of the text during the eleven years that this book has been used as a syllabus of the "Nonlinear Dynamics" course in the Department of Strength of Materials, at the Technical University of Catalonia, Spain.

I hope these notes will contribute to a better understanding of the nonlinear dynamics and encourage the reader to study this subject in greater depth.

Barcelona, May 2014

Sergio Oller

Contents

1 Introduction

Structural dynamics studies the structural equilibrium over time among external forces, elastic forces, mass forces and viscous forces for a discrete structural system with points that are internally linked to each other and all linked to a fixed reference system. These internal links between points describing the structural system may be elastic or not. If they are not elastic, the behavior of the system of points is non-conservative and therefore the structural material has a *nonlinear dissipative constitutive behavior*. Additionally to this nonlinear behavior, there is also a *nonlinear dissipative behavior due to the effects of the material viscosity* that leads to viscous forces dependent on the system velocity. In simpler cases, the damping non linearity is due to the development of viscous forces proportional to the velocity; however, in more complex cases the viscosity term may be time-dependent. Also, the system's non linearity can be observed in systems having large displacements and where the system works beyond its original geometric configuration, leading to a *nonlinear kinematic behavior*. Such non linearity is even more pronounced when large strain occurs along with large displacements, turning the solution of the structure's dynamic problem more complex.

All the above mentioned subjects will be thoroughly studied in this work; concepts are based on the *nonlinear dynamics of structures,* on the *mechanics of continuum media* and on *numerical techniques* such as the *finite element method.*

A nonlinear structural dynamics course may have different approaches to the content and development of concepts it must have and all of them are valid as long as the goals are achieved. This work deals with the required concepts to complete the basic training in structural nonlinear dynamics, in the mechanics of continuum media and in the finite element method. Accordingly, the topics included in this basic training in structures that are assumed to be already known by the reader will not be studied again.

A brief description of the book's contents follows: in **chapter 2**, an *introduction to the thermodynamical basis of the motion equation* is presented. This fundamental chapter contains the origins of the problem, which is set within a structured formulation that can address all remaining items in a consistent way. In **chapter 3**, the methods to solve the motion equation are described in detail; both the implicit and the explicit procedures and the advantages and drawbacks of each method are analyzed. In **chapter 4**, the stability concept of the solution of conservative systems is studied for different methods in order to solve the equation of movement. Once the basis of the solution stability of linear systems are established, an approximation to the nonlinear problem is made and criteria for the stability study are provided. Energy conservation here is a crucial requirement. This leads to the "formulation of conservative solution methods" currently being used in nonlinear dynamics. In **chapter 5,** once the basis of nonlinear dynamics are set, the time-independent constitutive formulation, such as plasticity and damage, is addressed showing also how the structural nonlinearity is affected by these behaviors. Similarly, in **chapter 6** the constitutive behavior of time-dependent materials, such as delayed elasticity and relaxation, is detailed, where the nonlinear damping is included in a natural way. This is emphasized because it is considered a part of the nonlinear dynamics where there is a conceptual gap.

2 Thermodynamic Basis of the Equation of Motion

2.1 Introduction

The thermodynamic basis defining the linear or nonlinear behavior of a solid during the mechanical process is introduced in this chapter. The synthesized concepts here help to understand the solid nonlinear behavior and to clearly set equilibrium at every time.

The kinematics of deformable solids is briefly reviewed to establish the notation to be used as well as the definitions of the mechanics of continuum media which are important to remember. A brief description of the *thermodynamics* is also presented to point out the most relevant aspects of the formulation of constitutive models for the nonlinear behavior of solids. Reference to the mechanics of continuous media and to thermodynamics[1,2,3] is highly recommended to deepen and broaden the concepts addressed here.

2.2 Kinematics of deformable bodies

In order to sustain the formulation of constitutive models, it is necessary to introduce the basic concepts describing the kinematics of a point in the space, the stress and the strain measurements as well as their relation in different configurations. The purpose of this chapter is to establish the notation and review some definitions. It is not intended to substitute any specific book of continuum mechanics. Therefore, reference to the sources[1,2,3] is recommended.

2.2.1 Basic definitions of tensors describing the kinematics of a point in the space

Let a continuous solid in three dimensions be considered, represented by the domain $\Omega_t \subset \mathbb{R}^3$ located in the space in its *current configuration* in time t, or by an image of this domain located in the space in an *intermediate configuration* $\overline{\Omega}_t \subset \mathbb{R}^3$ or by the domain $\Omega_0 \subset \mathbb{R}^3$ located in the *reference configuration* or *original configuration* (see Figure 2.1).

---

[1] Malvern, L. (1969). *Introduction to the mechanics of continuous medium.* Prentice Hall, Englewood Cliffs, NJ.

[2] Lubliner, J. (1990). *Plasticity theory.* MacMillan, New York.

[3] Maugin, G. A. (1992). *The thermomechanics of plasticity and fracture.* Cambridge University Press.

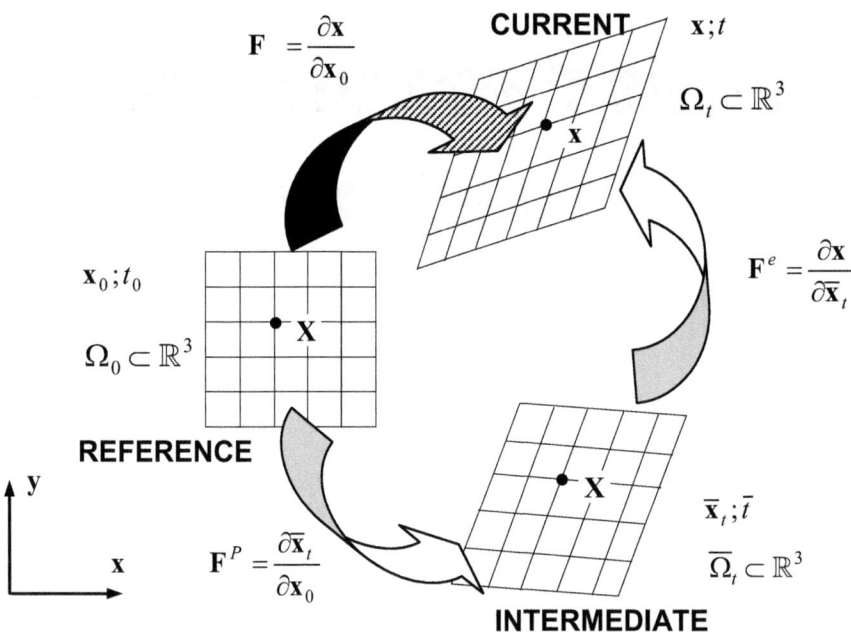

Figure 2.1 –Schematic representation of the kinematic configurations of a solid in the space.

A point $\mathbf{X} \in \Omega_0$, of coordinates $(\mathbf{x}_i)_0$, located at the *reference configuration*, to which one and only one of the points in the *intermediate configuration* corresponds, represented by $\mathbf{X} \in \overline{\Omega}_t$ of coordinates $(\overline{\mathbf{x}}_i)_t$, and similarly corresponds to it $\mathbf{X} \in \Omega_t$ with coordinates $(\mathbf{x}_i)_t$ corresponding to the *current configuration*. Thus, the body movement is described as a function of its position in the reference configuration and of the time,

$$\mathbf{x} = \mathbf{x}(\mathbf{X};t); \quad \mathbf{X} \in \Omega_0 \tag{2.1}$$

The *gradient of deformation tensor* is defined as the following transformation

$$\mathbf{F} = \mathbf{F}(\mathbf{X};t) = \nabla_0 \mathbf{x} = \frac{\partial \mathbf{x}}{\partial \mathbf{x}_0} = \mathbf{J} \tag{2.2}$$

where $\mathbf{J}$ is *the Jacobian matrix*. The remaining transformations shown in Figure 2.1 are obtained from the following definition,

$$\mathbf{F} = \frac{\partial \mathbf{x}}{\partial \mathbf{x}_0} = \frac{\partial \mathbf{x}}{\partial \overline{\mathbf{x}}_t} \frac{\partial \overline{\mathbf{x}}_t}{\partial \mathbf{x}_0} = \mathbf{F}^e \cdot \mathbf{F}^p \tag{2.3}$$

where

$$\mathbf{F}^e = \frac{\partial \mathbf{x}}{\partial \overline{\mathbf{x}}_t} \quad \text{Elastic transformation,}$$

$$\mathbf{F}^p = \frac{\partial \overline{\mathbf{x}}_t}{\partial \mathbf{x}_0} \quad \text{Plastic transformation,} \tag{2.4}$$

The change of the solid volume during a configuration change is obtained by the determinant of the Jacobian matrix commonly known as *Jacobian*. Thus,

$$J = |\mathbf{J}| = |\mathbf{F}| = \frac{dV}{dV_0} > 0 \tag{2.5}$$

dV and dV_0 are the infinitesimal volume in the configurations Ω_t and Ω_0, respectively. The strain gradient tensor can be decomposed as the following polar transformation,

$$\mathbf{F} = \mathbf{R} \cdot \mathbf{U} = \mathbf{V} \cdot \mathbf{R} \tag{2.6}$$

where $\mathbf{R}$ is the so called *orthogonal tensor*, which meets the following orthonormal condition $\mathbf{R} \cdot \mathbf{R}^T \equiv \mathbf{R}^T \cdot \mathbf{R} \equiv \mathbf{I}$, and both $\mathbf{U}$ and $\mathbf{V}$ are positive-defined symmetric tensors. The definition of the latter depends on the Cauchy-Green *right tensor* $\mathbf{C} = \mathbf{F}^T \cdot \mathbf{F}$, then the *right stretching tensor* is equal to $\mathbf{U} = \mathbf{C}^{1/2}$. The Cauchy Green *left tensor* is also defined as $\mathbf{B} = \mathbf{F} \cdot \mathbf{F}^T$, so that by substituting equation (2.6) into the latter $\mathbf{B} = \mathbf{F} \cdot \mathbf{F}^T \equiv \mathbf{R} \cdot \mathbf{U} \cdot \mathbf{U} \cdot \mathbf{R}^T = \mathbf{R} \cdot \mathbf{C} \cdot \mathbf{R}^T$ is then obtained and from here the *left stretching tensor* can also be defined as $\mathbf{V} = \mathbf{B}^{1/2}$, so it can be rewritten as $\mathbf{V} = \mathbf{R} \cdot \mathbf{U} \cdot \mathbf{R}^T = \mathbf{F} \cdot \mathbf{R}^T$. From here it is obvious that the gradient of deformation can also be written as $\mathbf{F} = \mathbf{V} \cdot \mathbf{R}$.

Following Noll's notation (see Lubliner[2]), $\vec{\mathcal{V}}_x$ is called an Euclidian generic spatial vector defined in any configuration $\mathbf{x}$; thus, $\vec{\mathcal{V}}_0$ and $\vec{\mathcal{V}}$ will be the vectors defined in the *reference* and *current* configuration, respectively. The linear continuous space is designated as $L(\mathbf{x};\mathbf{y})$ that transforms $\mathbf{x} \to \mathbf{y}$. Based on this criterion, these tensors are designated according to the origin and destination of the transformation they perform.

$$
\begin{aligned}
&\mathbf{C} \in L\left(\vec{\mathcal{V}}_0;\vec{\mathcal{V}}_0\right) \quad ; \quad \mathbf{U} \in L\left(\vec{\mathcal{V}}_0;\vec{\mathcal{V}}_0\right) \quad : \text{Reference tensors - Lagrangeans,} \\
&\mathbf{B} \in L\left(\vec{\mathcal{V}};\vec{\mathcal{V}}\right) \quad ; \quad \mathbf{V} \in L\left(\vec{\mathcal{V}};\vec{\mathcal{V}}\right) \quad : \text{Current Tensors - Eulerians,} \\
&\left.\begin{aligned}
\mathbf{F} \in L\left(\vec{\mathcal{V}}_0;\vec{\mathcal{V}}\right) \quad ; \quad \mathbf{R} \in L\left(\vec{\mathcal{V}}_0;\vec{\mathcal{V}}\right) \\
\mathbf{F}^e \in L\left(\vec{\mathcal{V}}_p;\vec{\mathcal{V}}\right) \quad ; \quad \mathbf{F}^p \in L\left(\vec{\mathcal{V}}_0;\vec{\mathcal{V}}_p\right)
\end{aligned}\right\} : \text{Bipunctual Tensors.}
\end{aligned}
\tag{2.7}
$$

Tensors $\mathbf{F}^e \in L\left(\vec{\mathcal{V}}_p;\vec{\mathcal{V}}\right)$ and $\mathbf{F}^p \in L\left(\vec{\mathcal{V}}_0;\vec{\mathcal{V}}_p\right)$ are also called *material tensors* and they are invariant under any Euclidian transformation.

2.2.2 Strain measurements

The strain in the reference configuration, also called *Lagrangian strain*, is defined as:

$$\mathbf{E}_n = \frac{1}{n}\left(\mathbf{U}^n - \mathbf{I}\right) \tag{2.8}$$

Then, the following strain measurements are obtained:

$$\mathbf{E}_n = \frac{1}{n}\left(\mathbf{U}^n - \mathbf{I}\right) \Rightarrow \begin{cases} \text{para}: n = 0 \Rightarrow \mathbf{E}_0 = \ln \mathbf{U} \;,\, \text{Def. natural} \\ \text{para}: n = 1 \Rightarrow \mathbf{E}_1 = \mathbf{U} - \mathbf{I} \\ \text{para}: n = 2 \Rightarrow \mathbf{E} = \mathbf{E}_2 = \frac{1}{2}\left(\mathbf{C} - \mathbf{I}\right) \;,\, \text{Def. de Green - St. Venant} \end{cases} \tag{2.9}$$

The *Eulerian strain* measured in the current configuration is expressed as the Almansi form. Then,

$$e = \frac{1}{2}\left(I - B^{-1}\right) \tag{2.10}$$

where B is the *Cauchy-Green left tensor* already defined, and B^{-1} is commonly called the *Finger tensor*. In case $|F - I| \ll 1$, all the strains previously defined coincide $E_n \cong e \cong \varepsilon$ and get closer to the *infinitesimal strain*,

$$\varepsilon = \nabla^S u = \frac{1}{2}\left(\nabla_0 u + \nabla_0^T u\right) \tag{2.11}$$

where $x = x_0 + u \implies u = x - x_0$ is satisfied, and x_0 is the coordinates of the point X in the reference configuration and u is the relative displacement of such a point. Thus, the gradient is obtained as,

$$\nabla_0 u = \frac{\partial u}{\partial x_0} = \left(\frac{\partial x}{\partial x_0} - I\right) = (F - I) = j \tag{2.12}$$

From the latter and equation (2.11), the *infinitesimal strain and the strain gradient* are obtained as

$$F = \frac{\partial x}{\partial x_0} = \frac{\partial}{\partial x_0}(x_0 + u) = \left(I + \frac{\partial u}{\partial x_0}\right) = I + j$$

$$\varepsilon = \nabla^S u = \frac{1}{2}\left(j + j^T\right) \tag{2.13}$$

2.2.3 Relationships among mechanical variables

Given the transformation of the *strain gradient* F, which can relate the position of a point in a particular configuration to its image in any other configuration, an equivalence relationship can be established among all the other mechanical variables in one configuration with respect to their images corresponding to any other configuration. Therefore, the following tensor transformations[1,2,34] are defined

---

[4] Marsden J. And Hughes T. (1983). *Mathematical foundations of elasticity.* Prentice Hall, Enlewood Cliffs.

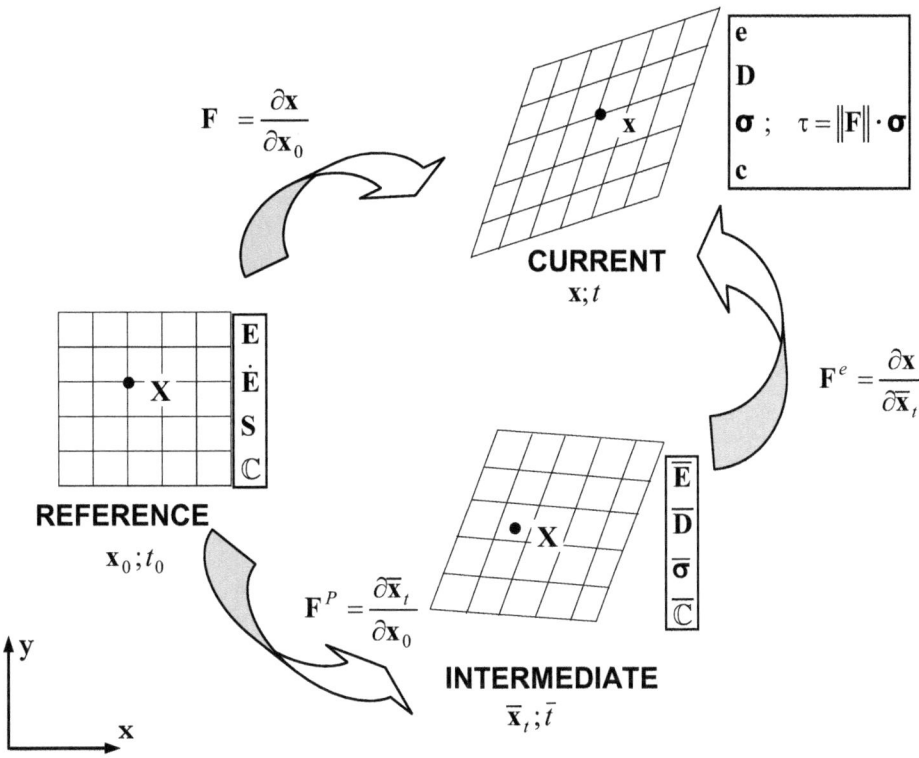

Figure 2.2 – Relationships among the mechanical variables in different configurations.

$$\text{Tranformations} \begin{cases} \text{Covariant} \begin{cases} \text{push forward} & :\underleftarrow{\varphi}\left(\mathbf{A}_{\#}\right) = \mathbf{F}^{-T} \cdot \left(\mathbf{A}_{\#}\right) \cdot \mathbf{F}^{-1} \\ \text{pull back} & :\underleftarrow{\varphi}\left(\mathbf{A}_{\#}\right) = \mathbf{F}^{T} \cdot \left(\mathbf{A}_{\#}\right) \cdot \mathbf{F} \end{cases} \\ \text{Contravariante} \begin{cases} \text{push forward} & :\vec{\varphi}\left(\mathbf{A}^{\#}\right) = \mathbf{F}^{T} \cdot \left(\mathbf{A}^{\#}\right) \cdot \mathbf{F} \\ \text{pull back} & :\bar{\varphi}\left(\mathbf{A}^{\#}\right) = \mathbf{F}^{-1} \cdot \left(\mathbf{A}^{\#}\right) \cdot \mathbf{F}^{-T} \end{cases} \end{cases} \qquad (2.14)$$

where the operators and their expression as a function of the *gradient of strain* are shown in Figure 2.2 and $\mathbf{A}_{\#}$ and $\mathbf{A}^{\#}$ are *co-variant* generic tensors of second-order (deformation tensor $\mathbf{E} \leftrightarrow \mathbf{e}$) and *contravariant* (stress tensor $\mathbf{S} \leftrightarrow \boldsymbol{\tau}(\boldsymbol{\sigma})$) respectively. Particularly, the following transformations are obtained for the transportation of the stress-deformation and constitutive tensors[1],

$\mathbf{e} = \underrightarrow{\phi}(\mathbf{E})$	$e_{ij} = F_{il}^{-T} E_{IJ} F_{jJ}^{-1}$	$\mathbf{e} = \mathbf{F}^{-T} \cdot \mathbf{E} \cdot \mathbf{F}^{-1}$
$\mathbf{E} = \underleftarrow{\phi}(\mathbf{e})$	$E_{IJ} = F_{il}^{T} e_{ij} F_{jJ}$	$\mathbf{E} = \mathbf{F}^{T} \cdot \mathbf{e} \cdot \mathbf{F}$
$\boldsymbol{\tau} = \underrightarrow{\phi}(\mathbf{S})$	$\tau_{ij} = F_{il} S_{IJ} F_{jJ}^{T}$	$\boldsymbol{\tau} = \mathbf{F} \cdot \mathbf{S} \cdot \mathbf{F}^{T}$
$\mathbf{S} = \underleftarrow{\phi}(\boldsymbol{\tau})$	$S_{IJ} = F_{il}^{-1} \tau_{ij} F_{jJ}^{-T}$	$\mathbf{S} = \mathbf{F}^{-1} \cdot \boldsymbol{\tau} \cdot \mathbf{F}^{-T}$
$\mathbf{c} = \underrightarrow{\phi}(\mathbb{C})$	$c_{ijkl} = F_{il} F_{jJ} F_{kK} F_{lL} C_{IJKL}$	
$\mathbb{C} = \underleftarrow{\phi}(\mathbf{c})$	$C_{IJKL} = F_{il}^{-1} F_{jJ}^{-1} F_{kK}^{-1} F_{lL}^{-1} c_{ijkl}$	

Table 2.1 Kinematics: relation among tensors of the current and reference configurations.

2.2.4 The objective derivative

One way to carry out an *objective time derivation* of a tensor variable over a particular configuration while its reference system is in movement is through the concept of *objective derivation of Lie*[1,3,4] $L_v(\bullet)$. The temporal derivation of a tensor variable in the current configuration while its reference system is in movement can be carried out by transporting this variable to a fixed reference system, reference configuration, then make the derivation with respect to time and finally send these variables back to the current configuration (Figure 2.3). The objectivity of the operation can be guaranteed based on the aforementioned transport operators. Thus,

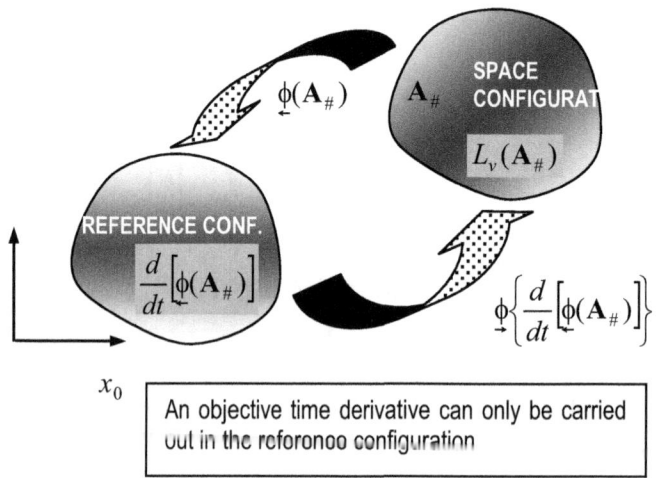

Figure 2.3 – Schematic representation of Lie's objective time derivative.

$$L_v\left(\mathbf{A}_{\#}\right) = \mathbf{F}^{-T} \cdot \left\{\frac{\partial}{\partial t}\left[\mathbf{F}^{T} \cdot \left(\mathbf{A}_{\#}\right) \cdot \mathbf{F}\right]\right\} \cdot \mathbf{F}^{-1}$$

$$L_v\left(\mathbf{A}^{\#}\right) = \mathbf{F} \cdot \left\{\frac{\partial}{\partial t}\left[\mathbf{F}^{-1} \cdot \left(\mathbf{A}^{\#}\right) \cdot \mathbf{F}^{-T}\right]\right\} \cdot \mathbf{F}^{T}$$

(2.15)

In the first expression above, the time derivation of *a covariant* tensor can be observed, whereas in the second one a *contravariant* tensor of a time derivation is observed.

2.2.5 Velocity

The velocity compute needs either an objective time derivative as defined before or the classic *objective material derivative*[1,2,3]. The velocity in the reference configuration or Lagrangian formulation is defined as $\dot{\mathbf{x}}_0 = \partial\mathbf{x}_0/\partial t$, and its corresponding image in the current configuration as $\mathbf{v} = \dot{\mathbf{x}}$. Now the *objective material derivative* of any variable is defined as

$$\phi(x_i; t) \xrightarrow{\substack{\text{derivada} \\ \text{material}}} \frac{D}{Dt}\phi = \frac{\partial\phi}{\partial t} + \mathbf{v}_i\frac{\partial\phi}{\partial x_i}$$

(2.16)

Such that the *velocity of the gradient of strain* is then,

$$\mathbf{F} = \boldsymbol{\nabla}_0\mathbf{x} = \frac{\partial\mathbf{x}}{\partial\mathbf{x}_0} \quad \Rightarrow \quad \dot{\mathbf{F}} = \frac{\partial^2\mathbf{x}}{\partial t\,\partial\mathbf{x}_0} = \frac{\partial\dot{\mathbf{x}}}{\partial\mathbf{x}_0} = \frac{\partial\mathbf{v}}{\partial\mathbf{x}_0} = \frac{\partial\mathbf{v}}{\partial\mathbf{x}}\frac{\partial\mathbf{x}}{\partial\mathbf{x}_0} = \mathbf{L}\cdot\mathbf{F}$$

(2.17)

where the *spatial gradient of velocities* can also be expressed as:

$$\mathbf{L} = \nabla \, \mathbf{v} = \frac{\partial \mathbf{v}}{\partial \mathbf{x}} = \dot{\mathbf{F}} \cdot \mathbf{F}^{-1} \qquad \in L(\vec{\mathcal{V}}; \vec{\mathcal{V}}) \tag{2.18}$$

Let the position of a point $\mathbf{x}_0$ in the reference configuration, the infinitesimal change of its position $\mathbf{x}_0 + d\mathbf{x}_0$ and the corresponding of this movement in the current configuration $\mathbf{x} \to \mathbf{x} + d\mathbf{x}$ be considered. On the other hand, if the differential movement can be expressed according to equation (2.17) as $d\mathbf{x} = \mathbf{F} d\mathbf{x}_0$, then the velocity in the current configuration is obtained once the corresponding objective differentiation is applied, $D\mathbf{x}/Dt = \dot{\mathbf{F}} d\mathbf{x}_0 = \mathbf{L} d\mathbf{x}$ (see also equations (2.17) and (2.18)).

The deformation velocity tensor is defined as,

$$\mathbf{D} = \frac{1}{2}\left(\mathbf{L} + \mathbf{L}^T\right) = \{\mathbf{L}\}_S \qquad \in L(\vec{\mathcal{V}}; \vec{\mathcal{V}}) \tag{2.19}$$

The vorticity tensor, which is the anti-symmetric part of the spatial gradient of velocities, is expressed as,

$$\mathbf{\Omega} = \frac{1}{2}\left(\mathbf{L} - \mathbf{L}^T\right) = \{\mathbf{L}\}_A$$
$$\mathbf{\Omega} = \frac{1}{2}\nabla \times \mathbf{v} \qquad \in L(\vec{\mathcal{V}}; \vec{\mathcal{V}}) \tag{2.20}$$

According to these last two equations, the velocity gradient tensor can be also written as,

$$\mathbf{L} = \mathbf{D} + \mathbf{\Omega} \tag{2.21}$$

The Lagrangian strain velocity can be expressed starting from equation (2.9) as,

$$\dot{\mathbf{E}} = \frac{\partial \mathbf{E}}{\partial t} = \frac{\partial}{\partial t}\left[\frac{1}{n}\left(\mathbf{U}^n - \mathbf{I}\right)\right] = \phi(\mathbf{D}) = \mathbf{F}^T \cdot (\mathbf{D}) \cdot \mathbf{F} \tag{2.22}$$

whose demonstration is formalized for *n=2*, where $\mathbf{U}^2 = \mathbf{C}$, and the following expression for the time change of the strain tensor is obtained from

$$\dot{\mathbf{E}} = \frac{1}{2}\dot{\mathbf{C}} = \frac{1}{2}\frac{\partial}{\partial t}\left(\mathbf{F}^T \cdot \mathbf{F}\right) = \frac{1}{2}\left(\dot{\mathbf{F}}^T \cdot \mathbf{F} + \mathbf{F}^T \cdot \dot{\mathbf{F}}\right) \tag{2.23}$$

But as $\dot{\mathbf{F}} = \mathbf{L} \cdot \mathbf{F}$ and $\dot{\mathbf{F}}^T = \mathbf{F}^T \cdot \mathbf{L}^T$, the following expression of the Lagrangian strain velocity is obtained,

$$\dot{\mathbf{E}} = \frac{1}{2}\left(\mathbf{F}^T \cdot \mathbf{L}^T \cdot \mathbf{F} + \mathbf{F}^T \cdot \mathbf{L} \cdot \mathbf{F}\right) = \frac{1}{2}\mathbf{F}^T \cdot \left(\mathbf{L}^T + \mathbf{L}\right) \cdot \mathbf{F} = \mathbf{F}^T \cdot (\mathbf{D}) \cdot \mathbf{F} = L_v(\mathbf{e}) \tag{2.24}$$

For $\mathbf{D} = 0 \iff \dot{\mathbf{E}} = 0$ the rigid body movement is obtained.

2.2.6 Stress measurements

Based on the true stress or *Cauchy stress* $\boldsymbol{\sigma}$, which is found in the current configuration, the *Kirchoff stress*, expressed as $\boldsymbol{\tau} = J\boldsymbol{\sigma}$, is also found in the current configuration but whose magnitude is not influenced by volume change $J = dV/dV_0 = \det|\mathbf{J}|$ while changing its configuration. The equivalent stress in the reference configuration can be obtained from the Kirchoff stress, also known as *Piola-Kirchoff second order tensor* $\mathbf{S} = \bar{\phi}(\boldsymbol{\tau}) = \mathbf{F}^{-1} \cdot \boldsymbol{\tau} \cdot \mathbf{F}$. This tensor represents a stress state on a non-deformed ideal solid[1,2,3], and in some calculations it is useful to have its definition to avoid constantly updating the geometry and getting a total equivalence to the process developed in the current configuration.

$$\tau = \vec{\phi} \mathbf{S} = \mathbf{F} \cdot \mathbf{S} \cdot \mathbf{F}^T \quad ; \quad \boldsymbol{\sigma} = \frac{1}{J} \mathbf{F} \cdot \mathbf{S} \cdot \mathbf{F}^T$$
$$\dot{\mathbf{S}} = f_{\text{REF}}\left(\dot{\mathbf{E}}; \mathbb{C}^T\right) \qquad \dot{\boldsymbol{\sigma}} = f_{\text{ACT}}\left(\dot{\mathbf{e}}; \mathbf{c}\right)$$

$$(2.25)$$

As observed in the equation above, two temporal variations of the constitutive relations that can be established similarly in any configuration are shown. For instance, in the reference configuration (REF) the Piola-Kirchoff stress is related to the Green-Lagrange strain and in the current configuration (CUR) the Cauchy stress is related to the Almansi strain. The three stress measurements coincide in small strains.

2.3 Thermodynamic basis

To sustain the formulation of a constitutive model it is also necessary to introduce the basic principles of thermodynamics[1,2,3]. Therefore, a brief review of the two laws of thermodynamics will be presented in this section as well as some of their consequences.

The purpose of this section is to review the essential elements of the thermodynamics of a continuum within the "small perturbation hypothesis" (SDH)[3].

2.3.1 First law of thermodynamics

The energy conservation principle is one of the most important general principles. This principle claims that energy can be neither created nor destroyed, it can only be changed from one form to another. For a closed system, the rate of the work done by external agents must be equal to the rate of total energy of the system. This principle is also known as the *first law of thermodynamics*. This law postulates energy balance, the conservation demand of the total internal energy W of the system. In other words, the first principle relates the power introduced into the system P_{in} and the amount of heat Q or caloric power existing in it, to the temporal change of internal energy, or global power $\partial W / \partial t$ in the system (see the schematic representation below).

Axiom. There exist:

1. An amount of own heat $Q_{own.}$ (own heat plus global heat transferred), governed by very specific physical laws (Newton, Fourier).
2. And an amount of global internal energy W, as a function of the physical state of the solid depending on state variable ω that represents the internal energy density or density of internal energy per unit mass ρ and volume V, such that the following relation is fulfilled,

$$\frac{d}{dt}\left[\int_V \rho\omega \, dV\right] = \frac{dW}{dt} = \dot{W} = Q_{own} + P_d$$

$$(2.26)$$

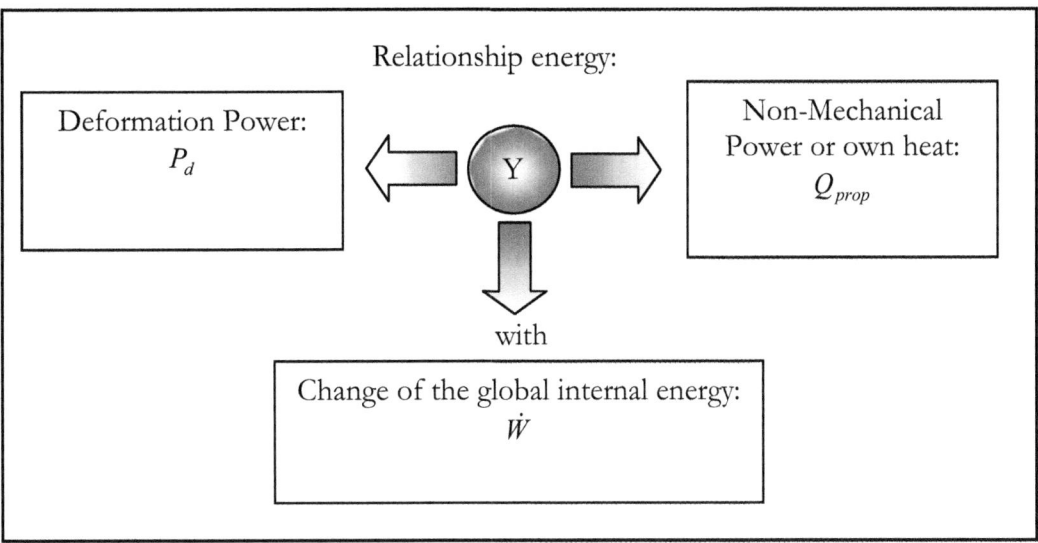

Obtaining the deformation power P_d:

In order to obtain the deformation power, we will start with the definition of mechanical power introduced P_{in} or external Power,

$$P_{int} = \oint_S \mathbf{t} \cdot \mathbf{v} \ dS + \int_V \rho \, \mathbf{b} \cdot \mathbf{v} \ dV = \oint_S t_i v_i \ dS + \int_V \rho \, b_i v_i \ dV \qquad (2.27)$$

where t_i is the surface force applied on the boundary S, (being $t_i = \sigma_{ij} n_j$, such that σ_{ij} is the Cauchy stress tensor and n_j is the normal vector to the surface S that envelops the solid[1,2,3]); b_i is the volume forces per unit mass; $\rho = \partial M / \partial V$ is the mass density, M is the mass and V is the volume; $v_i = du_i / dt = \dot{u}_i$ is the velocity field (if t=constant, then the velocity $v_i = \dot{u}_i$ is transformed into a time increment of the displacement field and the potential introduced is transformed into the time increase of work done).

Through Green's theorem[1,2,3], the surface integral of Figure (2.27) can be transformed into an integral over the volume of solid, leaving the introduced power with the following mathematical expression,

$$P_{int} = \int_V \left[v_i \left(\frac{\partial \sigma_{ij}}{\partial x_j} + \rho \, b_i \right) + \sigma_{ij} \frac{\partial v_i}{\partial x_j} \right] dV \qquad (2.28)$$

The term in parentheses is Cauchy's well-known equation or momentum balance per volume unit.

$$\frac{\partial \sigma_{ij}}{\partial x_j} + \rho \, b_i = \rho \frac{\partial v_i}{\partial t} \qquad (2.29)$$

If the acceleration is null or the problem is not time-dependent, Cauchy's classic equation of static equilibrium is obtained from expression (2.29),

$$\underbrace{\frac{\partial v_i}{\partial t} = 0}_{\text{quasistatic problem}} \quad \Rightarrow \quad \underbrace{\frac{\partial \sigma_{ij}}{\partial x_j} = -\rho b_i}_{\text{Cauchy equilibrium}} \qquad (2.30)$$

Substituting equation (2.29) into (2.28), the power introduced is obtained as a function of both the deformative power and the kinetic power,

$$P_{\text{int}} = \underbrace{\int_V \frac{\partial}{\partial t}\left[\frac{1}{2}\rho v_i v_i\right]dV}_{\text{Kinetic power}} + \underbrace{\int_V \sigma_{ij}\frac{\partial v_i}{\partial x_j}dV}_{\text{Deformative power}} = \dot{K} + P_d \tag{2.31}$$

where $\dot{K}$ is the kinetic power and P_d is the deformative power, such that this latter equation is,

$$P_{\text{int}} = \dot{K} + P_d \Rightarrow P_d = P_{\text{int}} - \dot{K} \tag{2.32}$$

By observing this last expression, three typical cases of the mechanics result:

- Rigid solid behavior problems: $\dfrac{\partial v_i}{\partial x_j} = 0 \Rightarrow P_{\text{int}} \equiv \dot{K}$

- Quasi-static behavior problems: $\dot{K} = 0 \Rightarrow P_{\text{int}} = P_0$

- Free vibration problems: $P_{\text{int}} = 0 \Rightarrow \dot{K} = -P_0$

Obtaining its own thermal power

To complete equation (2.26), it is necessary to define the Thermal Power or Own Heat Q_{own} as the Internal Heat r minus heat $q = q_i\, n_i$ lost by the borders

$$Q_{prop} = \int_V \rho\, r\, dV - \oint_S q_i n_i\, dS \tag{2.33}$$

where q is the heat flow that comes off the border, q_i is the heat flow vector, n_i is the normal vector to the surface around the border and r is the heat source per unit mass.

Substituting P_d and Q_{prop} into equation (2.26), the conservation equation for quasi-static problems is then obtained

$$\int_V \rho\dot{\omega}\, dV = \underbrace{\int_V \rho\, r\, dV - \oint_S q_i n_i\, dS}_{Q_{own}} + \underbrace{\int_V \sigma_{ij}\frac{\partial v_i}{\partial x_j}dV}_{P_d} \tag{2.34}$$

By transforming the surface integral into a volume integral (Green's Theorem), the Eulerian local form of the First law of thermodynamics is obtained. This is,

$$\rho\,\dot{\omega} = \rho\, r - \text{div}(q_i) + \sigma_{ij}\frac{\partial v_i}{\partial x_j} \tag{2.35}$$

Taking into consideration that the space strain gradient tensor can be expressed as a symmetric tensor plus another anti-symmetric, $\partial v_i/\partial x_j = L_{ij} = D_{ij} + \Omega_{ij}$, equation (2.35) is then written as,

$$\sigma_{ij}\frac{\partial v_i}{\partial x_j} = \sigma_{ij}L_{ij} = \sigma_{ij}D_{ij} + \underbrace{\sigma_{ij}\Omega_{ij}}_{0} \tag{2.36}$$

$$\rho\,\dot{\omega} = \left(\rho\, r - \text{div}(q_i)\right) + \sigma_{ij}D_{ij} \tag{2.37}$$

where D_{ij} and Ω_{ij} are respectively the strain velocity tensor (symmetric) and the vorticity tensor (antisymmetric).

2.3.2 Second law of thermodynamics

The second law of thermodynamics limits the sense of energy transformation. This energy can only be transformed from one way into another, which is not taken into account in the first law of thermodynamics. Consequently, a second law of thermodynamics must be defined. More specifically, the second law states the entropy balance η. The entropy is a conjugated state function of temperature θ, and it is related to power (usually thermal power).

Entropy density is defined as a function of the specific power and strain velocity,

$$\eta = \eta\left(\dot{\omega}; D_{ij}\right) \tag{2.38}$$

such that for $\dot{\eta} = 0$ there is an isentropic and adiabatic thermodynamic process (without heat loss by the borders) $\left(\rho r - div q_i = 0\right)$. For a reversible process, the law of the rate of entropy is defined as,

$$\dot{\eta} \overset{def}{=} \frac{1}{\theta}\dot{q} \quad \Rightarrow \quad \eta = \oint \dot{\eta}\, dt = \oint \left(\frac{1}{\theta}\dot{q}\, dt\right)_{rev} \tag{2.39}$$

The global entropy for the solid is then,

$$\Im \overset{def}{=} \int_V \rho\, \eta\, dV \tag{2.40}$$

An alternative way to present the second principle of thermodynamics is the Clausius-Duhem form, which is one of the most common forms in Cauchy's continuum. It is assumed that there are two conjugated scalar fields, temperature θ and entropy η, both local state functions such that

$$Q_{own} = \int_V \rho\, r\, dV - \oint_S q_i n_i\, dS = \dot{\Im}\theta \tag{2.41}$$

Or else

$$\dot{\Im} = \int_V \rho\frac{r}{\theta}\, dV - \oint_S \frac{q_i}{\theta} n_i dS \tag{2.42}$$

That represents the entropy change introduced by the heat transfer to the system.

Consequently, the second law of thermodynamics claims that in a continuum with some entropy, any change of entropy that may occur would be equal or greater than the amount of entropy introduced. From this inequality, the concept of energy loss or dissipation is introduced for the first time.

Entropy change introduced in the system

$$\frac{d}{dt}\Im \geq \dot{\Im}$$

Entropy increase in the system   Entropy increase introduced

By substituting equations (2.40) and (2.42) into this definition of the second law of thermodynamics, the so-called "Clausius-Duhem Inequality" is achieved,

$$\text{Clausius-Duhem} \atop \text{inequality} \left\{ \frac{d}{dt}\int_V \rho\,\eta\,dV \geq \int_V \rho\,\frac{r}{\theta}\,dV - \oint_S \frac{q_i}{\theta}n_i\,dS \right. \tag{2.43}$$

By transforming the surface integral into a volume integral (Green Theorem), the Eulerian local form of the second law of thermodynamics is obtained.

$$\dot{\eta} - \frac{r}{\theta} + \frac{1}{\rho}\mathrm{div}\left(\frac{q_i}{\theta}\right) \geq 0 \tag{2.44}$$

And solving the divergence of the ratio between the heat flow and the temperature, then

$$\rho\dot{\eta} - \rho\frac{r}{\theta} + \frac{1}{\theta}\mathrm{div}(q_i) - \frac{1}{\theta^2}q_i\nabla\theta \geq 0 \tag{2.45}$$

And reordering the terms, the rate of the thermo-mechanical dissipation can be obtained

$$\Xi = \theta\dot{\eta} - \left(r - \frac{1}{\rho}\mathrm{div}(q_i)\right) - \frac{1}{\rho\theta}q_i\nabla\theta \geq 0 \tag{2.46}$$

Combining the dissipation with the first principle of the Eulerian form, equation (2.35), the dissipation as a function of the deformative and caloric power is obtained,

$$\Xi = \theta\dot{\eta} - \dot{\omega} + \left(\frac{\sigma_{ij}D_{ij}}{\rho}\right) - \frac{1}{\theta\rho}q_i\nabla\theta \geq 0 \tag{2.47}$$

For thermo-mechanical processes where thermal problems can be decoupled from mechanical problems, by imposing the Coleman and Noll rational theory[1,2,3], the inequality above can be transformed into the known Clausius-Planck form,

$$\begin{cases} \Xi_m = \theta\dot{\eta} - \dot{\omega} + \left(\dfrac{\sigma_{ij}D_{ij}}{\rho}\right) \geq 0 & \rightarrow \quad \text{Mechanical dissipation} \\[2mm] \Xi_\theta = -q_i\nabla\theta \geq 0 & \rightarrow \quad \text{Thermal dissipation} \end{cases} \tag{2.48}$$

In the definition of constitutive models it is common to use dissipation in the Clausius-Planck form, also as a function of Helmholtz free energy, keeping in mind that the latter defines the capacity to produce work. The Helmholtz free energy Ψ is defined as:

$$\Psi = \Psi\left(e_{ij}; \theta; p_i\right) \overset{def}{=} \omega - \theta\eta \qquad \text{with} \quad p_i = \left\{F_{ij}^P; \alpha_i; d_i\right\} \tag{2.49}$$

where e_{ij} is the Almansi strain tensor and p_i is the set of internal variables such as in a plasticity and damage problem where the internal variables of each of the problems are considered as shown in equation (2.49). The time derivative of the Helmholtz free energy is then,

$$\dot{\Psi} = \dot{\omega} - \eta\dot{\theta} - \dot{\eta}\theta \quad \Rightarrow \quad -\dot{\omega} + \eta\dot{\theta} = -\dot{\Psi} - \dot{\eta}\theta \tag{2.50}$$

and substituting this equation into the equation of mechanical dissipation (2.48), mechanical dissipation by the local Eulerian form can be written as

$$\boxed{\Xi_m = -\dot{\Psi} - \dot{\theta}\eta + \frac{\sigma_{ij}D_{ij}}{\rho} \geq 0}$$ (2.51)

$$\Xi_m = \rho\left(-\dot{\Psi} - \dot{\theta}\eta\right) + \sigma_{ij}D_{ij} \geq 0$$ (2.52)

Substituting here the rate of free energy in a general form,

$$\dot{\Psi} = \frac{\partial\Psi}{\partial e_{ij}}D_{ij} + \frac{\partial\Psi}{\partial\theta}\dot{\theta} + \frac{\partial\Psi}{\partial p_i}\dot{p}_i$$ (2.53)

where e_{ij} and θ represent the thermal and mechanical free variables and p_i represents the internal variables of the nonlinear process. By considering this, the dissipation is then,

$$\Xi_m = \left(\sigma_{ij} - \rho\frac{\partial\Psi}{\partial e_{ij}}\right)D_{ij} - \rho\left(\frac{\partial\Psi}{\partial\theta} + \eta\right)\dot{\theta} - \rho\frac{\partial\Psi}{\partial p_i}\dot{p}_i \geq 0$$ (2.54)

D_{ij} and $\dot{\theta}$ are the temporal variations of the free variables, therefore in order to guarantee the Clausius-Duhem inequality, their multipliers must be identically null. In other words, from here the constitutive equations and the mechanical dissipation are obtained,

$$\left.\begin{aligned}
\left(\sigma_{ij} - \rho\frac{\partial\Psi}{\partial e_{ij}}\right) = 0 \Rightarrow \sigma_{ij} = \rho\frac{\partial\Psi}{\partial e_{ij}} \\
\left(\frac{\partial\Psi}{\partial\theta} + \eta\right) = 0 \Rightarrow \eta = -\frac{\partial\Psi}{\partial\theta}
\end{aligned}\right\} \text{ Constitutive equation}$$ (2.55)

$$\left.\Xi_m = -\rho\frac{\partial\Psi}{\partial p_i}\dot{p}_i \geq 0\right\} \text{ Dissipation}$$ (2.56)

2.3.3 Lagrangian local form of mechanical dissipation

To transform the dissipation expression into the Lagrangian local form, it is necessary to carry out the following transformation:

$$\boldsymbol{\sigma}:\mathbf{D} = \text{tr}(\boldsymbol{\sigma}:\mathbf{D}) \quad ; \quad \mathbf{D} = \mathbf{F}^{-T}\cdot\dot{\mathbf{E}}\cdot\mathbf{F}^{-1} \quad ; \quad \mathbf{F} = \frac{\partial\mathbf{x}}{\partial\mathbf{x}_0}$$ (2.57)

where $\mathbf{F}$ represents the deformation gradient which is a bi-punctual tensor relating a point of a reference configuration $\mathbf{x}_0$ with the same point in a current configuration $\mathbf{x}$, (this will be studied further and can be found in references[1,2,3]). By operating algebraically we obtain,

$$\boldsymbol{\sigma}:\mathbf{D} = \text{tr}\left(\boldsymbol{\sigma}\cdot\mathbf{F}^{-T}\cdot\dot{\mathbf{E}}\cdot\mathbf{F}^{-1}\right) = \text{tr}\left(\mathbf{F}^{-1}\cdot\boldsymbol{\sigma}\cdot\mathbf{F}^{-T}\cdot\dot{\mathbf{E}}\right) = \text{tr}\left(\frac{1}{J}\mathbf{S}:\dot{\mathbf{E}}\right)$$ (2.58)

where J is the determinant of the Jacobian matrix and $\mathbf{S}$ is the stress tensor in the reference configuration, which is also called Piola-Kirchoff tensor.

$$\boldsymbol{\sigma} : \mathbf{D} = \frac{1}{J} \mathbf{S} : \dot{\mathbf{E}} \quad ; \quad J = \det(\mathbf{F}) = \frac{dV}{dV_0} \tag{2.59}$$

By substituting this last expression into the Eulerian form of the dissipation, the Lagrangian form is then obtained, which is more suitable to solve some structural problems. Then,

$$\Xi_m = J\rho\left(-\dot{\Psi} - \dot{\theta}\eta\right) + S_{ij}\dot{E}_{ij} \geq 0 \tag{2.60}$$

where the Helmholtz free energy Ψ is defined now as,

$$\Psi = \Psi\left(E_{ij}; \theta; P_i\right) \overset{def}{=} \omega - \theta\eta \qquad con \quad P_i = \left\{F_{ij}^P; \alpha_i; d_i\right\} \tag{2.61}$$

where E_{ij} is the Green strain tensor and P is the set of internal variables that. For instance, in a plasticity and damage problem, the internal variables for each problem are the same variables shown in equation (2.49).

Knowing by the continuity equation that the density change between the reference configuration and the current configuration is given by the following expression $\rho_0 = \rho J = \dfrac{dM}{dV}\dfrac{dV}{dV_0}$, and defining a general form of free energy as $\Psi = \Psi(E_{ij}, \theta, P_i)$, its time variation is expressed as,

$$\dot{\Psi} = \frac{\partial \Psi}{\partial E_{ij}}\dot{E}_{ij} + \frac{\partial \Psi}{\partial \theta}\dot{\theta} + \frac{\partial \Psi}{\partial P_i}\dot{P}_i \geq 0 \tag{2.62}$$

where E_{ij} and θ are the free mechanic and thermal variables and P_i are the internal variables of the nonlinear process. Taking this into consideration, the dissipation is then,

$$\Xi_m = \left(S_{ij} - \rho_0\frac{\partial \Psi}{\partial E_{ij}}\right)\dot{E}_{ij} - \rho_0\left(\frac{\partial \Psi}{\partial \theta} + \eta\right)\dot{\theta} - \rho_0\frac{\partial \Psi}{\partial P_i}\dot{P}_i \geq 0 \tag{2.63}$$

$\dot{E}_{ij}$ and $\dot{\theta}$ are the rate variations of the free variables; therefore, to guarantee that the Clasius-Duhem inequality is met, their multipliers must be identically null and from here the constitutive equations and the dissipation in the Langrangean configuration can be obtained.

$$\left.\begin{array}{l} \left(S_{ij} - \rho_0\dfrac{\partial \Psi}{\partial E_{ij}}\right) = 0 \Rightarrow S_{ij} = \rho_0\dfrac{\partial \Psi}{\partial E_{ij}} \\[3mm] \left(\dfrac{\partial \Psi}{\partial \theta} + \eta\right) = 0 \Rightarrow \eta = -\dfrac{\partial \Psi}{\partial \theta} \end{array}\right\} \text{Constitutive equation} \tag{2.64}$$

$$\left.\Xi_m = \rho_0\frac{\partial \Psi}{\partial P_i}\dot{P}_i \geq 0\right\} \text{ Dissipation} \tag{2.65}$$

2.4 Internal variables

The origin of the thermodynamic internal variables $p_i = \mathbf{p}$ can be found firstly in the need to describe the kinetics of the physicochemical evolution of thermo-mechanical processes. However, the way it developed is related to rheological processes involved in the elasto-viscoplastic behavior of deformable solids.

These internal variables $\mathbf{p}$ provide a powerful alternative to characterize the continuum mechanical behavior and as with the free variable of the problem such as the stress $\boldsymbol{\sigma}$ or the strain $\boldsymbol{\varepsilon}$, they lead to the definition of the thermodynamic state of the solid.

The internal variables can be scalar, vector or tensor and they represent the average microscopic behavior of one or more physical effects. These variables normally represent a set of complex physical phenomena that cannot be split at a macroscopic level. In each constitutive problem to be formulated, we have to figure out and measure appropriately the suitable internal variable representing the macroscopic behavior, or having one or more physic phenomena at micro or medium scale level. Therefore, these variables can be measured but not controlled as their evolution law over time will only be established depending on the free variable and the remaining internal variables $\dot{\mathbf{p}} = f_p(\boldsymbol{\sigma}, \mathbf{p})$.

Thus, the definition of a state law or constitutive law will depend on the state variable or free variable $\boldsymbol{\sigma}$ or $\boldsymbol{\varepsilon}$ and the group of internal variables $\mathbf{p}$. This definition is solved if, additionally, the evolution law of the internal variables is established.

Then,

$$\dot{\boldsymbol{\sigma}} = f_\sigma\left(\dot{\boldsymbol{\varepsilon}}, \mathbf{X}, \dot{\mathbf{X}}, \dot{p}\right) \qquad \text{Constitutive law}$$

$$\dot{p} = f_p(\boldsymbol{\sigma}, p) \qquad\qquad \text{Evolution law of the internal variables} \tag{2.66}$$

In the equation above, $\mathbf{X}$ represents the variables defining the constitutive elastic behavior of the material.

2.5 Dynamic equilibrium equation for a discrete solid

The dynamic equilibrium equation of a discrete solid subjected to actions in time can be obtained directly from the first law of thermodynamics and previous knowledge of the finite element method the reader is supposed to have. From equations (2.26) and (2.32), the conservation law can be written as follows,

$$\frac{d}{dt}\int_V \rho\omega \, dV - Q_{prop} = P_d = P_{\text{int}} - \dot{K}$$

$$\underbrace{\int_V \rho\dot{\omega} \, dV - \left[\int_V \rho\, r \, dV - \oint_S q_i n_i \, dS\right]}_{\text{Mechanical power}} = \underbrace{\int_V \sigma_{ij} D_{ij} \, dV}_{\text{Deformative power}} = \underbrace{\oint_S t_i \dot{u}_i \, dS + \int_V \rho b_i \dot{u}_i dV}_{\text{Input power}} - \underbrace{\int_V \rho\dot{u}_i \frac{\partial \dot{u}_i}{\partial t} d}_{\text{Kinetic power}} \tag{2.67}$$

where the deformation velocity, which is now the temporal increment of deformation, can be written as $D_{ij} = \left\{L_{ij}\right\}_S = \left\{\nabla_i^S \dot{u}_j\right\}_S = \left\{\dot{F}_{ij} F_{kj}^{-1}\right\}_S$, which substituted into the previous equation leads to the power equilibrium in a continuum solid,

$$\int_V \sigma_{ij} \nabla_i^S \dot{u}_j \, dV = \oint_S t_i \dot{u}_i \, dS + \int_V \rho b_i \dot{u}_i dV - \int_V \rho \dot{u}_i \frac{\partial \dot{u}_i}{\partial t} dV \tag{2.68}$$

Following the concept of polynomial approximation of a continuous field of displacements $u_j(x,y,z)$ or velocities $\dot{u}_j(x,y,z)$, through a polynomial normalized local support function $N_{jk}(x,y,z)$ known as shape function[5].

$$u_j(x,y,z)\Big|_{\Omega^e} = N_{jk}(x,y,z) U_k\Big|_{\Omega^e} \quad \Rightarrow \quad \dot{u}_j(x,y,z)\Big|_{\Omega^e} = N_{jk}(x,y,z) \dot{U}_k\Big|_{\Omega^e} \tag{2.69}$$

This function, $N_{jk}(x,y,z)$, valid on the bounded domain Ω^e called finite element, allows the approximation within such domain of the displacement fields $u_k(x,y,z)$, velocities $\dot{u}_k(x,y,z)$ and accelerations $\ddot{u}_k(x,y,z)$ through the valuation of their respective magnitudes U_k, $\dot{U}_k$, $\ddot{U}_k$ in a finite number of points, called *nodes* belonging to the finite element domain Ω^e. Accordingly, the derived fields can be established from the displacements, such as the Almansi strain, among others $e_{ik} = \nabla_i^S u_k$. Then,

$$u_j(x,y,z)\Big|_{\Omega^e} = N_{jk}(x,y,z) U_k\Big|_{\Omega^e} \quad \Rightarrow \quad e_{ij}\Big|_{\Omega^e} = \nabla_i^S u_j\Big|_{\Omega^e} = \nabla_i^S N_{jk} U_k\Big|_{\Omega^e} \tag{2.70}$$

The finite element method[5] is called the numerical procedure that results from using this polynomial approximation for field functions. This approximation reduces the infinite unknowns of the field function to a finite number of unknowns, defined in certain points and pre-established as nodes of the finite element.

Substituting approximations (2.69) and (2.70) into equation (2.68), the potential equilibrium equation can be written from the following approximation as

$$\left[\int_{V^e} \sigma_{ij} \nabla_i^S N_{jk} \, dV\right]_{\Omega^e} \dot{U}_k\Big|_{\Omega^e} = \left[\oint_{S^e} t_i N_{ik} \, dS + \int_{V^e} \rho b_i N_{ik} dV - \int_{V^e} \rho N_{ki} N_{ij} \ddot{U}_j dV\right]_{\Omega^e} \dot{U}_k\Big|_{\Omega^e} \tag{2.71}$$

But this equation is fulfilled for any velocity $\dot{U}_k\Big|_{\Omega^e}$; therefore, the established equality in equation (2.71) does not depend on this velocity, and from here the following equation of dynamic equilibrium for a discrete solid is obtained.

$$\underbrace{\int_{V^e} \sigma_{ij} \underbrace{\nabla_i^S N_{jk}}_{\overline{B}_{ijk}\big|_{\Omega^e}} \, dV\Bigg|_{\Omega^e}}_{f_k^{int}\big|_{\Omega^e}} = \underbrace{\oint_{S^e} t_i N_{ik} \, dS + \int_{V^e} \rho b_i N_{ik} dV\Bigg|_{\Omega^e}}_{f_k^{ext}\big|_{\Omega^e}} - \underbrace{\overbrace{\int_{V^e} \rho N_{ki} N_{ij} dV}^{M_{kj}\big|_{\Omega^e}}\Bigg|_{\Omega^e} \cdot \ddot{U}_j\Big|_{\Omega^e}}_{f_k^{mas}\big|_{\Omega^e}} \tag{2.72}$$

where $f_k^{int}\Big|_{\Omega^e}$, $f_k^{mas}\Big|_{\Omega^e}$ and $f_k^{ext}\Big|_{\Omega^e}$ are the sets of internal, mass and external forces ordered in column matrices and developed at each point of the discrete system

---

[5] Zienkiewicz, O. and Taylor, R. (1989). The finite element method. McGraw-Hill, Vol I y II.

approaching the continuum $\ddot{U}_j\big|_{\Omega^e}$ the acceleration in such points, $M_{kj}\big|_{\Omega^e}$ the elemental mass and $\overline{B}_{ijk}\big|_{\Omega^e} = \nabla_i^S N_{jk}\big|_{\Omega^e}$ the strain compatibility tensor or symmetric gradient of the shape function.

Equation (2.72) represents the elemental equation of the *dynamic equilibrium on the current configuration* which expressed in the *reference configuration* adopts the following form,

$$\left(\int_{V_0^e} S_{ij}\nabla_i^S N_{jk}\,dV_0\right)_{\Omega_0^e} = \left(\oint_{S_0^e} t_i N_{ik}\,dS_0 + \int_{V_0^e}\rho_0 b_i N_{ik}\,dV_0\right)_{\Omega_0^e} - \left(\int_{V_0^e}\rho_0\,N_{ki}\,N_{ij}\,dV_0\right)_{\Omega_0^e}\ddot{U}_j\big|_{\Omega_0^e}$$

(2.73)

$$M_{kj}\ddot{U}_j + f_k^{\text{int}} = f_k^{\text{ext}} \qquad \in \Omega_0^e$$

where M_{kj} is the mass matrix, S_{ij} is the Piola Kirchoff stress tensor, ρ_0, V_0 and S_0 are the solid density, volume and body surface in the reference configuration (see Figure 2.1).

From a mechanical-numerical point of view, the nonlinearity in equations (2.72) or (2.73) may be due to diverse phenomena,

- **Constitutive nonlinearity**, resulting from the loss of linearity between the stress and strain fields σ_{ij}-e_{ij} (o S_{ij}-E_{ij} for the reference configuration), such as in plasticity, damage, etc. This nonlinearity is due to the property change of the material during its mechanical behavior and is reflected in its constitutive tensor $\mathbb{C}_{ijkl}$.

- **Large strain nonlinearity**, due to the nonlinear influence of the solid change of configuration in the strain field (see Figure 2.1 and equation (2.8)). This change of configuration also modifies the constitutive tensor $\mathbb{C}_{ijkl}$ (see Table 2.1), and therefore it establishes a nonlinear relation between stresses and strains (2.25)). Moreover, these changes of configuration are produced by large movements, translations and rotations, which also produce changes in the local reference system in the solid points, affecting then the strain compatibility tensor $\overline{B}_{ijk}$.

- **Large displacement nonlinearity**, unlike large strains, only affect the strain compatibility tensor $\overline{B}_{ijk}$, as changes only occur in the local reference system of the solid points due to large displacements.

These possible nonlinearities, which can occur either at the same time or separately, are summarized in a descriptive table as follows

σ_{ij} —
$\begin{cases}
\text{•Constitutive non-linearity :} \\
\quad \text{Non linear dependence between stresses and strains} \\
\quad \text{strains due to changes in the constitutive tensor } \mathbf{C}_{ijkl} \\
\text{•Large Strains:} \\
\quad \text{Non linear dependence among strains } e_{ij} \text{ or } E_{ij} \text{ and} \\
\quad \text{displacements.} \\
\quad \text{Non linear dependence between stresses and strains due to} \\
\quad \text{changes in the constitutive tensor } \mathbf{C}_{ijkl} \text{ , and configuration changes.} \\
\quad \text{Non linearity due to changes in the geometric configuration of} \\
\quad \text{the solid that is reflected in the tensor } \overline{\mathbf{B}}_{ijk}.
\end{cases}$

$\overline{\mathbf{B}}_{ijk}$ —
$\begin{cases}
\text{• Large Displacements:} \\
\quad \text{Only represents a part of the problem in large strains,} \\
\quad \text{because only affects the strain compatibility tensor } \overline{\mathbf{B}}_{ijk}.
\end{cases}$

Equation (2.72) (or (2.73) for the reference configuration) represents the equilibrium in the elemental domain Ω^e, and its contribution to the global domain Ω is carried out through the "assembly" of this equilibrium equation using the linear operator $\mathbf{A}$ that represents the sum between the force components , according to the position and direction of the local contributions[5].

2.5.1 Nonlinear problem – Linearization of the equilibrium equation

If there is a linear solid behavior, the following relation of global equilibrium is met and its expression is obtained from the equation assembly of the local equilibrium as represented in equation (2.72) or (2.73)

$$0 = \mathop{\mathbf{A}}_{\Omega^e}\left[f_k^{\text{mas}} + f_k^{\text{int}} - f_k^{\text{ext}} \right]_{\Omega^e} = \Delta f_k\big|_{\Omega} \tag{2.74}$$

The nonlinearity in the solid global behavior is expressed as a residual force $\Delta f_k\big|_{\Omega}$, produced by the unbalance of internal forces $f_k^{\text{int}}\big|_{\Omega}$, mass forces $f_k^{\text{mas}}\big|_{\Omega}$ and the external forces $f_k^{\text{ext}}\big|_{\Omega}$. This unbalance at some time "t" of the dynamic process can be eliminated by linearizing this residual force $\Delta f_k\big|_{\Omega}$ (2.74), in the surrounding of the current equilibrium state ($i+1$). Therefore, it is necessary to force the equilibrium in the current state ($i+1$) and express such condition through an expansion of Taylor's series considering only the first variation,

$$0 = \mathop{\mathbf{A}}_{\Omega^e}^{i+1} \left[\Delta f_k \right]_{\Omega^e}^t \cong \mathop{\mathbf{A}}_{\Omega^e}^i \left[\Delta f_k \right]_{\Omega^e}^t + \mathop{\mathbf{A}}_{\Omega^e} \left[{}^i\left[\frac{\partial(\Delta f_k)}{\partial U_r} \right]_{\Omega^e}^t \cdot {}^{i+1}\left[\Delta U_r \right]_{\Omega^e}^t \right]$$

$$0 = {}^{i+1}\left[\Delta f_k \right]_{\Omega}^t \cong {}^i\left[\Delta f_k \right]_{\Omega}^t + \mathop{\mathbf{A}}_{\Omega^e} \left[{}^i\left[M_{kj} \frac{\partial \ddot{U}_j}{\partial U_r} + \frac{\partial f_k^{\text{int}}}{\partial U_r} + \frac{\partial f_k^{\text{int}}}{\partial \dot{U}_j} \frac{\partial \dot{U}_j}{\partial U_r} - \frac{\partial f_k^{\text{ext}}}{\partial U_r} \right]_{\Omega^e}^t \cdot {}^{i+1}\left[\Delta U_r \right]_{\Omega^e}^t \right]$$

(2.75)

where the acceleration and the velocity must be expressed through a linear approximation in finite differences, as shown further in section B3.3.21, the Newmark method is an example of this approximation. By substituting into this equation the internal and mass forces expressed in equation (2.72), (the same procedure can be used in equation (2.73)) then,

$$0 = \mathop{\mathbf{A}}_{\Omega^e} {}^i\left[M_{kj} \ddot{U}_j + \int_{V^e} \sigma_{ij} \nabla_i^S N_{jk} \, dV - f_k^{\text{ext}} \right]_{\Omega^e}^t +$$

$$+ \mathop{\mathbf{A}}_{\Omega^e} {}^i\left[\left(\int_{V^e} \rho N_{ki} N_{ij} dV \right) \frac{\partial \ddot{U}_j}{\partial U_r} + \frac{\partial}{\partial U_r} \left(\int_{V^e} \sigma_{ij} \nabla_i^S N_{jk} \, dV \right) + \frac{\partial}{\partial \dot{U}_m} \left(\int_{V^e} \sigma_{ij} \nabla_i^S N_{jk} \, dV \right) \frac{\partial \dot{U}_m}{\partial U_r} - \frac{\partial f_k^{\text{ext}}}{\partial U_r} \right]_{\Omega^e}^t \cdot$$

$$\cdot \mathop{\mathbf{A}}_{\Omega^e} {}^{i+1}\left[\Delta U_r \right]_{\Omega^e}^t$$

$$0 = \mathop{\mathbf{A}}_{\Omega^e} {}^i\left[M_{kj} \ddot{U}_j + \int_{V^e} \sigma_{ij} \nabla_i^S N_{jk} \, dV - f_k^{\text{ext}} \right]_{\Omega^e}^t +$$

$$+ \mathop{\mathbf{A}}_{\Omega^e} {}^i\left[\left(\int_{V^e} \rho N_{ki} N_{ij} dV \right) \frac{\partial \ddot{U}_j}{\partial U_r} + \left(\int_{V^e} \frac{\partial \sigma_{ij}}{\partial e_{st}} \frac{\partial e_{st}}{\partial U_r} \nabla_i^S N_{jk} \, dV \right) + \left(\int_{V^e} \frac{\partial \sigma_{ij}}{\partial D_{st}} \frac{\partial D_{st}}{\partial \dot{U}_m} \nabla_i^S N_{jk} \, dV \right) \frac{\partial \dot{U}_m}{\partial U_r} - \frac{\partial f_k^{\text{ext}}}{\partial U_r} \right]_{\Omega^e}^t \cdot$$

$$\cdot \mathop{\mathbf{A}}_{\Omega^e} {}^{i+1}\left[\Delta U_r \right]_{\Omega^e}^t$$

Particularizing this dynamic equilibrium equation for a material having the following viscoelastoplastic constitutive law $\sigma_{ij} = \rho(\partial \Psi(e_{ij}, p_i) / \partial e_{ij}) = \mathbb{C}_{ijkl} : e_{kl}^e + \xi_{ijkl} : D_{kl}$ for a kinematic relation of the type $e_{ij} = \nabla_i^S u_j = \nabla_i^S N_{jk} U_k$, and $D_{ij} = \nabla_i^S \dot{u}_j = \nabla_i^S N_{jk} \dot{U}_k$, then,

$$0 = \mathop{\mathbf{A}}_{\Omega^e} {}^i\left[M_{kj} \ddot{U}_j + \int_{V^e} \sigma_{ij} \nabla_i^S N_{jk} \, dV - f_k^{\text{ext}} \right]_{\Omega^e}^t + \mathop{\mathbf{A}}_{\Omega^e} {}^i\left[\left(\int_{V^e} \rho N_{ki} N_{ij} dV \right) \frac{\partial \ddot{U}_j}{\partial U_r} + \right.$$

$$\left. + \left(\int_{V^e} \left(\nabla_s^S N_{tr} \right) \mathbb{C}_{ijst}^T \left(\nabla_i^S N_{jk} \right) dV \right) + \left(\int_{V^e} \left(\nabla_s^S N_{tr} \right) \xi_{ijst}^T \left(\nabla_i^S N_{jk} \right) dV \right) \frac{\partial \dot{U}_m}{\partial U_r} - \frac{\partial f_k^{\text{ext}}}{\partial U_r} \right]_{\Omega^e}^t \cdot$$

(2.76)

$$\mathop{\mathbf{A}}_{\Omega^e} {}^{i+1}\left[\Delta U_r \right]_{\Omega^e}^t$$

$$0 = {}^i\left[\Delta f_k \right]_{\Omega}^t + {}^i\left[J_{kr}^T \right]_{\Omega}^t \cdot {}^{i+1}\left[\Delta U_r \right]_{\Omega}^t$$

where $\xi_{ijst}^T = \xi^T$ is the tangent viscosity tensor and $J_{kr}^T = \mathbb{J}^T$ is the Jacobian tangent operator. This equation can also be presented in matrix form. The operators contributing to the Jacobian definition are detailed,

$$0 = {}^{i+1}[\Delta\mathbf{f}]_\Omega^t \cong {}^i[\Delta\mathbf{f}]_\Omega^t + \underbrace{{}^i\left[\mathbb{M}\frac{\partial\ddot{\mathbf{U}}}{\partial\mathbf{U}} + \mathbb{K}^T + \mathbb{D}^T\frac{\partial\dot{\mathbf{U}}}{\partial\mathbf{U}} - \frac{\partial\mathbf{f}^{\text{ext}}}{\partial\mathbf{U}}\right]_\Omega^t}_{{}^i\mathbb{J}_\Omega^t} \cdot {}^{i+1}[\Delta\mathbf{U}]_\Omega^t \tag{2.77}$$

As the latter is the linearized equilibrium equation where $[\mathbb{K}^T]_\Omega = \underset{\Omega^e}{\mathbf{A}}\int_{\mathbf{V}^e}(\nabla^S\mathbf{N}):\mathbf{C}^T:(\nabla^S\mathbf{N})dV$ represents the tangent stiffness matrix, $[\mathbb{M}]_\Omega = \underset{\Omega^e}{\mathbf{A}}\int_{\mathbf{V}^e}\rho\mathbf{N}:\mathbf{N}\,dV$ is the mass matrix, $[\mathbb{D}^T]_\Omega = \underset{\Omega^e}{\mathbf{A}}\int_{\mathbf{V}^e}(\nabla^S\mathbf{N}):\xi^T:(\nabla^S\mathbf{N})dV$ is the tangent damping matrix, all defined in the domain Ω, $\mathbb{C}_{ijst}^T$ is the tangent tensor corresponding to the constitutive law used in each point of the solid and the external force is expressed as $\mathbf{f}^{\text{ext}} = \underset{\Omega^e}{\mathbf{A}}\left[\oint_{S^e}\mathbf{N}:\mathbf{t}\,dS + \int_{\mathbf{V}^e}\rho\mathbf{N}:\mathbf{b}\,dV\right]$. The solid unbalanced force ${}^{i+1}[\Delta\mathbf{f}_k]_\Omega^t$ is eliminated following the Newton-Raphsons resolution[5] until this residue becomes negligible, a situation known as the convergence of the linearized process towards the exact solution (see Figure 2.4).

In Figure 2.4 the space equilibrium is described, the convergence is kept in time domain to be dealt with in the solution methods in the time domain of the dynamic equilibrium equation.

2.6 Different types of nonlinear dynamic problems

A brief description of the different types of behaviors is given in this section and the nonlinearity condition to the dynamic problem is introduced. Firstly, creep and stress relaxation problems will be studied as a consequence of the scope of this book on nonlinear dynamics. These problems lead to the so-called viscous damping which depends on velocity, plasticity and damage and contributes to energy dissipation regardless of the velocity. Although large strains also lead to nonlinearity in the dynamic problem, it will not be studied in depth like other nonlinear phenomena.

Let small strain problems be considered, in which the Jacobian matrix (2.5) meets the following condition,

$$J = |\mathbf{J}| = |\mathbf{F}| = \frac{dV}{dV_0} \cong 1 \tag{2.78}$$

from here the coincidence between the Cauchy and the Piola Kirchoff stresses are obtained $\sigma_{ij} = S_{ij}$, between the strain velocity in the current configuration and the corresponding infinitesimal magnitude $D_{ij} = \dot{\varepsilon}_{ij}$ and between the density in the different configurations $\rho = \rho_0$. Under these particular conditions, displacements and strains are negligible with respect to the solid dimensions and therefore the strain can be written as follows,

$$\boldsymbol{\varepsilon} = \frac{1}{2}\left(\mathbf{F}\,\mathbf{F}^{T} - \mathbf{I}\right) = \nabla^{S}\mathbf{u} = \frac{1}{2}\left[\left(\frac{\partial \mathbf{u}}{\partial \mathbf{x}}\right) + \left(\frac{\partial \mathbf{u}}{\partial \mathbf{x}}\right)^{T}\right] \tag{2.79}$$

Moreover, for materials with elastic behavior and small strains, the following coincidence occurs in the strain definitions ($\mathbf{e} = \mathbf{E} = \boldsymbol{\varepsilon}$), and the free energy is written in the following simplified manner as,

$$\Psi = \frac{1}{2\rho_0}\left(\boldsymbol{\varepsilon} : \mathbb{C} : \boldsymbol{\varepsilon}\right) \tag{2.80}$$

such that substituted in equations (2.55) or (2.64), the following constitutive law is obtained,

$$\boldsymbol{\sigma} = \rho_0 \frac{\partial \Psi}{\partial \boldsymbol{\varepsilon}} = \mathbb{C} : \boldsymbol{\varepsilon} \tag{2.81}$$

where the constitutive tensor $\mathbb{C}$ coincides exactly with the one obtained through Hooke's generalized law and its canonical expression is,

$$\mathbb{C}_{ijkl} = \lambda\,\delta_{ij}\,\delta_{kl} + \mu\left(\delta_{ik}\delta_{jl} + \delta_{il}\delta_{jk}\right) \tag{2.82}$$

where λ and μ are the constants of Lamé and δ_{ij} is the Kronecker tensor. Hooke's elasticity tensor is defined as positive and with the following symmetries

$$\mathbb{C}_{ijkl} = \mathbb{C}_{klij} = \mathbb{C}_{ijlk} = \mathbb{C}_{jilk} \tag{2.83}$$

Cauchy defined the elastic body as "that one in which strains in any of the solid's points were determined by their stress and temperature state". In contrast, a material will have an inelastic behavior when it is necessary to establish additional definitions to the classic elasticity theory and its formulation is related to the history of the material behavior. This situation makes it difficult to guarantee a biunivocal relation between the stress and the strain tensor; in other words, they are not invertible expressions

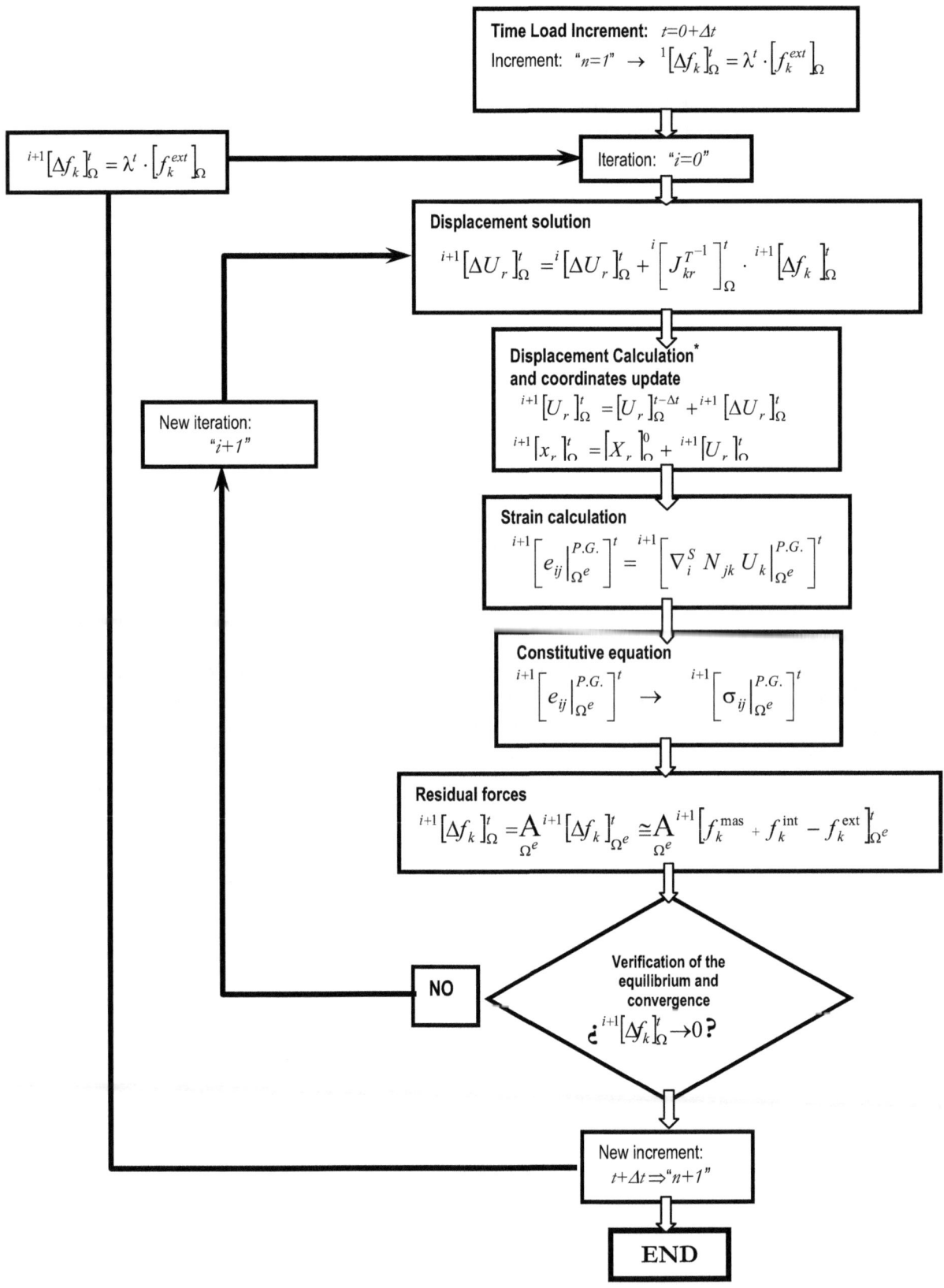

* **NOTE:** The velocities and acceleration fields must be calculated by the acceleration approximation model (see Newmark method as an example)

Figure 2.4 – Schematic representation of the solution to the non-linear problem through Newton-Raphson.

2.6.1 Material Non-linearities

Time produces irreversible behaviors in some solids. Basically, there are three types of non-linear time-dependent behaviors:

- **Delayed Elasticity or "Creep"**, where the strain increases while the stress applied is constant (see Figure 2.5).
- **Stresses Relaxation**, where the stress loss is produced while the strain level remains constant. This behavior, although not invertible, is the inversed implicit creep form (see Figure 2.5)
- **Visco-plasticity,** whose non-linear behavior is due to an increase of the inelastic strain field, but this occurs provided that the stress field surpasses some pre-established threshold. (See Figure 2.7)

There are materials whose non-linear behavior is time-independent and can be due to:

- **Plasticity or inelastic behavior with instantaneous flow.** This behavior can be established mathematically as a limit of the viscoplastic particular case, but the physics of the problem is qualitatively different. This concept will be described in detail in the following chapters. (See Figure 2.6)
- **Damage or stiffness degradation in which** a material strength loss is produced as a result of the elasticity degradation of the material.

These behaviors can occur either isolated or all in different degrees. Constitutive modeling and its influence in structure behavior will be further analysed.

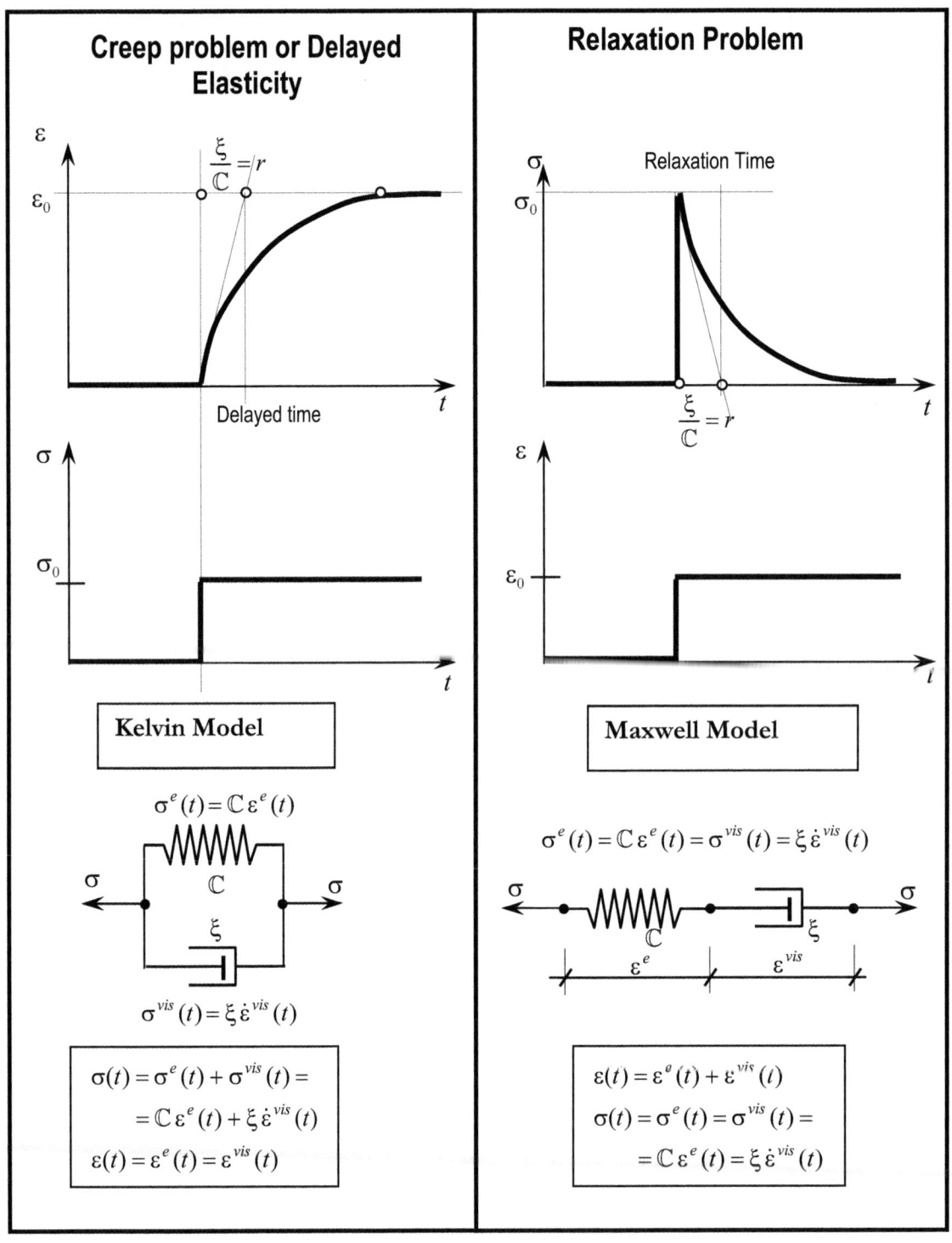

Figure 2.5 – Simplified ways to understand Kelvin and Maxwell viscous behavior.

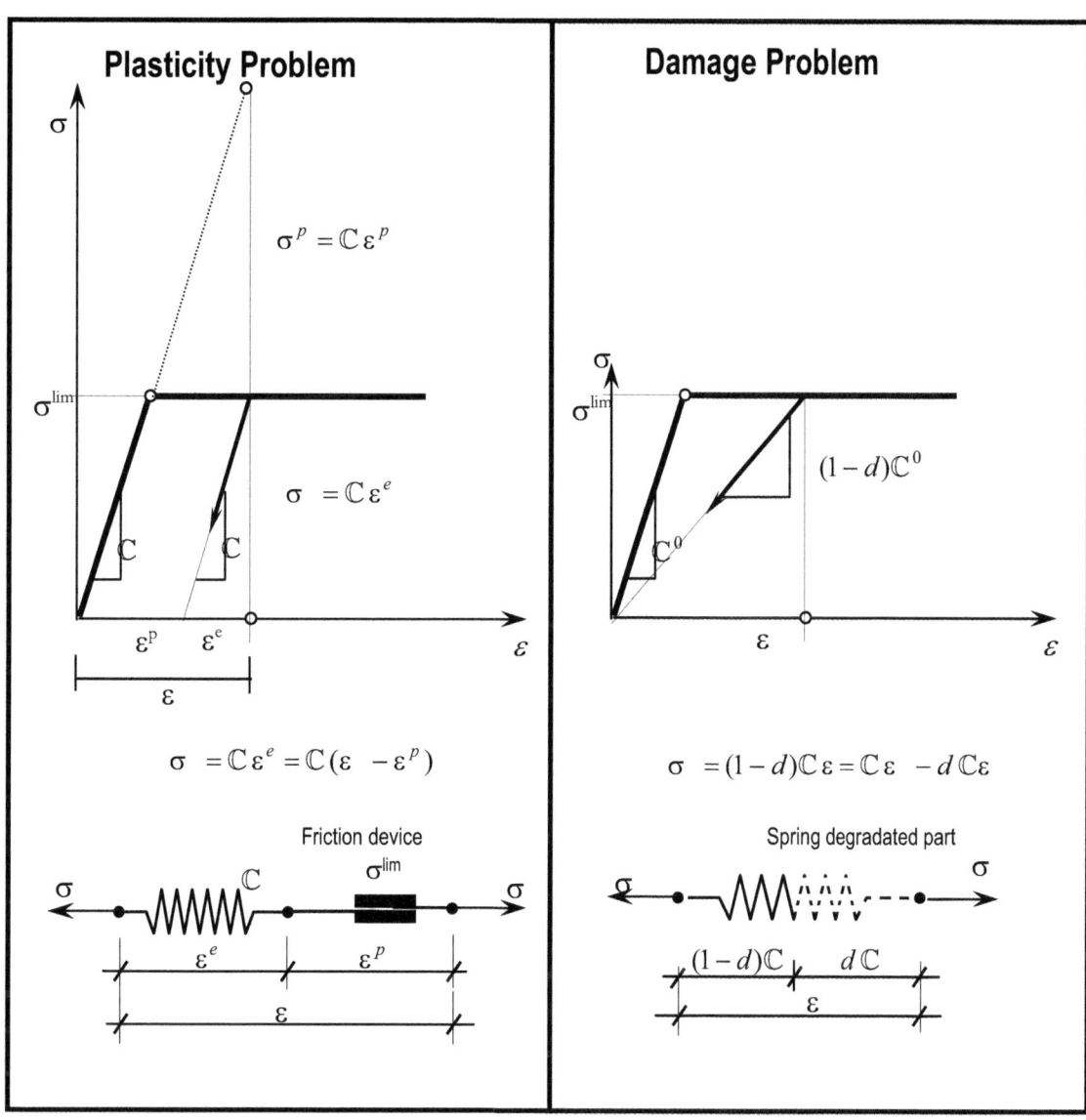

Figure 2.6 – Simplified ways to understand the elastoplastic and damage behavior.

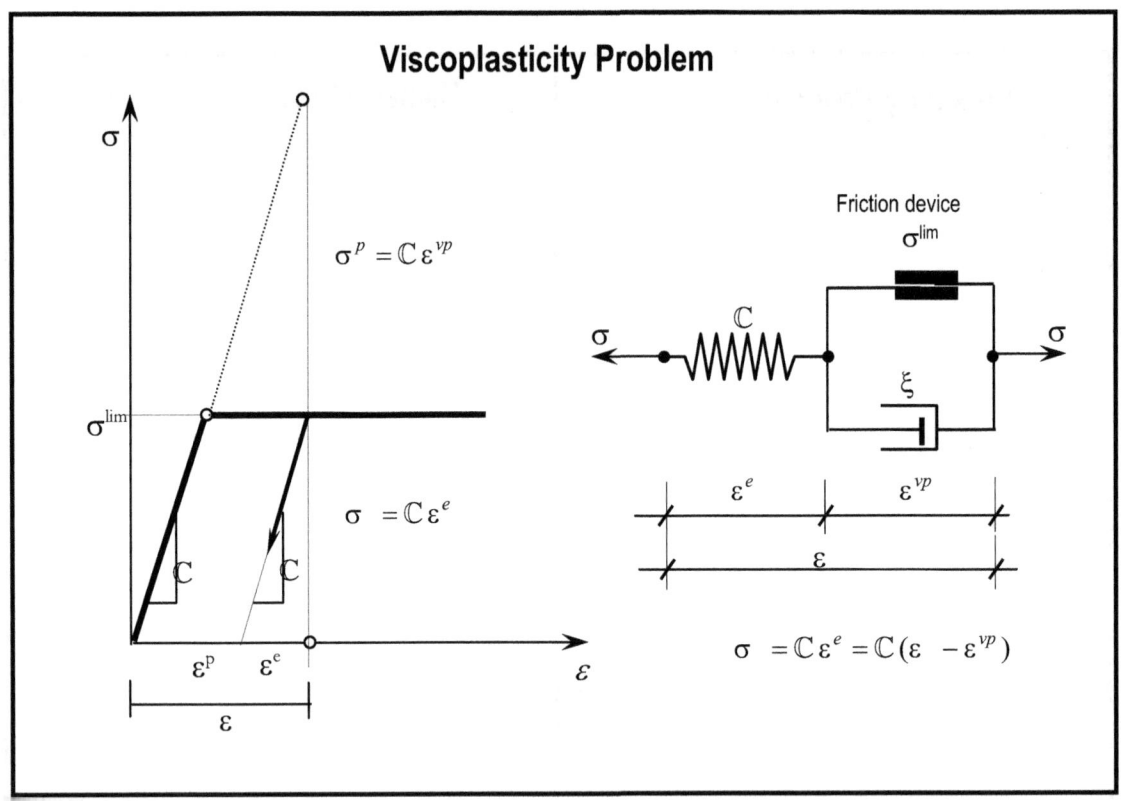

Figure 2.7 – Simplified ways to understand the viscoplastic behavior.

3 Solution of the Equation of Motion

3.1 Introduction

This chapter deals with the solution of the equation of motion in its semi-discrete form in the time domain (see equilibrium equation, section 2.5). Below is the assembly of equation 2.72 (or 2.73 if the equilibrium is achieved in the reference configuration) which defines the equilibrium in the solid at time $t + \Delta t$,

$$
\mathop{A}_{\Omega^e}\left[\int_{V^e}\sigma_{ij}\nabla_i^S N_{jk}\,dV\right]_{\Omega^e}^{t+\Delta t} = \mathop{A}_{\Omega^e}\left[\oint_{S^e}t_i N_{ik}\,dS + \int_{V^e}\rho\,b_i N_{ik}\,dV\right]_{\Omega^e}^{t+\Delta t} -\mathop{A}_{\Omega^e}\left[\int_{V^e}\rho\,N_{ki}N_{ij}\,dV\right]_{\Omega^e}^{t+\Delta t}\ddot{U}_j\Big|_{\Omega^e}^{t+\Delta t}
$$

and it can be written in the following compact form,

$$
\Delta f_k = M_{kj}\ddot{U}_j(t+\Delta t) + f_k^{int}(\dot{\mathbf{U}},\mathbf{U},t+\Delta t) - f_k^{ext}(t+\Delta t) = 0 \qquad \in \Omega
$$

$$
\Delta \mathbf{f} = \mathbb{M}\,\ddot{\mathbf{U}}(t+\Delta t) + \mathbf{f}^{int}(\dot{\mathbf{U}},\mathbf{U},t+\Delta t) - \mathbf{f}^{ext}(t+\Delta t) = 0 \qquad \in \Omega
$$

(3.1)

This system of ordinary differential equations contains the polynomial approximation of the displacement, velocity and acceleration fields defined in equation 2.69. The spatial discretization of the displacement, velocity and acceleration fields is represented by the particular values of these functions, $\mathbf{U}(t)$, $\dot{\mathbf{U}}(t)$ and $\ddot{\mathbf{U}}(t)$ respectively. These are defined only in some points of the continuous medium (nodes), where the local polinomial approximation is defined. It can also be called as shape function of a finite element. It must be noted that in the equation (3.1) the rate field is continuous and only the space functions were defined. Also, in this equation the mass matrix equation $\mathbb{M} = \int_{V^e}\rho\,N_{ki}N_{ij}\,dV$, the external forces $\mathbf{f}^{ext}(t+\Delta t) = \oint_{S^e}t_i N_{ik}\,dS + \int_{V^e}\rho\,b_i N_{ik}\,dV$ and the internal forces $\mathbf{f}^{int}(\dot{\mathbf{U}},\mathbf{U},t+\Delta t) = {} = \int_V \sigma_{ij}\nabla_i^S N_{jk}\,dV$ are represented, which contain the non-linear terms of the material behavior.

Among the possible nonlinearities of a point in a generic solid are only those produced by plasticity, damage and viscosity which will be considered in this analysis. Also, small strain and motion problems will be tackled in such a way that the strain field can be represented as $e_{ij} = E_{ij} = \varepsilon_{ij}$. Under these assumptions, the stress σ_{ij} can be written as (see equation 2.55 for the Eulerian description and 2.64 for the Lagrangian description),

$$\boldsymbol{\sigma} = \rho \frac{\partial \Psi}{\partial \boldsymbol{\varepsilon}} = \begin{cases} \mathbb{C}^S : \boldsymbol{\varepsilon}^e + \boldsymbol{\xi} : \dot{\boldsymbol{\varepsilon}} & \text{Elasto-plastic case with viscosity} \\ \mathbb{C}^0 : \boldsymbol{\varepsilon} & \text{Elastic case} \end{cases} \qquad (3.2)$$

For a linear elastic problem, the viscosity is introduced into equilibrium equation through a viscous forces term $\mathbf{f}^{\text{visc}}(\dot{\mathbf{U}}, t + \Delta t) = \mathbb{D}\,\dot{\mathbf{U}}(t + \Delta t)$, where $\mathbb{D} = \int_V (\nabla^S \mathbf{N})\boldsymbol{\xi}(\nabla^S \mathbf{N}) dV$ is the so-called damping matrix, usually taking the form given by the linear combination of Rayleigh[1] proportional damping. Thus, for linear behavior, the internal forces have the following form,

$$f_k^{\text{int}}(\dot{\mathbf{U}}, \mathbf{U}, t + \Delta t) = D_{kj}\,\dot{U}_j(t + \Delta t) + K_{kj}\,U_j(t + \Delta t) \qquad \in \Omega_0$$
$$\mathbf{f}^{\text{int}}(\dot{\mathbf{U}}, \mathbf{U}, t + \Delta t) = \mathbb{D}\,\dot{\mathbf{U}}(t + \Delta t) + \mathbb{K}\,\mathbf{U}(t + \Delta t) \qquad \in \Omega_0 \qquad (3.3)$$

The main philosophy for the solution of this classical problem in structural dynamics is based on the "concept of separation of variables" that assumes that spatial and temporal problems are independent from one another. Therefore, a different strategy is established for each solution to the problem . Thus, the finite element method is adopted for the solution of the spatial problem, while the finite differences is used for temporal problems. In other words, the semi-discrete equation (3.1) is solved at each time step t, which represents the spatial equilibrium in such time t (see figure 3.1).

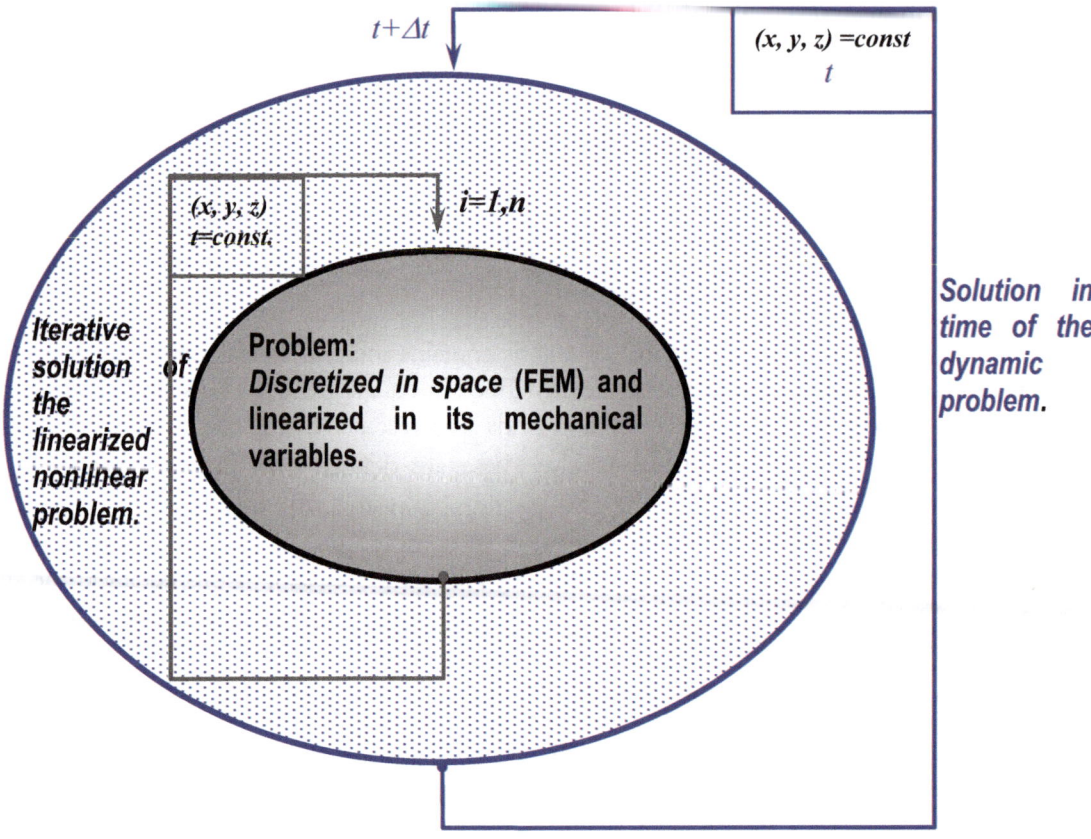

Figure 3.1 – Schematic representation of the nonlinear resolution of a dynamic problem.

3.2 Explicit-implicit solution

The time solution of the differential equation (3.3) can be obtained either through an implicit or explicit strategy . If the response at current time $(t + \Delta t)$ depends completely on the solution of the previous step (t), then we have an explicit solution, but if the current solution depends on the velocity and the acceleration at current time $(t + \Delta t)$, then we have an implicit solution. As an example, a simpler equation is adopted, such as the first order differential equation for the heat conduction phenomenon, $\mathbb{M} \, \dot{\mathbf{U}}(t) + \mathbf{f}^{\mathrm{int}}(\mathbf{U}, t) - \mathbf{f}^{\mathrm{ext}}(t) = 0$. Through this equation the concept of implict-explicit solution can be explained in a simple way as: given the solution in displacements and velocities in time t, the next solution is obtained in time $(t + \Delta t)$ as

$$\begin{cases} \mathbf{U}(t + \Delta t) = \mathbf{U}(t) + \Delta t \, \dot{\mathbf{U}}(t + \alpha \, \Delta t) \\ \dot{\mathbf{U}}(t + \Delta t) = (1 - \alpha) \, \dot{\mathbf{U}}(t) + \alpha \, \dot{\mathbf{U}}(t + \Delta t) \end{cases} \tag{3.4}$$

where α is a parameter to obtain a solution:

- **Explicit:** it states the equilibrium at time t, with $\alpha = 0$. The displacement in the next step is obtained depending on the velocity and displacement of the previous step.
- **Implicit:** it formulates the equilibrium at time $(t+\Delta t)$ with $\alpha = 1$. The displacement in the next step is obtained depending on the current time velocity and on the displacement of the previous step.

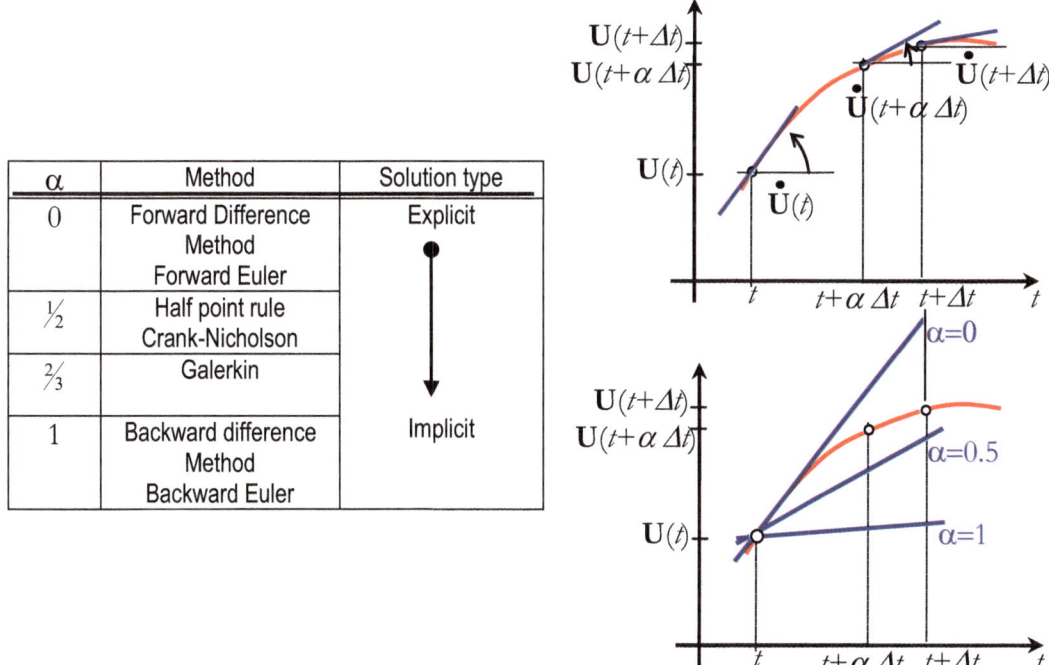

α	Method	Solution type
0	Forward Difference Method Forward Euler	Explicit
½	Half point rule Crank-Nicholson	
⅔	Galerkin	
1	Backward difference Method Backward Euler	Implicit

Figure 3.2 – Simplified representation of an explicit-implicit solution.

The following aspects are obtained by comparing an explicit solution with an implicit one, which must be analyzed depending on the problem,

Explicit time integration:

This requires small time analysis for each time step:
1. The solution algorithm is simple in terms of logic and structure and it allows carrying out a simple treatment of the different nonlinearities.
2. It requires less memory storage.
3. It does not need expensive tangent operators, usually found in implicit methods.
4. The explicit methods lead to reliable algorithms.
5. The solution time increment is bounded and it is usually very small. This makes harder the problems solution at very large time domains.

Implicit time integration:

These are very robust and stable methods:
1. The time increments can be much larger than in explicit methods, preserving the solution stability.
2. They allow more precise solutions, with lower error tolerances.
3. A relative drawback is the linearization of the solution through Newton-Raphson which requires tangent operators that are usually very difficult to obtain.
4. Another drawback is the large storage demand when using direct solution methods for the system of equations.

Because of the robustness of the implicit methods, related aspects will also be addressed. In all cases, we will admit a spatial problem which is approximated by a discrete form by the finite element method. We will only study the time problem solution, leaving the spatial problem for a specific study of time-independent problem.

3.3 Implicit solution

The implicit time integration method assumes that displacements $\mathbf{U}(t + \Delta t) = \mathbf{U}^{t+\Delta t}$ and velocities $\dot{\mathbf{U}}(t + \Delta t) = \dot{\mathbf{U}}^{t+\Delta t}$ at time $t + \Delta t$ are obtained using the following *linear approximation in differences,*

$$\begin{cases} \dot{\mathbf{U}}^{t+\Delta t} = k_V \ddot{\mathbf{U}}^{t+\Delta t} \Delta t + \boldsymbol{f}_V\left(\dot{\mathbf{U}}^t, \ddot{\mathbf{U}}^t, \cdots\right) \\ \mathbf{U}^{t+\Delta t} = k_U \ddot{\mathbf{U}}^{t+\Delta t} \Delta t^2 + \boldsymbol{f}_U\left(\mathbf{U}^t, \dot{\mathbf{U}}^t, \ddot{\mathbf{U}}^t, \cdots\right) \end{cases} \tag{3.5}$$

where Δt is the time step and k_U and k_V are coefficients to be determined when defining the particular solution method.

3.3.1   Equilibrium at time $(t + \Delta t)$

It is assumed that coefficients k_U and k_V are not null, accelerations $\ddot{\mathbf{U}}^{t+\Delta t}$ and velocities $\dot{\mathbf{U}}^{t+\Delta t}$ are functions of the displacement $\mathbf{U}^{t+\Delta t}$ at time $(t + \Delta t)$ and the residual force in such time $^i\left[\Delta \mathbf{f}\right]_\Omega^{t+\Delta t}$, linearized for each iteration "i", is defined over the domain Ω (equation 2.77), by the expression

$$0 = {}^{i+1}\left[\Delta\mathbf{f}\right]_\Omega^{t+\Delta t} \cong {}^{i}\left[\Delta\mathbf{f}\right]_\Omega^{t+\Delta t} + {}^{i}\mathbb{J}_\Omega^{t+\Delta t} \cdot {}^{i+1}\left[\Delta\mathbf{U}\right]_\Omega^{t+\Delta t}$$

$$0 = {}^{i+1}\Delta\mathbf{f}^{t+\Delta t} \cong {}^{i}\Delta\mathbf{f}^{t+\Delta t} + {}^{i}\mathbb{J}^{t+\Delta t} \cdot \left[{}^{i+1}\mathbf{U}^{t+\Delta t} - {}^{i}\mathbf{U}^{t+\Delta t}\right]$$

(3.6)

In this chapter, we will disregard the subscript Ω when referring the global equilibrium in order to simplify the notation. This global equilibrium in nonlinear behaviors can be obtained by using the Newton-Raphson iterative procedure to obtain the approximation of the solution in the surrounding of point "(i+1)", by means of the linearization described in equation (3.6). In this linearized solution, the tangent Jacobian operator has the following form,

$$^{i}\mathbb{J}^{t+\Delta t} = \mathbb{J}\left(^{i}\mathbf{U}^{t+\Delta t}\right) = {}^{i}\left[\frac{\partial\Delta\mathbf{f}}{\partial\mathbf{U}}\right]^{t+\Delta t} = {}^{i}\left[\mathbb{M}\frac{\partial\ddot{\mathbf{U}}}{\partial\mathbf{U}} + \frac{\partial\mathbf{f}^{\text{int}}}{\partial\mathbf{U}} + \frac{\partial\mathbf{f}^{\text{int}}}{\partial\dot{\mathbf{U}}}\frac{\partial\dot{\mathbf{U}}}{\partial\mathbf{U}} - \frac{\partial\mathbf{f}^{\text{ext}}}{\partial\mathbf{U}}\right]^{t+\Delta t}$$

(3.7)

where the tangent stiffness operator is $\mathbb{K}^T = \dfrac{\partial\mathbf{f}^{\text{int}}}{\partial\mathbf{U}}$, the tangent damping operator is $\mathbb{D}^T = \dfrac{\partial\mathbf{f}^{\text{int}}}{\partial\dot{\mathbf{U}}}$ and the derivatives $\dfrac{\partial\mathbf{f}^{\text{ext}}}{\partial\mathbf{U}}$ is the effect of the change of position of the external forces due to the successive configuration changes. This last term can be neglected under small displacements, as in this case changes of position of the loads are quasi negligible as compared to the model size. By substituting these tangent operators into equation (3.7), the Jacobian matrix becomes,

$$^{i}\mathbb{J}^{t+\Delta t} = \mathbb{J}\left(^{i}\mathbf{U}^{t+\Delta t}\right) = {}^{i}\left[\frac{\partial\Delta\mathbf{f}}{\partial\mathbf{U}}\right]^{t+\Delta t} = {}^{i}\left[\mathbb{M}\frac{\partial\ddot{\mathbf{U}}}{\partial\mathbf{U}} + \mathbb{K}^T + \mathbb{D}^T\frac{\partial\dot{\mathbf{U}}}{\partial\mathbf{U}} - \frac{\partial\mathbf{f}^{\text{ext}}}{\partial\mathbf{U}}\right]^{t+\Delta t}$$

(3.8)

In linear-elastic dynamic problems this operator is further simplified, becoming the following constant operator,

$$^{i}\mathbb{J}^{t+\Delta t} \equiv \mathbb{J}^0 = \mathbb{M}\frac{\partial\ddot{\mathbf{U}}}{\partial\mathbf{U}} + \mathbb{K}^0 + \mathbb{D}^0\frac{\partial\dot{\mathbf{U}}}{\partial\mathbf{U}}$$

(3.9)

And for the quasi-static problems, where $\ddot{\mathbf{U}} = \dot{\mathbf{U}} = \mathbf{0}$ and $\mathbf{f}^{\text{ext}} = cons$, the Jacobian operator tends to be the classical stiffness matrix.

$$^{i}\mathbb{J}^{t+\Delta t} = \mathbb{J}\left(^{i}\mathbf{U}^{t+\Delta t}\right) = {}^{i}\left[\frac{\partial\Delta\mathbf{f}}{\partial\mathbf{U}}\right]^{t+\Delta t} = {}^{i}\left[\mathbb{K}^T\right]^{t+\Delta t}$$

(3.10)

3.3.2 Equilibrium solution in time - Implicit methods

Once the equilibrium equation and its linearization are introduced, the fundamental problem to solve is the relation among the fields of acceleration, velocity and displacement ($\partial\ddot{\mathbf{U}}/\partial\mathbf{U}$; $\partial\dot{\mathbf{U}}/\partial\mathbf{U}$). This will lead to the Jacobian operator expressed in equations (3.7), (3.8) and (3.9). To reach this goal there are several ways to formulate an implicit integrator in time. Among the most common procedures are the *one step integration* method, which belongs to the Newmark´s family and its main characteristic is that it is unconditionally stable to solve linear dynamic problems. However, the solution stability in nonlinear problems cannot be guaranteed, but this subject will be discussed further.

Also there exist *multi-steps methods* such as the family of Houbolt methods, which offers more accuracy but demand more information stored and consequentely are more expensive in terms of computational calculus.

In order to show the reader both the one and two-steps methods, we will discuss the time integration procedures of Newmark and Houbolt.

3.3.2.1   The Newmark method

This is a one-step integration method, which is unconditionally stable in linear elastic dynamic problems. It is one of the most widely used methods because of the relation among computational cost, accuracy and simplicity in its numerical implementation.

A one-step formulation is characterized because the displacements and velocities at time $(t + \Delta t)$ are partially obtained from a system already known in the previous step (t). This is (equation (3.5)),

$$\begin{cases} \mathbf{U}^{t+\Delta t} = \boldsymbol{f}_1\left(\ddot{\mathbf{U}}^{t+\Delta t}, \mathbf{U}^t, \dot{\mathbf{U}}^t, \ddot{\mathbf{U}}^t\right) \\ \dot{\mathbf{U}}^{t+\Delta t} = \boldsymbol{f}_2\left(\ddot{\mathbf{U}}^{t+\Delta t}, \mathbf{U}^t, \dot{\mathbf{U}}^t, \ddot{\mathbf{U}}^t\right) \end{cases} \tag{3.11}$$

The implicit feature of this method comes from the dependence of the acceleration at current time.

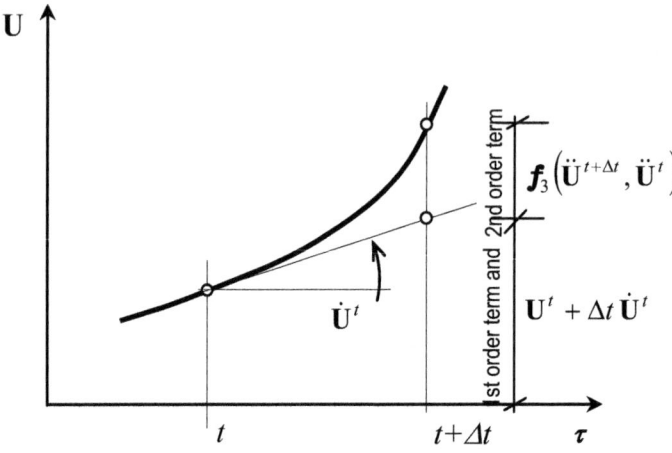

Figure 3.3 – Displacement approximation in time.

The main difference with the classical method of centered differences[1] is that the velocity and the displacement do not come from the instantaneous acceleration (a derivative), but from the integration of the acceleration which makes the solution more stable. Then, the velocity and the displacement are defined using the following procedure,

- **Velocity calculation:**

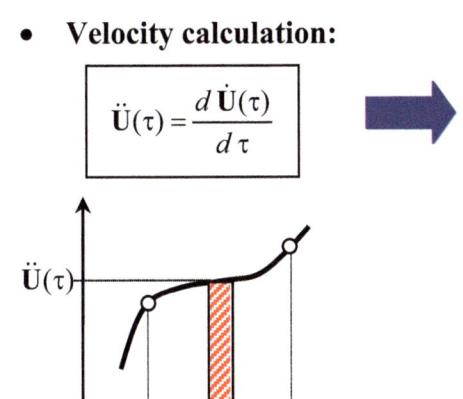

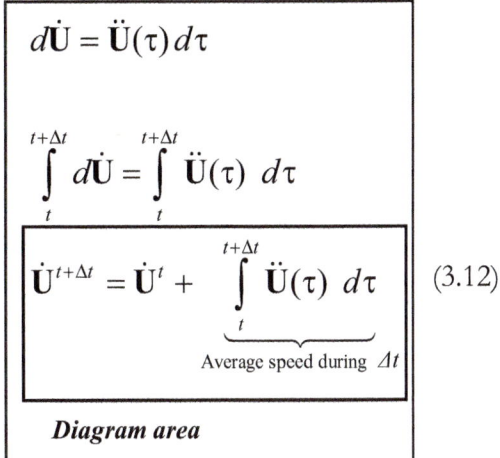

$$dU̇ = Ü(\tau)\,d\tau$$

$$\int_{t}^{t+\Delta t} dU̇ = \int_{t}^{t+\Delta t} Ü(\tau)\ d\tau$$

$$U̇^{t+\Delta t} = U̇^{t} + \underbrace{\int_{t}^{t+\Delta t} Ü(\tau)\ d\tau}_{\text{Average speed during } \Delta t} \qquad (3.12)$$

Diagram area

- **Displacement calculation:**

$$dU = U̇(\tau)\,d\tau$$

$$\int_{t}^{t+\Delta t} dU = \int_{t}^{t+\Delta t} U̇(\tau)\ d\tau$$

$$U^{t+\Delta t} = U^{t} + \int_{t}^{t+\Delta t} U̇(\tau)\ d\tau \qquad (3.13)$$

First order moment.
Figure (3.3)

Substituting the velocity $U̇^{t+\Delta t} = U̇^{t} + \int_{t}^{t+\Delta t} Ü(\tau)\ d\tau$, previously obtained, in this last expression, we get the displacement as a function of the acceleration,

$$U^{t+\Delta t} = U^{t} + \int_{t}^{t+\Delta t} \left[U̇(\tau) + \int_{0}^{\tau} Ü(\tau)\ d\tau \right] d\tau$$

$$U^{t+\Delta t} = U^{t} + \int_{t}^{t+\Delta t} U̇(\tau)\ d\tau + \int_{t}^{t+\Delta t} \left[\int_{0}^{\tau} Ü(\tau)\ d\tau \right] d\tau \qquad (3.14)$$

Transforming the first-order moment in another integral form, $\int_{0}^{\Delta t} \left[\int_{0}^{\tau} Ü(\tau)\ d\tau \right] d\tau = \int_{0}^{\Delta t} \left[(t+\Delta t) - \tau \right] Ü(\tau)\ d\tau$, the following expression for the displacement is obtained,

$$U^{t+\Delta t} = U^{t} + U̇^{t} \Delta t + \int_{t}^{t+\Delta t} \left[\int_{0}^{\tau} Ü(\tau)\ d\tau \right] d\tau = U^{t} + U̇^{t} \Delta t + \underbrace{\int_{t}^{t+\Delta t} \left[(t+\Delta t) - \tau \right] Ü(\tau) d\tau}_{\text{Average displacement during } \Delta t} \qquad (3.15)$$

To solve this problem the expression for the acceleration must be assumed. Newmark adopts a linear variation of acceleration in time of the following type

$$\ddot{U}(\tau) = \ddot{U}^t + f(\tau)\left(\ddot{U}^{t+\Delta t} - \ddot{U}^t\right) \tag{3.16}$$

where $f(\tau)$ is a shape function which has the following definition $f(\tau = t) = 0$ and $f(\tau = t + \Delta t) = 1$ (see Figure 3.4).

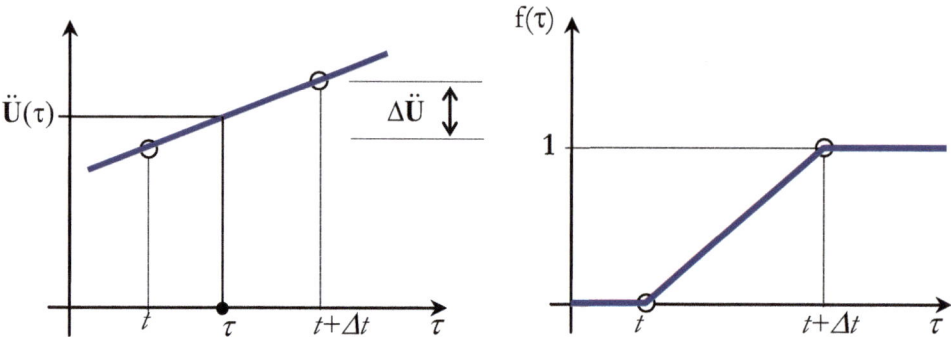

Figure 3.4 – Time approximation of displacement.

Substituting this linear variation into the velocity equation (3.12) and into the displacement equation (3.15), we will have the following system of equations,

$$\begin{cases} \dot{U}^{t+\Delta t} = \dot{U}^t + \int\limits_t^{t+\Delta t} \ddot{U}^t \, d\tau + \int\limits_t^{t+\Delta t} f(\tau)\left(\ddot{U}^{t+\Delta t} - \ddot{U}^t\right) d\tau \\[2mm] U^{t+\Delta t} = U^t + \dot{U}^t \Delta t + \int\limits_t^{t+\Delta t} \left[(t+\Delta t) - \tau\right] \cdot \left[\ddot{U}^t + f(\tau)\left(\ddot{U}^{t+\Delta t} - \ddot{U}^t\right)\right] d\tau \end{cases} \tag{3.17}$$

After some calculations this can be written as,

$$\begin{cases} \dot{U}^{t+\Delta t} = \dot{U}^t + \ddot{U}^t \Delta t + \left(\ddot{U}^{t+\Delta t} - \ddot{U}^t\right) \cdot \int\limits_0^{\Delta t} f(\tau) \, d\tau \\[2mm] U^{t+\Delta t} = U^t + \dot{U}^t \Delta t + \ddot{U}^t \dfrac{\Delta t^2}{2} + \int\limits_0^{\Delta t}\left[\left(\ddot{U}^{t+\Delta t} - \ddot{U}^t\right) \cdot \underbrace{\int\limits_0^{\tau} f(\tau)\, d\tau}_{g(\tau)} \right] d\tau \end{cases} \tag{3.18}$$

The result of the integral $g = \int\limits_t^{t+\Delta t} f(\tau)\, d\tau = k_U \Delta t = \gamma \Delta t$ must be interpreted as the area of the linear approximation function of the accelerations (Figure 3.4). Note that in this particular case, the coefficient is $\gamma = 0.5$ (shadow area in **Figure 3.5**).

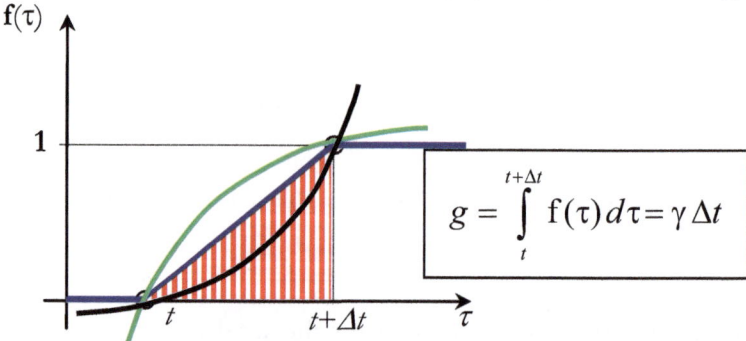

Figure 3.5 – Area of the approximation function of the accelerations.

Keeping in mind that,

$$\begin{cases} \displaystyle\int_{t}^{t+\Delta t} f(\tau)\,d\tau = k_{\mathrm{U}}\Delta t = \gamma\,\Delta t \\[4mm] \displaystyle\int_{t}^{t+\Delta t}\left[\int_{0}^{\tau} f(\tau)\,d\tau\right]d\tau = k_{\mathrm{V}}\Delta t^2 = \beta\,\Delta t^2 \end{cases} \tag{3.19}$$

and substituting these results into equation (3.18), the following equations for velocity and displacement result,

$$\begin{cases} \dot{\mathbf{U}}^{t+\Delta t} = \dot{\mathbf{U}}^{t} + \ddot{\mathbf{U}}^{t}\Delta t + \left(\ddot{\mathbf{U}}^{t+\Delta t} - \ddot{\mathbf{U}}^{t}\right)\gamma\,\Delta t \\[3mm] \mathbf{U}^{t+\Delta t} = \mathbf{U}^{t} + \dot{\mathbf{U}}^{t}\Delta t + \ddot{\mathbf{U}}^{t}\dfrac{\Delta t^2}{2} + \left(\ddot{\mathbf{U}}^{t+\Delta t} - \ddot{\mathbf{U}}^{t}\right)\beta\,\Delta t^2 \end{cases} \tag{3.20}$$

By reordering the terms of these equations, then:

$$\text{Predictor:}\quad \begin{cases} \dot{\mathbf{U}}^{t+\Delta t} = \dot{\mathbf{U}}^{t} + (1-\gamma)\ddot{\mathbf{U}}^{t}\,\Delta t + \gamma\,\ddot{\mathbf{U}}^{t+\Delta t}\,\Delta t \\[3mm] \mathbf{U}^{t+\Delta t} = \mathbf{U}^{t} + \dot{\mathbf{U}}^{t}\,\Delta t + \underbrace{\left(\dfrac{1}{2}-\beta\right)\ddot{\mathbf{U}}^{t}\,\Delta t^2 + \beta\,\ddot{\mathbf{U}}^{t+\Delta t}\,\Delta t^2}_{\Delta\mathbf{U}^{t+\Delta t}} \end{cases} \tag{3.21}$$

From the last two equations, Newmark's fundamental expressions (3.11) to describe the displacement, velocity and acceleration at the current time of the integration process are then obtained as,

$$\text{Corrector:}\quad \begin{cases} \ddot{\mathbf{U}}^{t+\Delta t} = \left(\dfrac{1}{\beta\,\Delta t^2}\right)\left(\mathbf{U}^{t+\Delta t} - \mathbf{U}^{t} - \dot{\mathbf{U}}^{t}\,\Delta t\right) - \left(\dfrac{1}{2\beta}-1\right)\ddot{\mathbf{U}}^{t} \\[3mm] \dot{\mathbf{U}}^{t+\Delta t} = \left(\dfrac{\gamma}{\beta\,\Delta t}\right)\Delta^{i}\mathbf{U}^{t} + \left(1-\dfrac{\gamma}{\beta}\right)\dot{\mathbf{U}}^{t} + (1-\dfrac{\gamma}{2\beta})\ddot{\mathbf{U}}^{t}\,\Delta t \\[3mm] \mathbf{U}^{t+\Delta t} = \mathbf{U}^{t} + \Delta^{i}\mathbf{U}^{t+\Delta t} \end{cases} \tag{3.22}$$

Coefficients $k_{\mathrm{U}} = \gamma$ and $k_{\mathrm{V}} = \beta$ determine the method's stability, as it will be shown later. Substituting these expressions into the linearized dynamic equilibrium equation (3.6) and into the Jacobian (3.8) at time $(t + \Delta t)$ iteration (i), then:

$$\mathbf{0} = {}^{i+1}\Delta\mathbf{f}^{t+\Delta t} \cong {}^{i}\Delta\mathbf{f}^{t+\Delta t} + {}^{i}\mathbf{J}^{t+\Delta t}\cdot\left[{}^{i+1}\mathbf{U}^{t+\Delta t} - {}^{i}\mathbf{U}^{t+\Delta t}\right]$$

with:

$$ {}^{i}\mathbb{J}^{t+\Delta t} = {}^{i}\left[\mathbb{M}\frac{\partial\ddot{\mathbf{U}}}{\partial\mathbf{U}} + \mathbb{K}^{T} + \mathbb{D}^{T}\frac{\partial\dot{\mathbf{U}}}{\partial\mathbf{U}} - \frac{\partial\mathbf{f}^{\mathrm{ext}}}{\partial\mathbf{U}}\right]^{t+\Delta t} = {}^{i}\left[\mathbb{M}\frac{1}{\beta\Delta t^2} + \mathbb{K}^{T} + \mathbb{D}^{T}\frac{\gamma}{\beta\Delta t} - \frac{\partial\mathbf{f}^{\mathrm{ext}}}{\partial\mathbf{U}}\right] \tag{3.23}$$

And substituted into equilibrium equation (3.1), the expression for the residual forces at time $(t + \Delta t)$ iteration (i), yields as

$$
\begin{aligned}
{}^{i}\Delta\mathbf{f}^{t+\Delta t} &= \mathbb{M}\ {}^{i}\ddot{\mathbf{U}}^{t+\Delta t} + {}^{i}\!\left[\mathbf{f}^{\text{int}}\right]^{t+\Delta t} - {}^{i+1}\!\left[\mathbf{f}^{\text{ext}}\right]^{t+\Delta t} = \\
&= \mathbb{M}\!\left[\left(\frac{1}{\beta\,\Delta t^{2}}\right)\!\left(\Delta^{i}\mathbf{U}^{t+\Delta t} - {}^{i}\dot{\mathbf{U}}^{t}\,\Delta t\right) - \left(\frac{1}{2\beta}-1\right){}^{i}\ddot{\mathbf{U}}^{t}\right] + \\
&\quad + \int_{V} {}^{i}\boldsymbol{\sigma}^{t+\Delta t}\!\left(\nabla^{S}\mathbf{N}\right)dV - {}^{i+1}\!\left[\mathbf{f}^{\text{ext}}\right]^{t+\Delta t}
\end{aligned}
\tag{3.24}
$$

It can be verified that the solution in time obtained by this method is unconditionally stable for values of $\gamma \geq 1/2$ and $\beta \geq 1/4\,(0.5+\gamma)^{2}$ (see chapter 4).

The integration scheme in time is carried out following the steps described below, after imposing the condition of zero residual force ($^{i+1}\Delta\mathbf{f}^{t+\Delta t}=\mathbf{0}$, equation (3.23)),

---

1.   Setting velocities and displacements, equation (3.21), from the initial condition of null acceleration at $(t+\Delta t)$,

$$\overset{\approx}{\mathbf{U}}{}^{t+\Delta t} = \mathbf{0}$$

$$\overset{\sim}{\dot{\mathbf{U}}}{}^{t+\Delta t} = \dot{\mathbf{U}}^{t} + (1-\gamma)\,\ddot{\mathbf{U}}^{t}\,\Delta t$$

$$\widetilde{\mathbf{U}}{}^{t+\Delta t} = \mathbf{U}^{t} + \dot{\mathbf{U}}^{t}\,\Delta t + \left(\frac{1}{2}-\beta\right)\ddot{\mathbf{U}}^{t}\,\Delta t^{2}$$

2.   Obtaining the correction $\Delta\mathbf{U}^{t+\Delta t}$ of the displacements from equation (3.23) of linearized equilibrium at time $(t+\Delta t)$

$${}^{i}\Delta\mathbf{f}^{t+\Delta t} = -{}^{i}\mathbb{J}^{t+\Delta t}\cdot\Delta^{i+1}\mathbf{U}^{t+\Delta t}$$

3.   Obtaining the corrected displacements, velocities and accelerations (3.22),

$${}^{i+1}\ddot{\mathbf{U}}^{t+\Delta t} = \left(\frac{1}{\beta\,\Delta t^{2}}\right)\Delta^{i+1}\mathbf{U}^{t+\Delta t}$$

$${}^{i+1}\dot{\mathbf{U}}^{t+\Delta t} = \overset{\sim}{\dot{\mathbf{U}}}{}^{t+\Delta t} + \left(\frac{\gamma}{\beta\,\Delta t}\right)\Delta^{i+1}\mathbf{U}^{t+\Delta t}$$

$${}^{i+1}\mathbf{U}^{t+\Delta t} = \widetilde{\mathbf{U}}{}^{t+\Delta t} + \Delta^{i+1}\mathbf{U}^{t+\Delta t}$$

4.   Calculation of the deformation field, resolution of the constitutive equation and convergence verification by calculating the residual force,

$${}^{i+1}\boldsymbol{\varepsilon}^{t+\Delta t} = \nabla^{S\,i+1}\mathbf{U}^{t+\Delta t} \quad;\quad {}^{i+1}\dot{\boldsymbol{\varepsilon}}^{t+\Delta t} = \frac{\left(\nabla^{S\,i+1}\mathbf{U}^{t+\Delta t}\right) - \left(\nabla^{S\,i+1}\mathbf{U}^{t}\right)}{\Delta t}$$

$${}^{i+1}\boldsymbol{\sigma}^{t+\Delta t} = {}^{i+1}\!\left[\mathbb{C}^{S}:\boldsymbol{\varepsilon}^{e} + \xi:\dot{\boldsymbol{\varepsilon}}\right]^{t+\Delta t}$$

$$
\begin{aligned}
{}^{i+1}\Delta\mathbf{f}^{t+\Delta t} &= \mathbb{M}\!\left[\left(\frac{1}{\beta\,\Delta t^{2}}\right)\!\left(\Delta^{i+1}\mathbf{U}^{t+\Delta t} - {}^{i+1}\dot{\mathbf{U}}^{t}\,\Delta t\right) - \left(\frac{1}{2\beta}-1\right){}^{i+1}\ddot{\mathbf{U}}^{t}\right] + \\
&\quad + \int_{V} {}^{i+1}\boldsymbol{\sigma}^{t+\Delta t}\!\left(\nabla^{S}\mathbf{N}\right)dV - {}^{i+1}\!\left[\mathbf{f}^{\text{ext}}\right]^{t+\Delta t}
\end{aligned}
$$

5.   If $\left|{}^{i+1}\Delta\mathbf{f}^{t+\Delta t}\right| > \text{TOL}$, then "GO TO 2"; or else "GO TO 1" and increase time $(t+\Delta t)$.

Table 3.1 Newmark's time integration procedure

3.3.2.2 The Houbolt method

It is a time integration method in two-steps, or four points, which means that to solve a problem in time $(t + \Delta t)$ information about the state of the variables at times (t), $(t - \Delta t)$ and $(t - 2\Delta t)$ are required. The main difference with the one-step method is the best accuracy, but it introduces more complexity in the calculation and also a higher demand for memory to store the variables in previous steps.

Acceleration is calculated from the following approximation,

$$\sum_{k=1}^{m} \left(\gamma_k \mathbf{U}^{(t+\Delta t)-k\cdot\Delta t} - \Delta t\, \delta_k\, \ddot{\mathbf{U}}^{(t+\Delta t)-k\cdot\Delta t} \right) = \mathbf{0} \tag{3.25}$$

where m represents the number of discrete points to be considered in time domain, γ_k and δ_k are coefficients to be determined. Particularly in the Houbolt case, $m = 4$, $\gamma_0 = 2$, $\gamma_1 = -5$, $\gamma_2 = 4$, $\gamma_3 = -1$, $\delta_0 = 1$, $\delta_k = 0$ are adopted for $(k = 1, 2, 3)$.

The expression for the residual forces takes now the following form,

$$^{i+1}\Delta\mathbf{f}^* = \frac{1}{\delta_0} \sum_{k=1}^{m} \delta_k \left[^{i+1}\Delta\mathbf{f}^{t+\Delta t} \right]_k =$$

$$= \frac{1}{\delta_0} \sum_{k=1}^{m} \delta_k \left[\mathbb{M}\, ^{i+1}\ddot{\mathbf{U}}^{(t+\Delta t)-k\cdot\Delta t} + \, ^{i+1}\left[\mathbf{f}^{\text{int}} \right]^{(t+\Delta t)-k\cdot\Delta t} - \left[\mathbf{f}^{\text{ext}} \right]^{(t+\Delta t)-k\cdot\Delta t} \right] = \mathbf{0} \tag{3.26}$$

According to the acceleration equation (3.25), the residual forces can be written as a function of the displacements only and, implicitly, as a function of the velocities. This is,

$$^{i+1}\Delta\mathbf{f}^* = \frac{1}{\delta_0} \sum_{k=1}^{m} \left[\frac{\gamma_k}{\Delta t^2} \mathbb{M}\, ^{i+1}\mathbf{U}^{(t+\Delta t)-k\cdot\Delta t} + \delta_k \left[^{i+1}\left[\mathbf{f}^{\text{int}} \right]^{(t+\Delta t)-k\cdot\Delta t} - \left[\mathbf{f}^{\text{ext}} \right]^{(t+\Delta t)-k\cdot\Delta t} \right] \right] = \mathbf{0} \tag{3.27}$$

Then the following expression for the Jacobian operator is,

$$^i\mathbb{J}^{t+\Delta t} = \mathbb{J}\left(^i\mathbf{U}^{t+\Delta t} \right) = \left[\frac{\partial \Delta\mathbf{f}^*}{\partial \mathbf{U}_{n+1}} \right]^i = \left[\mathbb{M}\frac{\gamma_0}{\delta_0\Delta t^2} + \mathbb{K}^T + \mathbb{D}^T\frac{\partial \dot{\mathbf{U}}}{\partial \mathbf{U}} - \frac{\partial \mathbf{f}^{\text{ext}}}{\partial \mathbf{U}} \right]^{t+\Delta t} \tag{3.28}$$

To solve the relation between velocity and displacement the following approximation is established, similarly to when the acceleration-displacement relation was introduced into equation (3.16)

$$\dot{\mathbf{U}}^{t+\Delta t} = \frac{\alpha_0}{\beta_0\Delta t}\mathbf{U}^{t+\Delta t} + \frac{1}{\beta_0\Delta t}\sum_{k=1}^{m}\left(\alpha_k\mathbf{U}^{(t+\Delta t)-k\Delta t} - \Delta t\,\beta_k\,\dot{\mathbf{U}}^{(t+\Delta t)-k\Delta t} \right) = \mathbf{0} \Rightarrow$$

$$\Rightarrow \left[\frac{\partial \dot{\mathbf{U}}}{\partial \mathbf{U}} \right]^{t+\Delta t} = \frac{\alpha_0}{\beta_0\Delta t} \tag{3.29}$$

Then the Jacobian operator is written as,

$$^i\mathbb{J}^{t+\Delta t} = \mathbb{J}\left(^i\mathbf{U}^{t+\Delta t} \right) = \left[\frac{\partial \Delta\mathbf{f}^*}{\partial \mathbf{U}_{n+1}} \right]^i = \left[\mathbb{M}\frac{\gamma_0}{\delta_0\Delta t^2} + \mathbb{K}^T + \mathbb{D}^T\frac{\alpha_0}{\beta_0\Delta t} - \frac{\partial \mathbf{f}^{\text{ext}}}{\partial \mathbf{U}} \right]^{t+\Delta t} \tag{3.30}$$

Thus, the equilibrium linearized equation is written as,

$$^{i+1}\Delta \mathbf{f}^* = -\,^i \mathbb{J}^{t+\Delta t} \cdot \Delta\,^{i+1}\mathbf{U}^{t+\Delta t} \tag{3.31}$$

The time integration scheme is carried out with the following steps,

after imposing the zero residual force ($^{i+1}\Delta \mathbf{f}^{t+\Delta t} = \mathbf{0}$), into equation (3.26),

1. Setting of velocities, equation (3.29), from t he initial condition of null displacement at $t + \Delta t$,

$$\widetilde{\mathbf{U}}^{t+\Delta t} = \mathbf{0}$$

$$\dot{\widetilde{\mathbf{U}}}^{t+\Delta t} = \frac{1}{\beta_0 \Delta t} \sum_{k=1}^{m} \left(\alpha_k \mathbf{U}^{(t+\Delta t)-k\cdot\Delta t} - \Delta t\,\beta_k\,\dot{\mathbf{U}}^{(t+\Delta t)-k\cdot\Delta t} \right)$$

2. Calculation of residual forces $^i\Delta \mathbf{f}^*$ according to equation (3.27)

3. Obtaining the correction factor $\Delta \mathbf{U}^{t+\Delta t}$ of the displacements from equilibrium linearized equation at time $t + \Delta t$

$$^i\Delta \mathbf{f}^* = -\,^i \mathbb{J}^{t+\Delta t} \cdot \Delta\,^{i+1}\mathbf{U}^{t+\Delta t}$$

4. Obtaining the corrected displacements and velocities,

$$^{i+1}\dot{\mathbf{U}}^{t+\Delta t} = \dot{\widetilde{\mathbf{U}}}^{t+\Delta t} + \left(\frac{\alpha_0}{\beta_0\,\Delta t} \right) \Delta\,^{i+1}\mathbf{U}^{t+\Delta t}$$

$$^{i+1}\mathbf{U}^{t+\Delta t} = \widetilde{\mathbf{U}}^{t+\Delta t} + \Delta\,^{i+1}\mathbf{U}^{t+\Delta t}$$

5. Obtaining the accelerations field from equation (3.25),

6. Calculation of the deformations field, solution of the constitutive equation and convergence verification for loads

$$^{i+1}\boldsymbol{\varepsilon}^{t+\Delta t} = \nabla^{S\,i+1}\mathbf{U}^{t+\Delta t} \quad ; \quad ^{i+1}\dot{\boldsymbol{\varepsilon}}^{t+\Delta t} = \frac{\left(\nabla^{S\,i+1}\mathbf{U}^{t+\Delta t} \right) - \left(\nabla^{S\,i+1}\mathbf{U}^t \right)}{\Delta t}$$

$$^{i+1}\boldsymbol{\sigma}^{t+\Delta t} = \,^{i+1}\left[\mathbf{C}^S : \boldsymbol{\varepsilon}^e + \boldsymbol{\xi} : \dot{\boldsymbol{\varepsilon}} \right]^{t+\Delta t}$$

$$^{i+1}\Delta \mathbf{f}^* = \frac{1}{\delta_0} \sum_{k=1}^{m} \left[\frac{\gamma_k}{\Delta t^2} \mathbf{M}\,^{i+1}\mathbf{U}^{(t+\Delta t)-k} + \delta_k \left[\,^{i+1}\left[\int_V \,^{i+1}\boldsymbol{\sigma}^{t+\Delta t} \left(\nabla^S \mathbf{N} \right) dV \right]^{(t+\Delta t)-k} - \,^{i+1}\left[\mathbf{f}^{\text{ext}} \right]^{(t+\Delta t)-k} \right] \right]$$

7. If $\left| ^{i+1}\Delta \mathbf{f}^* \right| > \text{TOL}$, then "GOTO 3"; else "GOTO 1" and increase time ($t + \Delta t$).

Table 3.2 Houbolt's time integration procedure.

3.3.3 Solution of the nonlinear-equilibrium equations system

A brief summary of some solution techniques of the linearized system of equations is presented in this section, which comes from either the Newmark or Houbolt integrations.

The reader is advised to check references[2,3] for a deeper understanding of the solution methods for systems of equations.

The solution of the following system of nonlinear equilibrium equations,

$$^i\Delta \mathbf{f}^{t+\Delta t} = -\,^i \mathbb{J}^{t+\Delta t} \cdot \Delta\,^{i+1}\mathbf{U}^{t+\Delta t} \tag{3.32}$$

can be carried out basically by using two different techniques[1] :

1. A linearization based on either the standard or modified *Newton-Raphson* techniques[1].
2. *Quasi-Newton* approximation techniques for the Jacobian matrix[2].

Both techniques are mainly different in the way the residual is expanded $^{i}\Delta\mathbf{f}^{t+\Delta t} = -\,^{i}\mathbb{J}^{t+\Delta t}\cdot\Delta^{i+1}\mathbf{U}^{t+\Delta t}$ in the surrounding of the previous solution and in the way to calculate the Jacobian operator.

3.3.3.1 The Newton-Raphson method

This is the fastest convergence method to solve nonlinear equations systems using the linearization technique. This technique assumes that the solution is within the *"attractive zone"*, in other words, the solution is convergent in the surrounding of it. In this case, the convergence ratio is quadratic.

In this iterative method, it is assumed that the equilibrium equation has the general form,

$$\boxed{\Delta\mathbf{f} = \mathbb{M}\,\ddot{\mathbf{U}}(t+\Delta t) + \mathbf{f}^{\text{int}}(\dot{\mathbf{U}},\mathbf{U},t+\Delta t) - \mathbf{f}^{\text{ext}}(t+\Delta t) = \mathbf{0}}$$
(3.33)

It can also be written, as stated in equation (B2.77), through a Taylor approximation series truncated at second term,

$$0 = \,^{i+1}\!\left[\Delta\mathbf{f}\right]_{\Omega}^{t+\Delta t} \cong \,^{i}\!\left[\Delta\mathbf{f}\right]_{\Omega}^{t+\Delta t} + \,^{i}\!\left(\frac{\partial\Delta\mathbf{f}}{\partial\mathbf{U}}\right)_{\Omega}^{t+\Delta t}\cdot\,^{i+1}\!\left[\Delta\mathbf{U}\right]_{\Omega}^{t+\Delta t} =$$

$$= \,^{i}\!\left[\Delta\mathbf{f}\right]_{\Omega}^{t+\Delta t} + \underbrace{\,^{i}\!\left[\mathbb{M}\frac{\partial\ddot{\mathbf{U}}}{\partial\mathbf{U}} + \mathbb{K}^{T} + \mathbb{D}^{T}\frac{\partial\dot{\mathbf{U}}}{\partial\mathbf{U}} - \frac{\partial\mathbf{f}^{\text{ext}}}{\partial\mathbf{U}}\right]_{\Omega}^{t+\Delta t}}_{^{i}\mathbb{J}_{\Omega}^{t+\Delta t}}\cdot\,^{i+1}\!\left[\Delta\mathbf{U}\right]_{\Omega}^{t+\Delta t}$$
(3.34)

In this equation (i) represents the iteration counter and (t) is the current time. Its solution comes from the inversion of the Jacobian operator $^{i}\mathbb{J}_{\Omega}^{t+\Delta t}$. This is,

$$\boxed{\,^{i+1}\!\left[\Delta\mathbf{U}\right]_{\Omega}^{t+\Delta t} = -\left[\,^{i}\mathbb{J}_{\Omega}^{t+\Delta t}\right]^{-1}\cdot\,^{i}\!\left[\Delta\mathbf{f}\right]_{\Omega}^{t+\Delta t}}$$
(3.35)

Thus, the displacement at the end of the linearized process (or when converged), will be (see Figure 3.6),

$$^{i+1}\!\left[\mathbf{U}\right]_{\Omega}^{t+\Delta t} = \,^{i}\!\left[\mathbf{U}\right]_{\Omega}^{t+\Delta t} + \,^{i+1}\!\left[\Delta\mathbf{U}\right]_{\Omega}^{t+\Delta t}$$
(3.36)

Despite the fast convergence of this method, some drawbacks can be observed such as,
1. It always needs a tangent Jacobian operator, which is not easy to obtain in all cases,
2. The convergence velocity is very low when the solution is far away,
3. In some cases, the problem needs the solution of equation (3.35) for anti-symmetric Jacobian operators. This turns difficult its inversion,
4. The method usually finds local minimum which are very hard to leave afterwards. Then, other auxiliary techniques are required, such as displacement-control methods (arc-length), which will be discussed later.

---

[1] Zienkiewicz, O. C. and Taylor, R. L. (2000)– *The Finite element method* – Vol. 1 y 2 –Butterworth-Heinemannl-CIMNE.

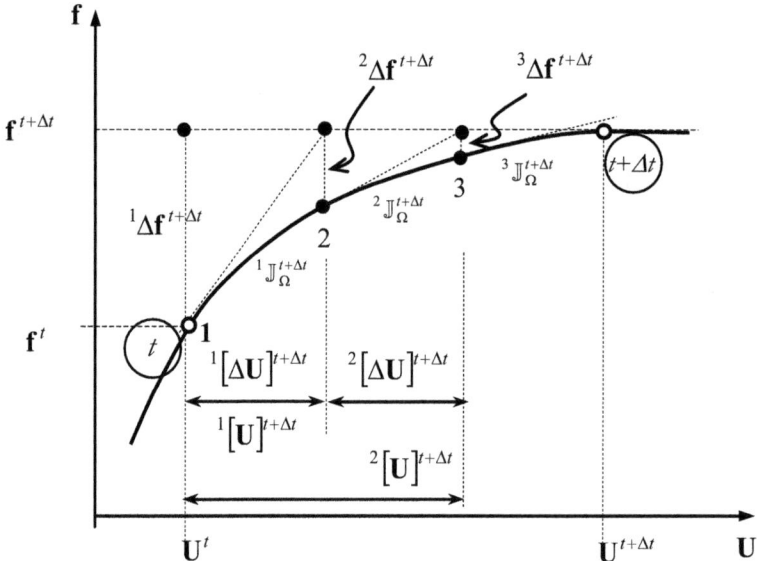

Figure 3.6 – Newton-Raphson method.

3.3.3.2 The modified Newton-Raphson method

This method solves equation (3.35) with the Jacobian operator defined in several forms:

1. **Initial stiffness method** (Figure 3.7). This numerical technique forces the Jacobian operator to be constant from the beginning to the end of the process.
2. **Updating time increment method** (Figure 3.8). The Jacobian operator is updated each time the load increases; in other words, while time t is constant there will be no changes in the Jacobian operator.

The Jacobian operator can be updated in other different ways based on other criteria, but it is not relevant here.

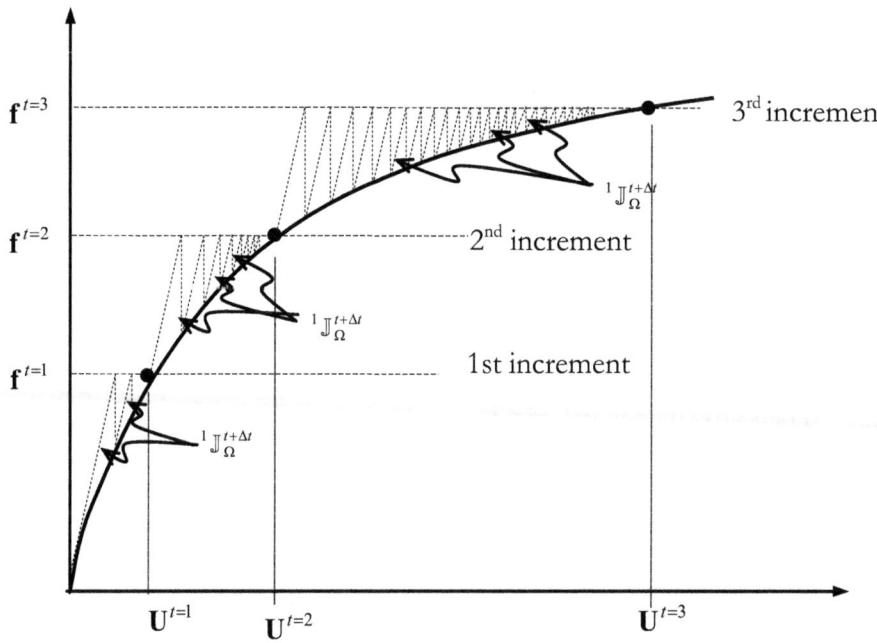

Figure 3.7 – The Newton-Raphson initial stiffness.

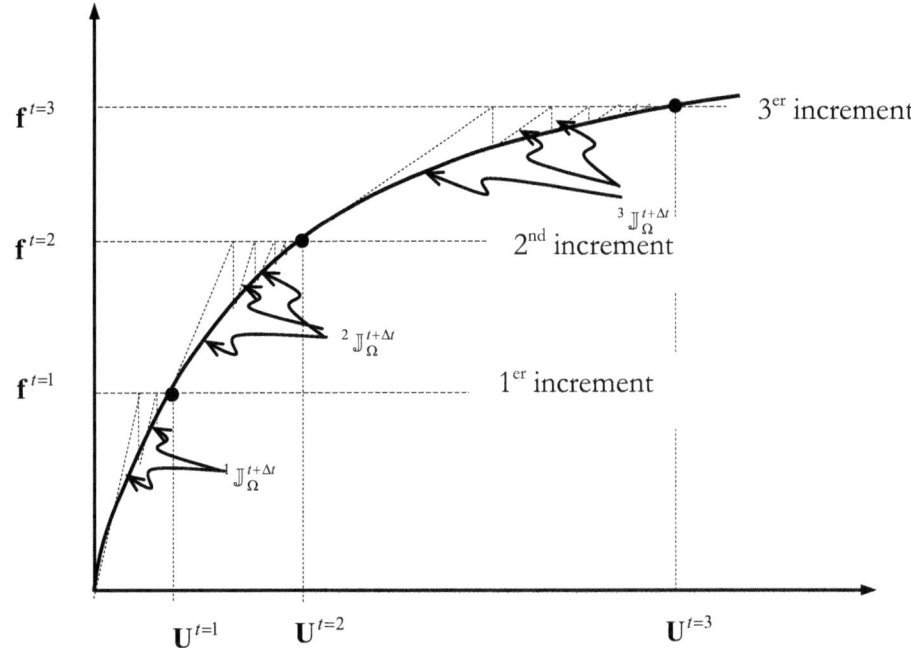

Figure 3.8 – Updated Newton-Raphson at each time increment.

The essence of the technique in any of the "modified Newton" methods, which is the convergence of speed, is missed and therefore these methods are only used in very particular cases. An example of such cases is when the Jacobian operator has no definition, or when this operator is ill-conditioned. In these cases a decreased convergence speed is generally accepted to obtain at least one solution to the problem.

3.3.3.3   Convergence accelerators

These numerical procedures help to improve the convergence and are based on the displacement field updating through its linear extrapolation. It can be said that convergence accelerators are the conceptual basis of the "Secant Newton" or "Quasi-Newton" methods. The updating of the displacement field has the following form:

Without accelerator:

$$^{i+1}[\mathbf{U}]_\Omega^{t+\Delta t} = {}^i[\mathbf{U}]_\Omega^{t+\Delta t} + {}^{i+1}[\Delta\mathbf{U}]_\Omega^{t+\Delta t} \tag{3.37}$$

where $^{i+1}[\Delta\mathbf{U}]_\Omega^{t+\Delta t}$ is the displacement increment obtained by equation (3.32) by using any algorithm.

With accelerator:

$$\boxed{^{i+1}[\mathbf{U}]_\Omega^{t+\Delta t} = {}^i[\mathbf{U}]_\Omega^{t+\Delta t} + \mathbf{A} \cdot {}^{i+1}[\Delta\mathbf{U}]_\Omega^{t+\Delta t}} \tag{3.38}$$

where $\mathbf{A}$ represents the "acceleration matrix" at time (t). Among the different ways to define this acceleration matrix there is the Aitken accelerator[2], shown below.

3.3.3.4   Aitken's accelerator or extrapolation algorithm

The general idea is to obtain the coefficients of this acceleration matrix from the extrapolation based on the proportional relation between the displacement components already consolidated in time (t) and the different iterative, not consolidated, values of the displacement at time ($t+\Delta t$),

$$
\mathbf{A} = \begin{bmatrix} \alpha_1 & & & \mathbf{0} \\ & \cdot & & \\ & & \cdot & \\ & & & \cdot \\ \mathbf{0} & & & \alpha_n \end{bmatrix}; \quad \text{with} \quad \alpha_n = \frac{{}^{i+1}\left(\Delta U_n\right)^{t+\Delta t}}{\left(\Delta U_n\right)^{t} - {}^{i+1}\left(\Delta U_n\right)^{t+\Delta t}} \tag{3.39}
$$

where ΔU_n represents the nth component of the displacement increment.

The bases of this technique are the computation of the changes in the residual forces between iteration (i) and the next ($i+1$). Their residual values can be written from a unique secant Jacobian operator $\left[\mathbb{J}_{\Omega}^{t+\Delta t}\right]^{*}$, for two successive iterations. This is,

$$
{}^{i-1}\left[\Delta \mathbf{f}\right]_{\Omega}^{t+\Delta t} = -\left[\mathbb{J}_{\Omega}^{t+\Delta t}\right]^{*} \cdot {}^{i}\left[\Delta \mathbf{U}\right]_{\Omega}^{t+\Delta t} \quad \text{and} \quad {}^{i}\left[\Delta \mathbf{f}\right]_{\Omega}^{t+\Delta t} = -\left[\mathbb{J}_{\Omega}^{t+\Delta t}\right]^{*} \cdot {}^{i+1}\left[\Delta \mathbf{U}\right]_{\Omega}^{t+\Delta t}
$$

Now, by establishing the difference between both levels of unbalanced forces, then

$$
{}^{i-1}\left[\Delta \mathbf{f}\right]_{\Omega}^{t+\Delta t} - {}^{i}\left[\Delta \mathbf{f}\right]_{\Omega}^{t+\Delta t} = -\left[\mathbb{J}_{\Omega}^{t+\Delta t}\right]^{*} \cdot \left[{}^{i}\left[\Delta \mathbf{U}\right]_{\Omega}^{t+\Delta t} - {}^{i+1}\left[\Delta \mathbf{U}\right]_{\Omega}^{t+\Delta t}\right] \tag{3.40}
$$

which can also be written using a unique secant operator (see Figure 3.9)

$$
{}^{i-1}\left[\Delta \mathbf{f}\right]_{\Omega}^{t+\Delta t} - {}^{i}\left[\Delta \mathbf{f}\right]_{\Omega}^{t+\Delta t} = -\left[\overline{\mathbb{J}}_{\Omega}^{t+\Delta t}\right] \cdot {}^{i}\left[\Delta \mathbf{U}\right]_{\Omega}^{t+\Delta t} \tag{3.41}
$$

Equating the first members of expressions (3.40) and (3.41), the following is obtained

$$
\left[\mathbb{J}_{\Omega}^{t+\Delta t}\right]^{*} \cdot \left[{}^{i}\left[\Delta \mathbf{U}\right]_{\Omega}^{t+\Delta t} - {}^{i+1}\left[\Delta \mathbf{U}\right]_{\Omega}^{t+\Delta t}\right] = \left[\overline{\mathbb{J}}_{\Omega}^{t+\Delta t}\right] {}^{i}\left[\Delta \mathbf{U}\right]_{\Omega}^{t+\Delta t}
$$

and from this last expression the acceleration (or extrapolation) matrix is defined as

$$
\mathbf{A} = \left[\mathbb{J}_{\Omega}^{t+\Delta t}\right]^{*} \cdot \left[\overline{\mathbb{J}}_{\Omega}^{t+\Delta t}\right]^{-1} \Rightarrow \mathbf{A} \cdot \left[{}^{i}\left[\Delta \mathbf{U}\right]_{\Omega}^{t+\Delta t} - {}^{i+1}\left[\Delta \mathbf{U}\right]_{\Omega}^{t+\Delta t}\right] = {}^{i}\left[\Delta \mathbf{U}\right]_{\Omega}^{t+\Delta t} \tag{3.42}
$$

substituting this acceleration matrix into the equation of changes in the residual forces between two successive iterations (equation (3.40)), we get the following linearized equilibrium equation

$$
{}^{i}\left[\Delta \mathbf{f}\right]_{\Omega}^{t+\Delta t} = -\left[\mathbb{J}_{\Omega}^{t+\Delta t}\right]^{*} \cdot {}^{i+1}\left[\Delta \mathbf{U}\right]_{\Omega}^{t+\Delta t} = -\mathbf{A} \cdot \left[\overline{\mathbb{J}}_{\Omega}^{t+\Delta t}\right] {}^{i+1}\left[\Delta \mathbf{U}\right]_{\Omega}^{t+\Delta t} \tag{3.43}
$$

Thus, the system is always solved with a secant operator, whose expression is summarized in the following equation,

$$
\boxed{\left[\overline{\mathbb{J}}_{\Omega}^{t+\Delta t}\right] = \mathbf{A}_n^{-1} \cdot \left[\mathbb{J}_{\Omega}^{t+\Delta t}\right]^{*}} \tag{3.44}
$$

The different ways to obtain the acceleration operator $\mathbf{A}$ lead to different algorithms.

3.3.3.5 B.F.G.S. algorithms

This algorithm was developed by **B**roydon-**F**letcher-**G**oldfarb-**S**hanno[2]. A brief presentation of the most important expressions to describe the algorithm follows. Also, the method's main ideas, based on an inverse approximation of the Jacobian operator from equality (3.40), will be shown. Therefore, after some mathematical steps, the following updating formula for the inverse of the Jacobian operator is reached.

where

$$
\begin{aligned}
{}^{i}\mathbf{V} &= \left\{1 - \frac{{}^{i-1}[\Delta\mathbf{U}]_{\Omega}^{t+\Delta t}\cdot\left({}^{i-1}[\Delta\mathbf{f}]_{\Omega}^{t+\Delta t} - {}^{i}[\Delta\mathbf{f}]_{\Omega}^{t+\Delta t}\right)}{{}^{i-1}[\Delta\mathbf{U}]_{\Omega}^{t+\Delta t}\cdot{}^{i-1}[\Delta\mathbf{f}]_{\Omega}^{t+\Delta t}}\right\}\cdot{}^{i-1}[\Delta\mathbf{f}]_{\Omega}^{t+\Delta t} - {}^{i}[\Delta\mathbf{f}]_{\Omega}^{t+\Delta t} \\
{}^{i}\mathbf{W} &= \frac{{}^{i-1}[\Delta\mathbf{U}]_{\Omega}^{t+\Delta t}}{{}^{i-1}[\Delta\mathbf{U}]_{\Omega}^{t+\Delta t}\left({}^{i-1}[\Delta\mathbf{f}]_{\Omega}^{t+\Delta t} - {}^{i}[\Delta\mathbf{f}]_{\Omega}^{t+\Delta t}\right)}
\end{aligned}
\tag{3.46}
$$

Among the advantages of this secant Newton method is the time-saving in calculations because the initial stiffness matrix must be inverted only once. Also, the way the Jacobian matrix is obtained by using equation (3.45) maintains the symmetry of the original matrix. On the other hand, this updating does not keep the distribution of the matrix elements or topology (in *"sparsed"* form). Therefore, it is convenient to recover the original topology of the Jacobian matrix ${}^{1}\mathbb{J}_{\Omega}^{t+\Delta t}$ at each iteration and to reformulate equation (3.45) as follows:

$$
{}^{i}\mathbf{b} = \left[{}^{1}\mathbb{J}_{\Omega}^{t+\Delta t}\right]\cdot\prod_{j=2}^{i}\left({}^{j}\mathbf{A}\cdot{}^{j}[\Delta\mathbf{f}]_{\Omega}^{t+\Delta t}\right) \quad\Rightarrow\quad {}^{i}[\Delta\mathbf{U}]_{\Omega}^{t+\Delta t} = \left[\prod_{j=0}^{i-j}\left({}^{i-j}\mathbf{A}^{T}\right)\right]\cdot{}^{i}\mathbf{b}
\tag{3.47}
$$

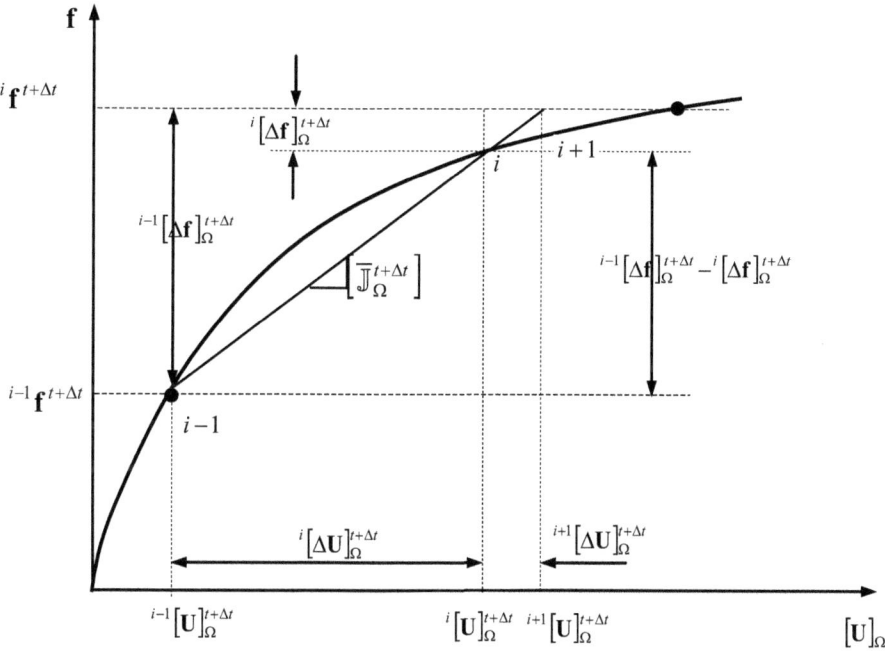

Figure 3.9 – Quasi-Newton – Aitken algorithm.

This numerical procedure requires storing vectors ${}^{i}\mathbf{V}$ and ${}^{i}\mathbf{W}$ at every iteration to set up ${}^{i}\mathbf{A}$ matrix.

3.3.3.6 The Secant-Newton algorithms

In this algorithm, the change of the Jacobian operator $\Delta\left[{}^{i}\mathbb{J}_{\Omega}^{t+\Delta t}\right]$, during a time increment, is not obtained as a function of its magnitude in the previous time increment $\Delta\left[{}^{i-1}\mathbb{J}_{\Omega}^{t+\Delta t}\right]$, but from the tangent operator of the first iteration of the increment $\left[{}^{1}\mathbb{J}_{\Omega}^{t+\Delta t}\right]$, and its form is the following

$$\left[{}^{i}\mathbb{J}_{\Omega}^{t+\Delta t}\right]^{-1} = \left[{}^{1}\mathbb{J}_{\Omega}^{t+\Delta t}\right]^{-1} + \frac{\mathbf{b}\cdot\mathbf{b}^{T}}{\mathbf{b}^{T}\left[{}^{i}[\Delta\mathbf{f}]_{\Omega}^{t+\Delta t} - {}^{i-1}[\Delta\mathbf{f}]_{\Omega}^{t+\Delta t}\right]} \tag{3.48}$$

being,

$$\mathbf{b} = [\Delta\mathbf{U}]_{\Omega}^{t} + {}^{i}[\Delta\mathbf{U}]_{\Omega}^{t+\Delta t} - {}^{i-1}[\Delta\mathbf{U}]_{\Omega}^{t+\Delta t} \tag{3.49}$$

such that the displacement increments in this equation are obtained according to the following form,

$${}^{i}[\Delta\mathbf{U}]_{\Omega}^{t+\Delta t} = -\left[{}^{i}\mathbb{J}_{\Omega}^{t+\Delta t}\right]^{-1}\cdot{}^{i}[\Delta\mathbf{f}]_{\Omega}^{t+\Delta t} \quad , \quad {}^{i-1}[\Delta\mathbf{U}]_{\Omega}^{t+\Delta t} = -\left[{}^{i}\mathbb{J}_{\Omega}^{t+\Delta t}\right]^{-1}\cdot{}^{i-1}[\Delta\mathbf{f}]_{\Omega}^{t+\Delta t} \tag{3.50}$$

3.3.3.7 "Line-Search" algorithms

This method improves the searching of the solution and can be applied to help any of the methods previously presented. It can be applied jointly with the Newton or Quasi-Newton algorithms.

Then, we propose here the solution of the system of nonlinear equations (3.35) or (3.41), either by using Newton or Quasi-Newton, which must be complemented with the minimization of a functional $\Pi(\mathbf{U})$ (see Figure 3.10). This involves a minimum search in the energy space,

$$\delta\Pi_{\Omega}(\mathbf{U}) = 0 = \left(\frac{\partial\Pi}{\partial\mathbf{U}}\right)_{\Omega}\delta\mathbf{U} \xrightarrow{\delta\mathbf{U}\neq 0} \left(\frac{\partial\Pi}{\partial\mathbf{U}}\right)_{\Omega} = \mathbf{0} = [\Delta\mathbf{f}]_{\Omega} \tag{3.51}$$

The method consists in obtaining ${}^{i}[\Delta\mathbf{U}]_{\Omega}^{t+\Delta t}$ by using any of the procedures mentioned before. This displacement increment will not be considered as an increment itself, but as the search direction. With this new search direction, the directional minimum P (which is not the absolute minimum) by minimizing $\mu(s) = \Pi[\mathbf{U}(s)]$ is obtained, and then the minimum position s^{*} is obtained. As proceeded with the Aitken accelerator in equation (3.38), the displacement increment is written as,

$${}^{i+1}[\mathbf{U}]_{\Omega}^{t+\Delta t} = {}^{i}[\mathbf{U}]_{\Omega}^{t+\Delta t} + {}^{i}[s^{*}]^{t+\Delta t}\cdot{}^{i+1}[\Delta\mathbf{U}]_{\Omega}^{t+\Delta t} \tag{3.52}$$

To find the value of s^{*} from the functional of energy Π is minimized in the search direction s or by applying the virtual works principle in such direction. This is,

$${}^{i+1}[\mathbf{U}]_{\Omega}^{t+\Delta t} = {}^{i}[\mathbf{U}]_{\Omega}^{t+\Delta t} + {}^{i}[s]^{t+\Delta t}\cdot{}^{i+1}[\Delta\mathbf{U}]_{\Omega}^{t+\Delta t} \implies \frac{\partial{}^{i+1}[\mathbf{U}]_{\Omega}^{t+\Delta t}}{\partial s} = {}^{i+1}[\Delta\mathbf{U}]_{\Omega}^{t+\Delta t} \tag{3.53}$$

Also, the minimum of this energy functional with respect to the displacement field is the residual force,

$${}^{i+1}\left(\frac{\partial\Pi}{\partial\mathbf{U}}\right)_{\Omega}^{t+\Delta t} = {}^{i+1}[\Delta\mathbf{f}]_{\Omega}^{t+\Delta t} \tag{3.54}$$

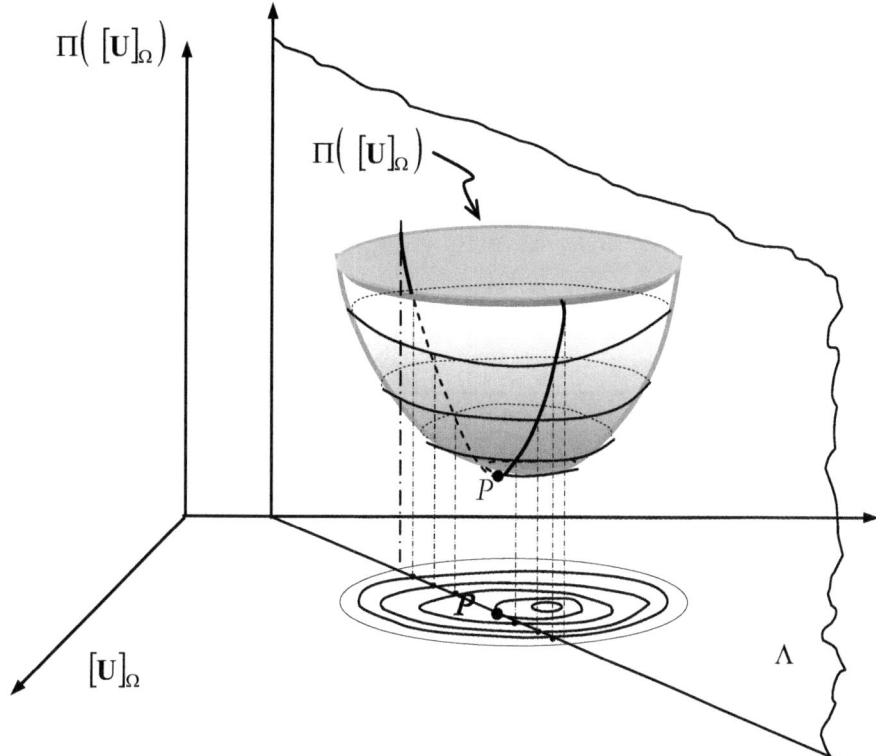

Figure 3.10 – Minimum search in the energy space using "Line-Search".

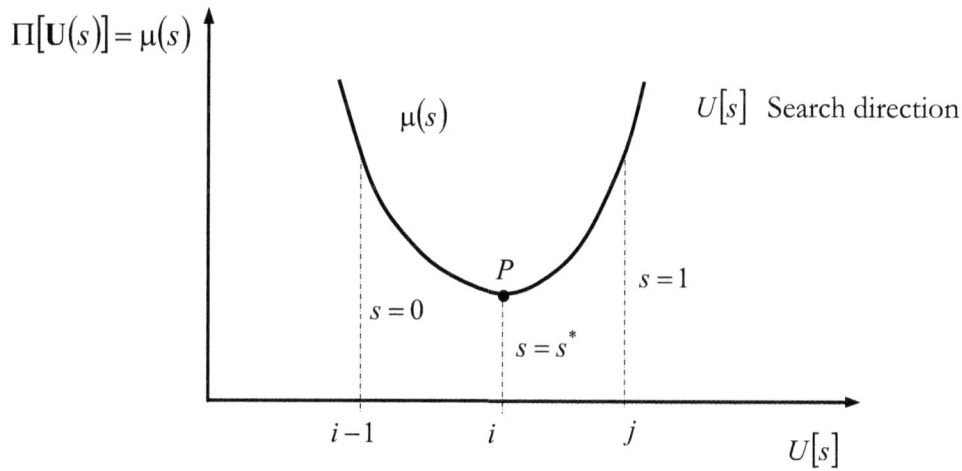

Figure 3.11 – Plane Λ - Search plane (see Figure (3.10).

Considering now the two previous equations, we get the following equation as a function of the coordinate (s),

$$\frac{\partial\Pi\left({}^{i}[\mathbf{U}]_{\Omega}^{t+\Delta t}+{}^{i}[s]^{t+\Delta t}\cdot{}^{i+1}[\Delta\mathbf{U}]_{\Omega}^{t+\Delta t}\right)}{\partial s}={}^{i+1}\left[\frac{\partial\Pi}{\partial\mathbf{U}}\frac{\partial\mathbf{U}}{\partial s}\right]^{t+\Delta t}={}^{i+1}[\Delta\mathbf{f}]_{\Omega}^{t+\Delta t}\cdot{}^{i+1}[\Delta\mathbf{U}]_{\Omega}^{t+\Delta t}=G(s)=0 \qquad (3.55)$$

Thus, we try to solve an equation $G(s)$ nonlinear in (s), whose solution is s^* (see Figure 3.12).

Figure 3.12 – Plane Λ for searching the zero of function $G(s)$.

To obtain the solution of this nonlinear equation we can use methods such as "Regula falsi"[2], "Illinois algorithm"[2], etc.

3.3.3.8 Solution control algorithms – "Arc-Length"

Unstable behaviours occur for various problems in structural mechanics, for which it is difficult to find a solution. To avoid this, a system of equilibrium equations with restrictions of type $\Delta \mathbf{f}(\ddot{\mathbf{U}}, \dot{\mathbf{U}}, \mathbf{U}, \lambda)$ are usually used, where the magnitude of external force $\lambda \, \mathbf{f}^{\text{ext}}$ is an unknown conditioned by an additional equation $c(\mathbf{U}, \lambda)$. Thus, a system of equilibrium equations enlarged by this restriction equation[2] is obtained,

$$
\begin{cases}
{}^{i+1}\left[\Delta\mathbf{f}(\ddot{\mathbf{U}},\dot{\mathbf{U}},\mathbf{U},\lambda)\right]_{\Omega}^{t+\Delta t} = \mathbf{M} \; {}^{i+1}\left[\ddot{\mathbf{U}}\right]_{\Omega}^{t+\Delta t} + {}^{i+1}\left[\mathbf{f}^{\text{int}}(\dot{\mathbf{U}},\mathbf{U})\right]_{\Omega}^{t+\Delta t} - {}^{i+1}\left[\lambda \; \mathbf{f}^{\text{ext}}\right]_{\Omega}^{t+\Delta t} = \mathbf{0} \\
{}^{i+1}\left[c(\mathbf{U},\lambda)\right]_{\Omega}^{t+\Delta t} = 0
\end{cases}
\tag{3.56}
$$

Developing the residual forces in Taylor series, in the surrounding of the solution at time $(t+\Delta t)$, and considering that both velocities and accelerations depend on the displacement field at this time, we have[2]

$$
\Delta\mathbf{f}\left({}^{i}[\mathbf{U}] + {}^{i+1}[\delta\mathbf{U}] , \; {}^{i}[\lambda] + {}^{i+1}[\delta\lambda]\right) = \mathbf{0} = \Delta\mathbf{f}\left({}^{i}[\mathbf{U}], {}^{i}[\lambda]\right) + {}^{i}\!\left(\frac{\partial\Delta\mathbf{f}}{\partial\mathbf{U}}\right) \cdot {}^{i+1}[\delta\mathbf{U}]_{\Omega}^{t+\Delta t} + {}^{i}\!\left(\frac{\partial\Delta\mathbf{f}}{\partial\lambda}\right) \cdot {}^{i+1}[\delta\lambda]
\tag{3.57}
$$

In this latter equation we can identify the following relations (see equation (3.34)),

$$
{}^{i}\!\left(\frac{\partial\Delta\mathbf{f}}{\partial\mathbf{U}}\right) = {}^{i}\mathbb{J} \quad ; \quad {}^{i}\!\left(\frac{\partial\Delta\mathbf{f}}{\partial\lambda}\right) = -\mathbf{f}^{\text{ext}}
\tag{3.58}
$$

which, substituted into equation (3.57), allows to rewrite the system of equilibrium equations with restrictions as,

$$
\begin{cases}
\mathbf{0} = \Delta\mathbf{f}\left({}^{i}\mathbf{U}, {}^{i}\lambda\right) + {}^{i}\mathbb{J} \cdot {}^{i+1}[\delta\mathbf{U}] - \mathbf{f}^{\text{ext}} \cdot {}^{i+1}[\delta\lambda] \\
0 = {}^{i+1}[c(\mathbf{U},\lambda)]
\end{cases}
\tag{3.59}
$$

[2] NOTE: To simplify the presentation, the superscript $(t+\Delta t)$ will be hereafter omitted in all the developments.

And from here we obtain the desired displacement field,

$$^{i}\mathbb{J}\cdot^{i+1}[\delta\mathbf{U}] = -\Delta\mathbf{f}\left(^{i}\mathbf{U},^{i}\lambda\right) + \mathbf{f}^{\text{ext}}\cdot^{i+1}[\delta\lambda]$$

$$^{i}\mathbb{J}\cdot^{i+1}[\delta\mathbf{U}] = -\underbrace{\left[\mathbb{M}\ ^{i}[\ddot{\mathbf{U}}] + ^{i}\left[\mathbf{f}^{\text{int}}(\dot{\mathbf{U}},\mathbf{U})\right] - ^{i}\left[\lambda\ \mathbf{f}^{\text{ext}}\right]\right]}_{^{i}\mathbb{J}\cdot^{i+1}[\delta\hat{\mathbf{U}}]} + ^{i}\mathbb{J}\cdot^{i+1}[\mathbf{U}]_{\text{TOT}}\cdot^{i+1}[\delta\lambda] \tag{3.60}$$

Since the Jacobian operator is the same in both members of the previous equation, the displacement increment can be written as,

$$^{i+1}[\delta\mathbf{U}] = ^{i+1}[\delta\hat{\mathbf{U}}] + ^{i+1}[\mathbf{U}]_{\text{TOT}}\cdot^{i+1}[\delta\lambda] \tag{3.61}$$

being in this expression:

$$\begin{cases} ^{i}\mathbb{J}\cdot^{i+1}[\delta\hat{\mathbf{U}}] = -\left[\mathbb{M}\ ^{i}[\ddot{\mathbf{U}}] + ^{i}\left[\mathbf{f}^{\text{int}}(\dot{\mathbf{U}},\mathbf{U})\right] - ^{i}\left[\lambda\ \mathbf{f}^{\text{ext}}\right]\right] \\ ^{i}\mathbb{J}\cdot^{i+1}[\mathbf{U}]_{\text{TOT}} = \mathbf{f}^{\text{ext}} \\ ^{i+1}[\lambda] = ^{i}[\lambda] + ^{i+1}[\delta\lambda] \end{cases} \tag{3.62}$$

where $^{i+1}[\mathbf{U}]_{\text{TOT}}$ is the total displacement obtained with the last (or maximum) value of the external force , $^{i+1}[\delta\hat{\mathbf{U}}]$ is the solution of the system of equations without correction and $\delta^{i+1}\lambda$ is the change of the application factor of the load.

Once the general equations of the method are defined, it is necessary to define the shape of the restriction equation. This will be introduced in the following section.

Displacement control equation - Spherical path

One of the mostly used displacement control equations is the "spherical control equation". It requires that a certain norm of increment of the displacement be contained inside an hyper-sphere of the displacement space. This is,

$$^{i+1}[c(\mathbf{U},\lambda)] = ^{i+1}[\Delta\mathbf{U}]^{T}\cdot^{i+1}[\Delta\mathbf{U}] - \Delta\ell^{2} = 0 \tag{3.63}$$

in such a way that the displacement and its increment are obtained as

$$^{i+1}[\mathbf{U}]^{t+\Delta t} = ^{1}[\mathbf{U}]^{t} + ^{i+1}[\Delta\mathbf{U}]^{t+\Delta t} \quad ; \quad ^{i+1}[\Delta\mathbf{U}]^{t+\Delta t} = ^{i}[\Delta\mathbf{U}]^{t+\Delta t} + ^{i+1}[\delta\mathbf{U}]^{t+\Delta t} \tag{3.64}$$

Substituting equation (3.61) into (3.64) and the resulting equation into (3.63), the following control equation at time $t + \Delta t$ is obtained,

$$\left\{^{i}[\Delta\mathbf{U}] + ^{i+1}[\delta\hat{\mathbf{U}}] + ^{i+1}[\mathbf{U}]_{\text{TOT}}\cdot^{i+1}[\delta\lambda]\right\}^{T}\cdot\left\{^{i}[\Delta\mathbf{U}] + ^{i+1}[\delta\hat{\mathbf{U}}] + ^{i+1}[\mathbf{U}]_{\text{TOT}}\cdot^{i+1}[\delta\lambda]\right\} - \Delta\ell^{2} = 0$$

$$\Downarrow \tag{3.65}$$

$$C_{1}\ ^{i+1}[\delta\lambda]^{2} + C_{2}\ ^{i+1}[\delta\lambda] + C_{3} = 0$$

where the coefficients of this second-degree equation have the following expression,

$$C_1 = {}^{i+1}[\mathbf{U}]_{TOT}^T \cdot {}^{i+1}[\mathbf{U}]_{TOT}$$

$$C_2 = 2\left[{}^{i}[\Delta\mathbf{U}] + {}^{i+1}[\delta\hat{\mathbf{U}}]\right]^T \cdot {}^{i+1}[\mathbf{U}]_{TOT}$$

$$C_3 = \left[{}^{i}[\Delta\mathbf{U}] + {}^{i+1}[\delta\hat{\mathbf{U}}]\right]^T \cdot \left[{}^{i}[\Delta\mathbf{U}] + {}^{i+1}[\delta\hat{\mathbf{U}}]\right] - \Delta\ell^2$$

(3.66)

By solving the second-degree equation in ${}^{i+1}[\delta\lambda]$, the load factor correction is then obtained (see Figure 3.13),

$${}^{i+1}[\delta\lambda] = -\frac{C_2}{2C_1} \pm \frac{\left(C_2^2 - 4C_1C_2\right)^{1/2}}{2C_1} \quad \Rightarrow \quad \begin{cases} {}^{i+1}[\delta\lambda]_1 \\ {}^{i+1}[\delta\lambda]_2 \end{cases}$$

(3.67)

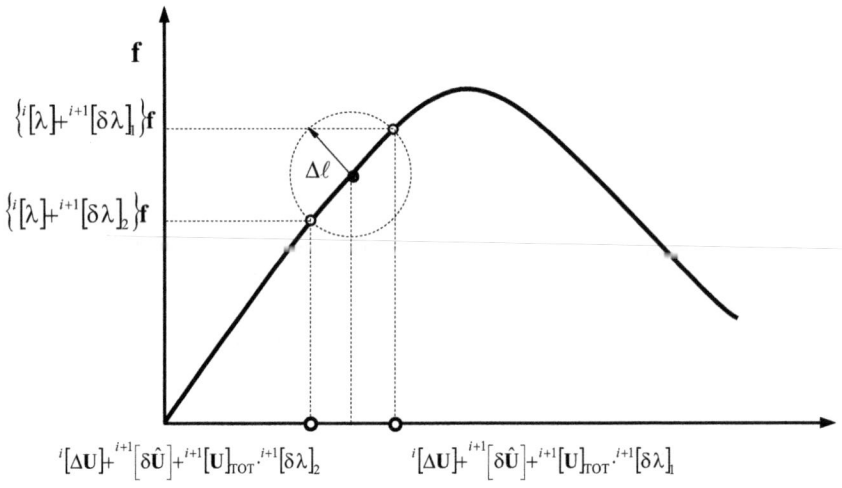

Figure 3.13 – Spherical path "Arc-Length" – Details in the search of the solution

Once the two load factors are obtained ${}^{i+1}[\delta\lambda]_{1,2}$, it is then necessary to find out which the right one is. To do this, the right *advance direction* ${}^{i+1}[\delta\mathbf{U}] = {}^{i+1}[\delta\hat{\mathbf{U}}] + {}^{i+1}[\mathbf{U}]_{TOT} \cdot {}^{i+1}[\delta\lambda]_{1,2}$ is explored. It is assumed that the right direction is the one with the angle with the displacement increment, in the previous step ${}^{i}[\Delta\mathbf{U}]$ is maximum (see Figure 3.14).

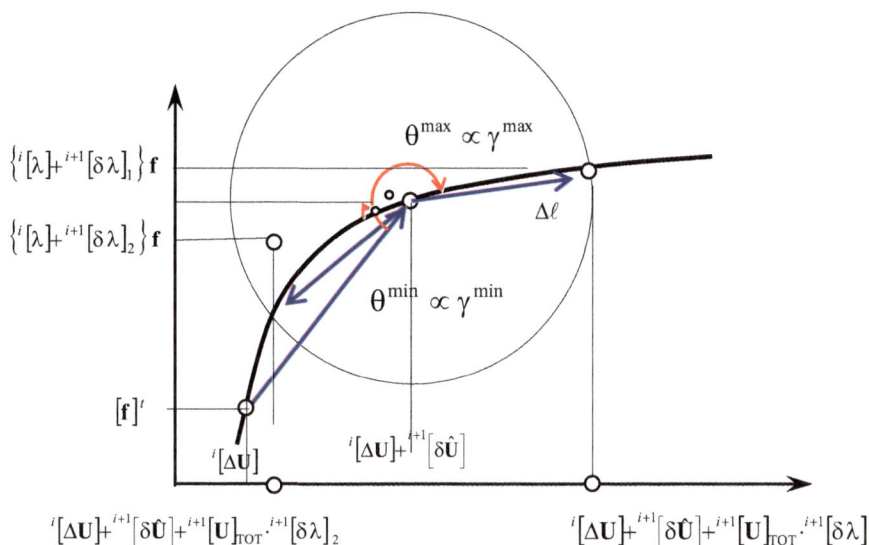

Figure 3.14 – Spherical path "Arc-Length" – Detail of the advance.

This direction follows from the scalar product,

$$\gamma^1 = {}^{i+1}\left[\Delta \mathbf{U}\right]_1^T \cdot {}^{i}\left[\Delta \mathbf{U}\right]$$
$$\gamma^2 = {}^{i+1}\left[\Delta \mathbf{U}\right]_2^T \cdot {}^{i}\left[\Delta \mathbf{U}\right]$$

(3.68)

then selecting ${}^{i+1}\left[\delta \lambda\right]_{1,2}$ that leads to ${}^{i+1}\left[\Delta \mathbf{U}\right]_{1,2}$ which maximizes γ_j.

4 Convergence Analysis of the Dynamic Solution

4.1 Introduction

In the first part of this chapter the dynamic equation (3.1) is particularized for linear problems to study the convergence of the solution for different numerical methods in the time domain. Strictly speaking, the concept of convergence cannot be guaranteed in the second-order nonlinear differential equations as studied in the solution shown in chapter B3 because the convergence involves stability in the solution and this cannot be guaranteed. Nevertheless, the "linearized stability" concept will be studied. It is the most commonly used concept. It can only guarantee the minimum stability conditions, although not enough.

The *stability analysis* is important when studying the dynamic solution in time for a problem because, together with the *consistency* solution, they determine the *convergence*[*] of the solution algorithm. In other words, when the stability and consistency conditions are met in a linear dynamic system, the solution convergence is guaranteed.

There are several ways to study the stability of a solution algorithm of the dynamic equilibrium differential equation, but Fourier's spectral study is the most widely used and it will be presented in this chapter.

4.2 Reduction to the linear elastic problem

The solution of the linear elastic problem through the modal analysis and the method of separation of variables will be briefly described in this section. Through the concept of diagonalization this technique can write a system of equations uncoupled to a one degree-of-freedom[1,2,3,4,5]. It will be assumed that the dynamic equilibrium differential equation in time *(t)* is a constant coefficient differential equation as,

$$\mathbb{M}\ \ddot{\mathbf{U}}(t) + \mathbb{D}\ \dot{\mathbf{U}}(t) + \mathbb{K}\ \mathbf{U}(t) - \mathbf{f}^{\text{ext}}(t) = 0 \qquad \in \Omega \qquad (4.1)$$

In this equation the mass matrix $\mathbb{M}$ must be symmetric and positive defined, while the

---

[*] NOTE: Convergence = Stability + Consistency

[1] Clough R. and Penzien J. (1977). Dynamics of Structures. Mc Graw-Hill - N. York.

[2] Paz M. (1992). Dinámica estructural. Reverté - Barcelona.

[3] Car, E., López F., Oller S. (2000). Estructuras sometidas a acciones dinámicas – CIMNE – Barcelona.

[4] Barbat A., Miquel J. (1994). Estructuras sometidas a acciones sísmicas - CIMNE -Barcelona.

[5] Barbat A., Oller S. (1997). Conceptos de cálculo de estructuras en las normativas de diseño sismo resistente - CIMNE IS-24 – Barcelona.

damping matrix $\mathbb{D}$ and the stiffness matrix $\mathbb{K}$ must be symmetric and positive semi-defined. As in the previous chapter, this ordinary differential equation system with constant coefficients contains the spatial discretization, such that the displacement field $\mathbf{U}(t)$, velocities $\dot{\mathbf{U}}(t)$ and accelerations $\ddot{\mathbf{U}}(t)$ are defined in only some points of the continuous medium (nodes), more specifically at those points where the local approximation polynomial (also known as shape function of the finite element) is defined (see section 2.5). Moreover, in equation (4.1) the time field is continuous. The following magnitudes are approximated by polynomial functions within the linear elastic discrete space, represented as (see equation 2.77),

$$
\begin{cases}
\textbf{Mass matrix:} \qquad \mathbb{M} = \mathbf{A}_{\Omega^e} \int_{V^e} \rho \, \mathbf{N} : \mathbf{N} \, dV \quad \text{, being } \rho \text{ the density} \\[4pt]
\textbf{Internal forces (see eq. 3.2):} \quad \mathbf{f}^{\text{int}}(\dot{\mathbf{U}}, \mathbf{U}, t+\Delta t) = \mathbf{A}_{\Omega^e} \int_{V^e} \boldsymbol{\sigma}\left(\nabla^S \mathbf{N}\right) dV = \\[4pt]
= \left[\mathbf{A}_{\Omega^e} \int_{V^e} \left(\nabla^S \mathbf{N}\right) : \xi : \left(\nabla^S \mathbf{N}\right) dV \right] : \dot{\mathbf{U}}(t+\Delta t) + \left[\mathbf{A}_{\Omega^e} \int_{V^e} \left(\nabla^S \mathbf{N}\right) : \mathbb{C}^0 : \left(\nabla^S \mathbf{N}\right) dV \right] : \mathbf{U}(t+\Delta t) = \qquad (4.2) \\[4pt]
= \mathbb{D} \ \dot{\mathbf{U}}(t+\Delta t) + \mathbb{K} \ \mathbf{U}(t+\Delta t) \\[4pt]
\textbf{External forces:} \qquad \mathbf{f}^{\text{ext}}(t+\Delta t) = \mathbf{A}_{\Omega^e} \left[\oint_{S^e} \mathbf{N} \cdot \mathbf{t} \, dS + \int_{V^e} \rho \, \mathbf{N} \cdot \mathbf{b} \, dV \right]
\end{cases}
$$

The magnitudes of this equation have been defined in previous chapters.

The main philosophy to solve this classical problem in linear dynamics is based on the assumption of the "concept of separation of variables", in which it is admitted that both time and spatial problems are independent of each other. Therefore, as mentioned before, a different solution strategy is established for each problem. Then for the spatial problem solution the finite element method is used, while the time problem is usually solved by finite procedures. In other words, the semi-discrete equation (4.1) representing the spatial equilibrium is solved at each time t.

By using the *modal superposition* technique, the displacement field is written as,

$$
\mathbf{U}(t) = \sum_{h=1}^{n-\text{modes}} \mathbf{U}_h(t) \tag{4.3}
$$

where $\mathbf{U}_h(t)$ is the displacement vector of the "hth" mode, which describes the "shape of *movement*" of the "n" Lagrangean normal coordinates $\varphi_h(t)$, or degrees-of-freedom system , when the "h^{th}" degree is perturbed. Through the method of separation of variables the displacement of the "h^{th}" mode is written as the product between its vibration shape $\mathbf{U}_h$ and the time amplitude of its normal coordinate $\varphi_h(t)$,

$$
\mathbf{U}_h(t) = \mathbf{U}_h \cdot \varphi_h(t) \tag{4.4}
$$

Thus, equation (4.3) can be rewritten as,

$$
\mathbf{U}(t) = \sum_{h=1}^{n-\text{modes}} \mathbf{U}_h(t) = \sum_{h=1}^{n-\text{modes}} \mathbf{U}_h \cdot \phi_h(t)
$$

$$\mathbf{U}(t) = \mathbb{U} \cdot \mathbf{\Phi}(t) = \begin{bmatrix} \mathbf{U}_1 & \cdots & \mathbf{U}_h & \cdots & \mathbf{U}_n \end{bmatrix} \cdot \begin{Bmatrix} \varphi_1(t) \\ \vdots \\ \varphi_h(t) \\ \vdots \\ \varphi_n(t) \end{Bmatrix} \tag{4.5}$$

where $\mathbb{U}$ is the modal matrix containing the n modal vectors $\mathbf{U}_h$ normalized with respect to mass and $\mathbf{\Phi}(t)$ is the vector of normal coordinates $\varphi_h(t)$, used to represent the behavior in time of all the system's normal coordinates.

Substituting equation (4.5) into equation (4.1), the following dynamic equilibrium equation[1,2,3,4,5] is obtained,

$$\mathbb{M} \, \mathbb{U} \cdot \ddot{\mathbf{\Phi}}(t) + \mathbb{D} \, \mathbb{U} \cdot \dot{\mathbf{\Phi}}(t) + \mathbb{K} \, \mathbb{U} \cdot \mathbf{\Phi}(t) - \mathbf{f}^{\,\text{ext}}(t) = 0 \tag{4.6}$$

Pre-multiplying the above equation by the modal matrix $\mathbb{U}^T$, and using the orthogonality conditions of the eigenvectors[*], the following uncoupled system of second-order differential equations with constant coefficients is obtained,

$$\ddot{\mathbf{\Phi}}(t) + \mathbb{N}\,\dot{\mathbf{\Phi}}(t) + \Lambda\,\mathbf{\Phi}(t) - \mathbb{U}^T \cdot \mathbf{f}^{\,\text{ext}}(t) = 0$$

$$\begin{bmatrix} \ddot{\varphi}_1(t) \\ \vdots \\ \ddot{\varphi}_h(t) \\ \vdots \\ \ddot{\varphi}_n(t) \end{bmatrix} + \begin{bmatrix} 2\xi_1\omega_1 & & & \\ & \ddots & & \\ & & 2\xi_h\omega_h & \\ & & & \ddots & \\ & & & & 2\xi_n\omega_n \end{bmatrix} \cdot \begin{bmatrix} \dot{\varphi}_1(t) \\ \vdots \\ \dot{\varphi}_h(t) \\ \vdots \\ \dot{\varphi}_n(t) \end{bmatrix} +$$

$$+ \begin{bmatrix} \omega_1^2 & & & \\ & \ddots & & \\ & & \omega_h^2 & \\ & & & \ddots & \\ & & & & \omega_n^2 \end{bmatrix} \begin{Bmatrix} \varphi_1(t) \\ \vdots \\ \varphi_h(t) \\ \vdots \\ \varphi_n(t) \end{Bmatrix} - \begin{bmatrix} U_1^1 & \cdots & U_1^h & \cdots & U_1^n \\ \vdots & \ddots & \vdots & \ddots & \vdots \\ U_h^1 & \cdots & U_h^h & \cdots & U_h^n \\ \vdots & \ddots & \vdots & \ddots & \vdots \\ U_n^1 & \cdots & U_n^h & \cdots & U_n^n \end{bmatrix} \begin{Bmatrix} f_1^{\text{ext}}(t) \\ \vdots \\ f_h^{\text{ext}}(t) \\ \vdots \\ f_n^{\text{ext}}(t) \end{Bmatrix} = \begin{Bmatrix} 0 \\ \vdots \\ 0 \\ \vdots \\ 0 \end{Bmatrix} \tag{4.7}$$

Each of these uncoupled equations $\ddot{\varphi}_h(t) + 2\xi_h\omega_h\dot{\varphi}_h(t) + \omega_h^2\varphi_h(t) - [a_h^{\text{ext}}(t)] = 0$, with $[a_h^{\text{ext}}(t)] = \sum_{i=1}^{n-GL} \mathbf{U}_h^i \cdot f_i^{\text{ext}}(t)$, represents the movement of an *equivalent one degree-of-freedom oscillator*, where the time solution techniques for the one degree-of-freedom dynamic equilibrium differential equation can be applied. It should be remembered that the natural pulsations $\omega_h = \sqrt{\lambda_h}$, also known as angular frequencies, result from the eigenvalues λ_h from the algebraic equation $\det[\mathbb{K} - \lambda_h \mathbb{M}] = 0$, of degree n-GL (number of degrees-of-freedom). The corresponding eigenvectors $\mathbf{U}_h$ are obtained from the solution of the system of equations $[\mathbb{K} - \lambda_h \mathbb{M}] \cdot \mathbf{U}_h = 0$.

---

[*] Note: The eigenvectors' orthogonality properties establish that,

$$\mathbf{U}_i^T \mathbb{M} \, \mathbf{U}_j = \begin{cases} 0 & \forall \; i \neq j \\ 1 & \forall \; i = j \end{cases} \quad ; \quad \mathbf{U}_i^T \mathbb{K} \, \mathbf{U}_j = \begin{cases} 0 & \forall \; i \neq j \\ \omega_i^2 = \lambda_i & \forall \; i = j \end{cases}$$

as long as these eigenvectors are normal to the mass.

The alternative way to the modal method is the use of "one-step" or "multiple-step" algorithms, described in the previous chapter, which can be used in nonlinear problems. Through both, the "modal superposition" method and the "step by step" time solution algorithms, the time solution of "n-degrees-of-freedom (n-DOF)" linear dynamic oscillators can be reached. In Figure 4.1 both paths for the differential equation are presented.

4.3 Solution of second-order linear symmetric systems

As observed in the previous section, there are two methods to solve the second-order linear differential equation system with constant coefficients. One is based on "modal decomposition" and the other on the time direct integration of the differential equation of movement through the "one or more steps algorithms". In the event of modal decomposition the system is diagonalized ant therefore it becomes an uncoupled system of differential equations, and it is faesible to solve one-by-one the differential equations for each degree-of-freedom. As for the total coupled system, the solution can be obtained by means of numerical techniques of one or more steps (see Figure 4.1). Then, the simplest path and the stability of these algorithms applied to a one degree-of-freedom dynamics system will be studied in this section.

To explain the stability study, let a generic equation corresponding to a degree-of-freedom "h" of the diagonalized system of equations (4.7) be considered,

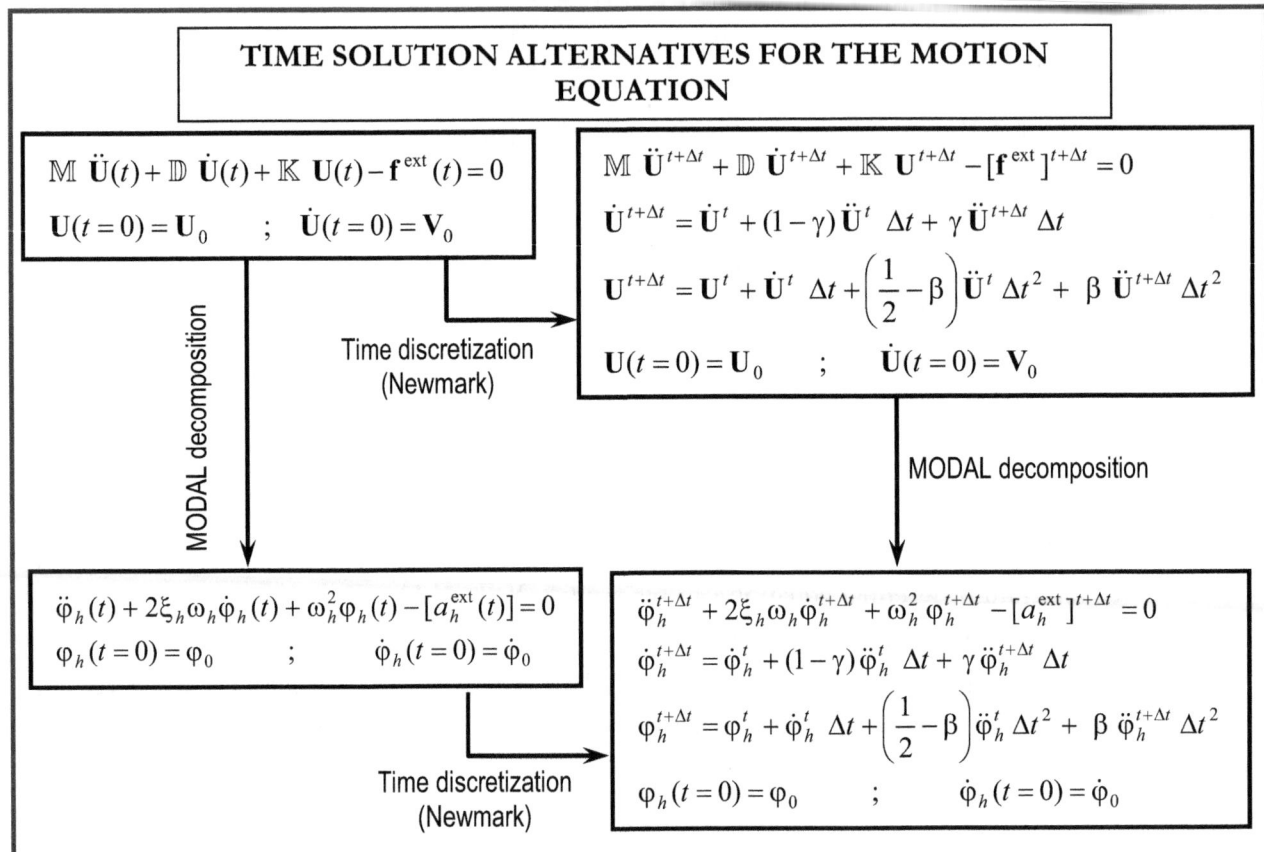

Figure 4.1 – Time solution alternatives for the motion equation in an elastic linear problem.

$$\ddot{\varphi}_h(t) + 2\xi_h\omega_h\dot{\varphi}_h(t) + \omega_h^2\varphi_h(t) - [a_h^{\text{ext}}(t)] = 0 \quad , \quad \forall \quad 1 \le h \le \text{n} - \text{GL} \tag{4.8}$$

The solution for this system of equations, through a one-step direct numerical integration technique, has the following form, known as *recurrent* formula[6],

$$\mathbf{X}^{t+\theta\Delta t} = \mathbf{A}^\theta \cdot \mathbf{X}^t + \mathbf{L}^{t+\theta\Delta t} \quad , \text{with} \quad 0 \le \theta \le \infty \tag{4.9}$$

where $\mathbf{X} = \{\ddot{\varphi}_h, \dot{\varphi}_h, \varphi_h\}$ and the matrix $\mathbf{A}$ and the vector $\mathbf{L}$ are the operators of integration and force, respectively. Each of these operators must be defined for each kind of resolution algorithm used. Once the time discretization is carried out by the Newmark method (see Figure 4.1) an algebraic system is then obtained, which can also be written as the general form of equation (4.9).

4.4 The dynamic equilibrium equation and its convergence-consistency and stability

Given a dynamic equilibrium equation in the form of a diagonalized second-order linear differential equation system (written according to (4.9)), it can be stated that: *the numerical solution procedure will be convergent if both conditions of stability and convergence are met simultaneously*[7]. In other words, the convergence condition of a solution algorithm requires that both consistency and stability conditions are met.

The **stability condition**[a] requires that the numerical solution is bonded for any arbitrary initial condition. It is obtained from the compliance of some algebraic conditions in the amplification matrix $\mathbf{A}$ (equation (4.9)) and will be studied in the next section.

This section will deal with the **consistency condition**[b] in detail, which requires truncation error limit of the solution as a consequence of the finite difference method used for the time discretization of the ordinary differential equation of the dynamic equilibrium.

Assuming an approximated solution of the dynamic problem $^*\mathbf{X}^{t+\theta\Delta t}$, in time $(t + \theta\Delta t)$, the round-off error in this solution is obtained from the general solution of the movement differential equation (4.9). This is,

$$^*\mathbf{X}^{t+\theta\Delta t} - \mathbf{A}^\theta \cdot {^*\mathbf{X}^t} - \mathbf{L}^{t+\theta\Delta t} = \boldsymbol{\tau}^t \quad , \text{ with } 0 \le \theta \le \infty \tag{4.10}$$

where $\boldsymbol{\tau}^t$ is the local error matrix and the asterisk (*) shows that it is an approximate solution. The simple solution algorithm meets the *consistency condition* if the matrix $\mathbf{A}$ is spectrally stable and at time $t + \theta\Delta t$ with $\theta = 1$, the following condition is satisfied,

$$\|\mathbf{e}^t\| = t \cdot c \cdot \Delta t^{K+1} \quad \forall \quad t \in [0, T] \tag{4.11}$$

where c is a constant independent of the time increment Δt and $K > 0$ is the convergence relation.

The error equation in the solution obtained by using finite differences comes from the difference between the right solution and the approximated one. Thus, substracting both solutions we have,

---

[6] Belytschko T. and Hughes T. (1983). *Computational methods for transient analysis*. North-Holland.
[7] Greenberg M. (1978). *Foundation of applied mathematics*. Prentice-Hall
[a] **NOTE: Stability condition**: bounded solution for any initial condition.
[b] **NOTE: Consistency condition**: if the matrix A is spectrally stable.

$$\begin{cases} \mathbf{X}^{t+\theta\Delta t} - \mathbf{A}^{\theta} \cdot \mathbf{X}^{t} - \mathbf{L}^{t+\theta\Delta t} = \mathbf{0} \\ {}^{*}\mathbf{X}^{t+\theta\Delta t} - \mathbf{A}^{\theta} \cdot {}^{*}\mathbf{X}^{t} - \mathbf{L}^{t+\theta\Delta t} - \boldsymbol{\tau}^{t} = \mathbf{0} \end{cases} \qquad (4.12)$$

Then the error expression at current time, i.e. at time $t + \theta\Delta t$, will be,

$$\left(\mathbf{X}^{t+\theta\Delta t} - {}^{*}\mathbf{X}^{t+\theta\Delta t}\right) - \mathbf{A}^{\theta} \cdot \left(\mathbf{X}^{t} - {}^{*}\mathbf{X}^{t}\right) - \boldsymbol{\tau}^{t} = \mathbf{0}$$
$$\mathbf{e}^{t+\theta\Delta t} - \mathbf{A}^{\theta} \cdot \mathbf{e}^{t} - \boldsymbol{\tau}^{t} = \mathbf{0} \qquad (4.13)$$

Following the same conceptual idea, the error in the previous time step (t) is then obtained,

$$\mathbf{e}^{t} - \mathbf{A}^{\theta} \cdot \mathbf{e}^{t-\theta\Delta t} - \boldsymbol{\tau}^{t-\theta\Delta t} = \mathbf{0} \qquad (4.14)$$

By substituting the last equation representing the error in the previous step (4.14) into the error equation at current time (4.13), in order to delete $\mathbf{e}^{t}$ from this last equation, then,

$$\mathbf{e}^{t+\theta\Delta t} - \mathbf{A}^{\theta} \cdot \left[\mathbf{A}^{\theta} \cdot \mathbf{e}^{t-\theta\Delta t} + \boldsymbol{\tau}^{t-\theta\Delta t}\right] - \boldsymbol{\tau}^{t} = \mathbf{0}$$
$$\mathbf{e}^{t+\theta\Delta t} - \left[\mathbf{A}^{\theta}\right]^{2} \cdot \mathbf{e}^{t-\theta\Delta t} - \left[\mathbf{A}^{\theta}\right] \cdot \boldsymbol{\tau}^{t-\theta\Delta t} - \boldsymbol{\tau}^{t} = \mathbf{0} \qquad (4.15)$$

By repeating now the deductive process to delete the error $\mathbf{e}^{t-\theta\Delta t}$ from the above equation then,

$$\mathbf{e}^{t+\theta\Delta t} - \left[\mathbf{A}^{\theta}\right]^{3} \cdot \mathbf{e}^{t-2\theta\Delta t} - \left[\mathbf{A}^{\theta}\right]^{2} \cdot \boldsymbol{\tau}^{t-2\theta\Delta t} - \left[\mathbf{A}^{\theta}\right] \cdot \boldsymbol{\tau}^{t-\theta\Delta t} - \boldsymbol{\tau}^{t} = \mathbf{0} \qquad (4.16)$$

And so on, following the accumulated error until the current time $(t_{n} = n \cdot \Delta t)$ can be written. Thus, during the whole solution process,

$$\mathbf{e}^{t+\theta\Delta t} - \left[\mathbf{A}^{\theta}\right]^{n+1} \cdot \mathbf{e}^{t=0} - \sum_{i=1}^{n}\left[\mathbf{A}^{\theta}\right]^{i} \cdot \boldsymbol{\tau}^{t-i\theta\Delta t} = \mathbf{0} \qquad (4.17)$$

The second term can be neglected as it is much smaller than the accumulation of errors in the rest, $\left[\mathbf{A}^{\theta}\right]^{n+1} \cdot \mathbf{e}^{t=0} << \sum_{i=1}^{n}\left[\mathbf{A}^{\theta}\right]^{i} \cdot \boldsymbol{\tau}^{t-i\theta\Delta t}$. Then, we can directly obtain the accumulated error in (t) instead of $(t + \theta \cdot \Delta t)$, and its norm must be bounded in the following form,

$$\left\|\mathbf{e}^{t}\right\| = \left\|\sum_{i=1}^{n-1}\left[\mathbf{A}^{\theta}\right]^{i} \cdot \boldsymbol{\tau}^{t-(i-1)\theta\Delta t}\right\| \leq t \cdot c \cdot \Delta t^{K} \qquad (4.18)$$

This conclusion, also known as "Equivalence Theorem of Lax" for initial value problems[7], leads to an obvious fact, when the time increment tends to zero the error in the finite-differences approximate solution dissapears ($\left\|\mathbf{e}^{t}\right\| \to 0$, si $\Delta t \to 0$).

4.5 Solution stability of second-order linear symmetric systems

4.5.1 Stability analysis procedure

The stability and consistency analysis of a time solution method of a second-order ordinary differential equation with constant coefficients will help to chose the suitable time increment that guarantees a convergent time solution, regardless of the initial conditions of the problem.

The stability study of the method is carried out by inspecting the numerical behavior of the solution and requiring that it is bounded for any arbitrary initial condition. Generally, the stability of a time integration method is studied on homogeneous equation systems, free of external actions $[a_h^{ext}]^{t+\Delta t} = 0$ and, in this particular case, it is a free vibration study. Thus, the solution of the dynamic equilibrium equation (4.9) is written as,

$$\boldsymbol{X}^{t+\theta\Delta t} = \boldsymbol{A}^\theta \cdot \boldsymbol{X}^t \quad ,\text{with} \quad 0 \le \theta \le \infty \tag{4.19}$$

To get a stable solution the matrix $\mathbf{A}^\theta$ must be bounded for $\theta \to \infty$. For this, the maximum eigenvalue $\max_k |\mu_k|$ of matrix $\mathbf{A}$ (or spectral ratio $\rho(\mathbf{A})$) must be less or equal than the norm of this matrix $\rho(\mathbf{A}) \le \|\mathbf{A}\|$. Then, the stablity is related to the growing relationship, or decreasing of $\mathbf{A}^\theta$ to the power of θ, also called as the spectral stability condition. To prove the convergence it is then necessary that the following condition be satisfied,

$$\left\|\mathbf{A}^\theta\right\| \le \text{const.} \quad , \quad \forall\, \theta \tag{4.20}$$

This requirement is met if:
1. The spectral radius is less than the unit $\rho(\mathbf{A}) \le 1$; the spectral radius of matrix $\mathbf{A}$ is the magnitude of its maximum eigenvalue. Thus,

$$\rho(\mathbf{A}) = \max_k |\mu_k| \quad \text{with} \quad k = 1,2,3 \tag{4.21}$$

2. The moduli of the eigenvalues of $\mathbf{A}$ must be less than the unit (contained in the above condition).

A matrix $\mathbf{A}$ that meets these two conditions is stable or spectrally stable.

4.5.2 Determination of $\mathbf{A}$ and $\mathbf{L}$ for Newmark

With the Newmark method the equilibrium is achieved in $(t + \Delta t)$ and therefore equation (4.8) is written as,

$$\ddot{\varphi}_h^{t+\Delta t} + 2\xi_h\omega_h\dot{\varphi}_h^{t+\Delta t} + \omega_h^2\,\varphi_h^{t+\Delta t} - [a_h^{ext}]^{t+\Delta t} = 0 \tag{4.22}$$

so that its solution in displacements and velocities has the following form (see chapter B3 and Figure 4.1),

$$\begin{cases} \dot{\varphi}_h^{t+\Delta t} = \dot{\varphi}_h^t + (1-\gamma)\,\ddot{\varphi}_h^t\,\Delta t + \gamma\,\ddot{\varphi}_h^{t+\Delta t}\,\Delta t \\[2mm] \varphi_h^{t+\Delta t} = \varphi_h^t + \dot{\varphi}_h^t\,\Delta t + \left(\dfrac{1}{2}-\beta\right)\ddot{\varphi}_h^t\,\Delta t^2 + \beta\,\ddot{\varphi}_h^{t+\Delta t}\,\Delta t^2 \end{cases} \tag{4.23}$$

where γ and β are parameters to control the method's stability, so this becomes unconditionally stable for values of $\gamma = 0.5$ and $\beta = 0.25$. These parameters are also consistent with the linear acceleration method (B3.3.2.1).

The elimination of acceleration $\ddot{\varphi}_h^t$ and $\ddot{\varphi}_h^{t+\Delta t}$ through the substitution of equation (4.23) into (4.22) enables a compact expression for matrix $\mathbf{A}$ and vector $\mathbf{L}$ to be found, such that the solution's acceleration is implicit and only expressed as a function of the velocity and displacement $\mathbf{X} = \{\dot{\varphi}_h, \varphi_h\}$. This enables the spectral radius of a (2x2) matrix to be found instead of a (3x3) matrix, and the solution is relatively simpler[*]. Therefore, we have,

$$\mathbf{A} = \mathbf{A}_1^{-1} \cdot \mathbf{A}_2 \quad , \quad \mathbf{L}^t = \mathbf{A}_1^{-1} \left\{ \begin{array}{c} \dfrac{\Delta t^2}{2}\left[(1-2\beta)\,[a_h^{ext}]^t + 2\beta[a_h^{ext}]^{t+\Delta t} \right] \\ \Delta t\left[(1-\gamma)\,[a_h^{ext}]^t + \gamma[a_h^{ext}]^{t+\Delta t} \right] \end{array} \right\} \tag{4.24}$$

where the matrices that make up the integration matrix $\mathbf{A}$ have the following expressions,

$$\mathbf{A}_1 = \begin{bmatrix} 1 + \Delta t^2 \beta \omega_h^2 & 2\Delta t^2 \beta \xi_h\,\omega_h \\ \Delta t\,\gamma\,\omega_h^2 & 1 + 2\Delta t\,\gamma\,\xi_h\,\omega_h \end{bmatrix}$$

$$\mathbf{A}_2 = \begin{bmatrix} 1 - \dfrac{\Delta t^2}{2}(1-2\beta)\omega_h^2 & \Delta t\left[1 - \Delta t(1-2\beta)\xi_h\omega_h \right] \\ -\Delta t(1-\gamma)\omega_h^2 & 1 - 2\Delta t(1-\gamma)\xi_h\,\omega_h \end{bmatrix} \tag{4.25}$$

Thus, the eigenvalues of $\mathbf{A}$ can be obtained from the following characteristic equation,

$$0 = \det(\mathbf{A} - \mu\mathbf{I}) = \mu^2 - (A_{11} + A_{22})\mu + (A_{11}A_{22} - A_{12}A_{21}) \tag{4.26}$$

Then, the roots of this equation are,

$$\mu_{1,2} = \left(\frac{A_{11} + A_{22}}{2} \right) \pm \sqrt{ \left(\frac{A_{11} + A_{22}}{2} \right)^2 - (A_{11}A_{22} - A_{12}A_{21}) } = a_1 \pm \sqrt{a_1^2 - a_2} \tag{4.27}$$

with $a_1 = (A_{11} + A_{22})/2$ and $a_2 = (A_{11}A_{22} - A_{12}A_{21})$. From this last equation and from the stability condition $\rho(\mathbf{A}) \le 1$ the domain where the solution is stable and whose boundary is delimited by doing $\mu = 1$ and adopting the negative sign in equation (4.27) can be graphically established (see Figure 4.2). From this, the following relation results,

$$0 = 1 - 2a_1 + a_2 \tag{4.28}$$

Now, graphing the possible solutions in the space a_1-a_2, it is observed that there is stability in the solution of this algorithm when equation (4.27) has the roots within the gray zone of Figure

---

[*] NOTE: An alternative way may be not to eliminate the acceleration and substitute equations (4.23) into (4.8), where the acceleration $\ddot{\varphi}_h^{t+\Delta t}$ is obtained and then substituting again the latter into (4.23), from where $\varphi_h^{t+\Delta t}$ and $\dot{\varphi}_h^{t+\Delta t}$ are obtained. By reordering these equations and substituting into (4.9), for $\theta = 1$, then,

$$\left\{ \begin{array}{c} \ddot{\varphi}_h^{t+\Delta t} \\ \dot{\varphi}_h^{t+\Delta t} \\ \varphi_h^{t+\Delta t} \end{array} \right\} = \begin{bmatrix} \left[-\left(\frac{1}{2}-\beta\right)x - 2(1-\gamma)\,y \right] & [-x-2y]\frac{1}{\Delta t} & -\frac{x}{\Delta t^2} \\ \Delta t\left[1 - \gamma - \left(\frac{1}{2}-\beta\right)\gamma\,x - 2(1-\gamma)\,\gamma\,y \right] & [1 - x\,\gamma - 2\,\gamma\,y] & -\frac{x\,\gamma}{\Delta t} \\ \Delta t^2\left[\frac{1}{2} - \beta - \left(\frac{1}{2}-\beta\right)\beta \right]x - 2(1-\gamma)\,\beta\,y & \Delta t[1 - x\beta - 2\beta y] & (1-\beta\,x) \end{bmatrix} \cdot \left\{ \begin{array}{c} \ddot{\varphi}_h^t \\ \dot{\varphi}_h^t \\ \varphi_h^t \end{array} \right\} + \left\{ \begin{array}{c} \dfrac{\beta}{\omega_h^2\Delta t^2} \\ \dfrac{\beta\,\gamma}{\omega_h^2\,\Delta t} \\ \dfrac{\beta\,x}{\omega_h^2} \end{array} \right\} \cdot [a_h^{ext}]^{t+\Delta t}$$

where $x = \left(\dfrac{1}{\omega_h^2\Delta t^2} + \dfrac{2\,\xi_h\,\gamma}{\omega_h\Delta t} + \beta \right)^{-1}$ and $y = \dfrac{\xi_h\,x}{\omega_h\Delta t}$. The problem with this procedure is that the spectral radius is obtained from a cubic equation, while it is easier to use a procedure that involves eliminating the acceleration and enables the quadratic equation to be obtained for the spectral radius.

4.2. Thus, to be unconditionally stable the following relation that defines the vertices of Figure 4.2 must be satisfied,

$$-\frac{a_2+1}{2} \le a_1 \le \frac{a_2+1}{2} \tag{4.29}$$

Substituting $2a_1 = (A_{11} + A_{22})$ and $a_2 = (A_{11}A_{22} - A_{12}A_{21})$ into equation (4.28) for the coefficients of matrix $\mathbf{A}$ (equation (4.24)), the magnitudes γ and β that guarantee a stable solution with the Newmark algorithm result. The variation ends of these coefficients meet the following inequality,

$$2\beta \ge \gamma \ge \tfrac{1}{2} \tag{4.30}$$

An example of the numerical solution of this problem can be found in "Appendix 1" of this chapter. Here a FORTRAN program is described to obtain the spectral radius and to verify the condition stated in equation (4.30).

Time equilibrium for Newmark with: $\beta = \tfrac{1}{4}$ and $\gamma = \tfrac{1}{2}$:

The stability study of the Newmark method shows that is unconditionally stable for $\beta = \tfrac{1}{4}$ and $\gamma = \tfrac{1}{2}$. Substituting these coefficients into equation (4.23) the following results,

$$\begin{cases}
\varphi_h^{t+\Delta t} = \varphi_h^t + \dot{\varphi}_h^t\,\Delta t + \dfrac{\Delta t^2}{4}\left[\ddot{\varphi}_h^t + \ddot{\varphi}_h^{t+\Delta t}\right] \\[2mm]
\dot{\varphi}_h^{t+\Delta t} = \dot{\varphi}_h^t + \dfrac{\Delta t}{2}\left[\ddot{\varphi}_h^t + \ddot{\varphi}_h^{t+\Delta t}\right] \\[2mm]
\ddot{\varphi}_h^{t+\Delta t} = \ddot{\varphi}_h^t + \dfrac{4}{\Delta t^2}\left[\Delta\varphi_h^{t+\Delta t} - \dot{\varphi}_h^t\,\Delta t\right]
\end{cases} \tag{4.31}$$

Substituting the second one into the the first one the displacement obtained from the average velocity at time increment Δt can be written as,

$$\varphi_h^{t+\Delta t} = \varphi_h^t + \frac{\Delta t}{2}\left[\dot{\varphi}_h^t + \dot{\varphi}_h^{t+\Delta t}\right] \tag{4.32}$$

Substituting these magnitudes into the equilibrium equation (4.22), the following results

$$\ddot{\varphi}_h^{t+\Delta t} + 2\xi_h\omega_h\dot{\varphi}_h^{t+\Delta t} + \omega_h^2\,\varphi_h^{t+\Delta t} - [a_h^{\text{ext}}]^{t+\Delta t} = 0$$

$$\left\{\ddot{\varphi}_h^t + \frac{4}{\Delta t^2}\left[\Delta\varphi_h^{t+\Delta t} - \dot{\varphi}_h^t\,\Delta t\right]\right\} + 2\xi_h\omega_h\left\{\dot{\varphi}_h^t + \frac{\Delta t}{2}\left[\ddot{\varphi}_h^t + \ddot{\varphi}_h^t + \frac{4}{\Delta t^2}\left[\Delta\varphi_h^{t+\Delta t} - \dot{\varphi}_h^t\,\Delta t\right]\right]\right\} + $$

$$+ \omega_h^2\left\{\varphi_h^t + \dot{\varphi}_h^t\,\Delta t + \frac{\Delta t^2}{4}\left[\ddot{\varphi}_h^t + \ddot{\varphi}_h^t + \frac{4}{\Delta t^2}\left[\Delta\varphi_h^{t+\Delta t} - \dot{\varphi}_h^t\,\Delta t\right]\right]\right\} - [a_h^{\text{ext}}]^{t+\Delta t} = 0 \tag{4.33}$$

Now, ordering the terms the following equation is obtained for the solution that guarantees the stable solution of the linear elastic dynamic problem

$$0 = \hat{a}_h^{t+\Delta t} + J_h^{t+\Delta t} \cdot \Delta \varphi_h^{t+\Delta t}$$

$$J_h^{t+\Delta t} = \left[\frac{4}{\Delta t^2} + 4\xi_h \omega_h + \omega_h^2 \right]_h^{t+\Delta t}$$

$$\hat{a}_h^{t+\Delta t} = \left\{ \left[2\xi_h \omega_h \Delta t + 2\omega_h^2 \Delta t^2 \right] \ddot{\varphi}_h^t - \left[\frac{4}{\Delta t} + 2\xi_h \omega_h \Delta t^2 \right] \dot{\varphi}_h^t \right\} +$$

$$+ \underbrace{\ddot{\varphi}_h^t + 2\xi_h \omega_h \dot{\varphi}_h^t + \omega_h^2 \varphi_h^t}_{a_h^t} - [a_h^{\text{ext}}]^{t+\Delta t}$$

(4.34)

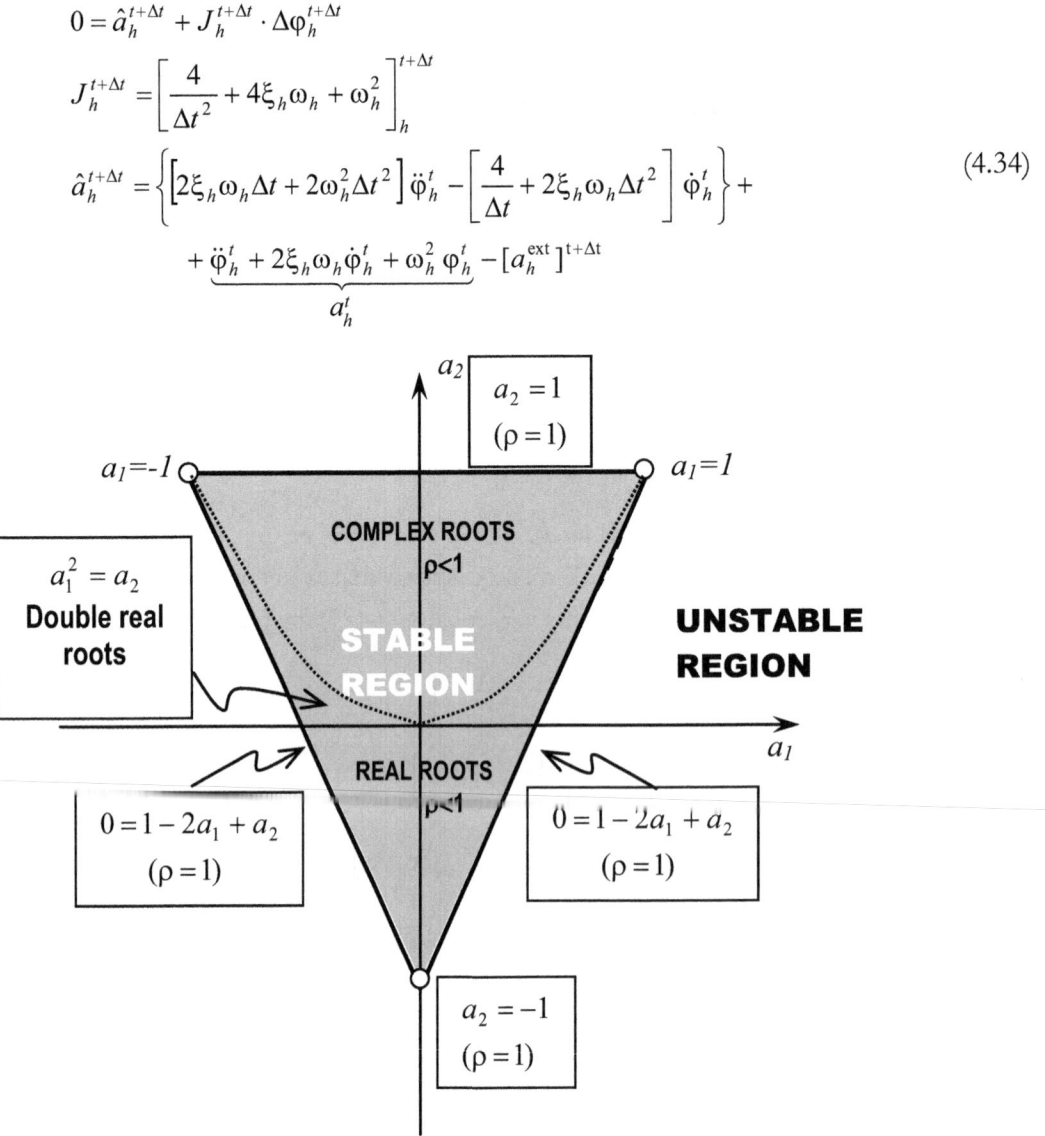

Figure 4.2 – Stability regions in plane $a_1 - a_2$.

The system of equations for the linear elastic equilibrium could also be obtained by eliminating the previous process of diagonalization from equation (4.1), or by using an analogous procedure as followed in chapter B3 with equations (4.23) and (4.24),

$$\mathbf{0} = \hat{\mathbf{f}}^{t+\Delta t} + \mathbb{J}^{t+\Delta t} \cdot \Delta \mathbf{U}^{t+\Delta t}$$

$$\text{with:} \quad \mathbb{J}^{t+\Delta t} = \left[\mathbb{M} \frac{4}{\Delta t^2} + \mathbb{D}^T \frac{2}{\Delta t} + \mathbb{K}^T \right]^{t+\Delta t}$$

$$\text{and} \quad : \quad \hat{\mathbf{f}}^{t+\Delta t} = \mathbb{M} \left[\frac{4}{\Delta t^2} \left(\Delta \mathbf{U}^{t+\Delta t} - \dot{\mathbf{U}}^t \Delta t \right) - \ddot{\mathbf{U}}^t \right] +$$

$$+ \left[\mathbf{f}^{\text{int}} \right]^{t+\Delta t} - \left[\mathbf{f}^{\text{ext}} \right]^{t+\Delta t}$$

(4.35)

4.5.3 Determination of A and L for central differences-Newmark's explicit form

The constants β and γ allow the Newmark method to become a totally explicit method such as the *central differences*[8], instead of an implicit method as discussed in chapter 3. A very versatile methodology fitting the solution for different particular problems is achieved by the changes made into Newmark's constants.

The following dynamic equilibrium in time (t) is established for the movement equation (4.8) for free oscillation of the h degree-of-freedom,

$$\ddot{\varphi}_h^t + 2\xi_h\omega_h\dot{\varphi}_h^t + \omega_h^2\,\varphi_h^t = 0 \qquad (4.36)$$

Then the acceleration and velocity expressed in terms of central differences in time (*t*) results[1,2,4,8],

$$\begin{cases} \ddot{\varphi}_h^t = \dfrac{1}{\Delta t^2}\left(\varphi_h^{t+\Delta t} - 2\varphi_h^t + \varphi_h^{t-\Delta t}\right) \\[2mm] \dot{\varphi}_h^t = \dfrac{1}{2\,\Delta t}\left(\varphi_h^{t+\Delta t} - \varphi_h^{t-\Delta t}\right) \end{cases} \qquad (4.37)$$

substituting equations (4.37) into equation (4.36) the displacement $\varphi_h^{t+\Delta t}$ is obtained,

$$\frac{1}{\Delta t^2}\left(\varphi_h^{t+\Delta t} - 2\varphi_h^t + \varphi_h^{t-\Delta t}\right) + 2\xi_h\omega_h\frac{1}{2\,\Delta t}\left(\varphi_h^{t+\Delta t} - \varphi_h^{t-\Delta t}\right) + \omega_h^2\varphi_h^t = 0$$

$$\varphi_h^{t+\Delta t} = \left[\frac{2-\omega_h^2\,\Delta t^2}{1+\xi_h\omega_h\Delta t}\right]\cdot\varphi_h^t - \left[\frac{1-\xi_h\omega_h\Delta t}{1+\xi_h\omega_h\Delta t}\right]\cdot\varphi_h^{t-\Delta t} \qquad (4.38)$$

which can be written in the form (4.19), for $\theta = 1$, as follows

$$\mathbf{X}^{t+\Delta t} = \mathbf{A}\cdot\mathbf{X}^t \quad \Rightarrow \quad \begin{Bmatrix} \varphi_h^{t+\Delta t} \\ \varphi_h^t \end{Bmatrix} = \begin{bmatrix} \dfrac{2-\omega_h^2\,\Delta t^2}{1+\xi_h\omega_h\Delta t} & -\dfrac{1-\xi_h\omega_h\Delta t}{1+\xi_h\omega_h\Delta t} \\ 1 & 0 \end{bmatrix}\cdot\begin{Bmatrix} \varphi_h^t \\ \varphi_h^{t-\Delta t} \end{Bmatrix} \qquad (4.39)$$

In such a way that the eigenvalues μ_k of the characteristic equation of the amplification matrix $\mathbf{A}$, for a non-damping problem $\xi_h = 0$, are obtained from the following expression

$$\det\left\{\begin{bmatrix} 2-\omega_h^2\,\Delta t^2 & -1 \\ 1 & 0 \end{bmatrix} - \mu_k\begin{bmatrix} 1 & 0 \\ 0 & 1 \end{bmatrix}\right\} = 0$$

$$\mu_k^2 - \mu_k\left(2-\omega_h^2\,\Delta t^2\right) + 1 = 0 \quad \Rightarrow \quad \mu_{1,2} = \frac{\left(2-\omega_h^2\,\Delta t^2\right)}{2} \pm \sqrt{\frac{\left(2-\omega_h^2\,\Delta t^2\right)^2}{4} - 1} \qquad (4.40)$$

Imposing the condition of spectral radius less than or equal to the unity (section 4.5.1) to the eigenvalues already obtained, the time increment Δt can be obtained which guarantees the stability of the solution for the central finite difference method with null damping $\xi_h = 0$. Thus,

$$\rho(\mathbf{A}) = \max|\mu_1, \mu_2| \le 1 \quad \Rightarrow \quad \Delta t \le \frac{2}{\omega_h} \quad \text{or also} \quad \frac{\Delta t}{T} \le \frac{1}{\pi} \qquad (4.41)$$

These central difference equations can be obtained by considering $\beta = 0$ and $\gamma = \frac{1}{2}$ in the Newmark method, followed by some mathematical manipulation. With this more general

---

[8] Clough R. and Penzien J. (1977). Dynamics of Structures. Mc Graw Hill – N. York.

approach the stability of this method can be studied by using all the concepts described in section 4.5.2. After substituting $\beta = 0$ and $\gamma = \frac{1}{2}$ into (4.23) then,

$$\begin{cases} \dot{\phi}_h^{t+\Delta t} = \dot{\phi}_h^t + \ddot{\phi}_h^t \dfrac{\Delta t}{2} + \ddot{\phi}_h^{t+\Delta t} \dfrac{\Delta t}{2} \\[2mm] \phi_h^{t+\Delta t} = \phi_h^t + \dot{\phi}_h^t \Delta t + \dfrac{1}{2} \ddot{\phi}_h^t \Delta t^2 \end{cases} \tag{4.42}$$

Using equations (4.42) and (4.22), the acceleration terms can be deleted. The general solution as a condensed expression for the matrix $\mathbf{A}$ and the vector $\mathbf{L}$ can be witten. Thus, the solution will only be expressed as a function of the velocity and displacement $\mathbf{X} = \{\dot{\phi}_h, \phi_h\}$, leading the following expression of the matrix $\mathbf{A}$

$$\mathbf{A} = \begin{bmatrix} \dfrac{\Delta t}{2}\left(1 - \dfrac{x}{2} - y\right) & \left(1 - \dfrac{x}{2} - y\right) \\[4mm] \dfrac{\Delta t^2}{2} & \Delta t \end{bmatrix} \tag{4.43}$$

where $x = \left(\dfrac{1}{\omega_h^2 \Delta t^2} + \dfrac{\xi_h}{\omega_h \Delta t}\right)^{-1}$ and $y = \dfrac{\xi_h x}{\omega_h \Delta t}$.

The matrix $\mathbf{A}$ could also be obtained by using central differences and substituting coefficients $\beta = 0$ and $\gamma = \frac{1}{2}$ into equation (4.25). Then,

$$\left. \begin{aligned} \mathbf{A}_1 &= \begin{bmatrix} 1 & 0 \\[1mm] \dfrac{\Delta t}{2}\omega_h^2 & 1 + \Delta t \xi_h \omega_h \end{bmatrix} \\[4mm] \mathbf{A}_2 &= \begin{bmatrix} 1 - \dfrac{\Delta t^2}{2}\omega_h^2 & \Delta t[1 - \Delta t \xi_h \omega_h] \\[3mm] -\dfrac{\Delta t}{2}\omega_h^2 & 1 - \Delta t \xi_h \omega_h \end{bmatrix} \end{aligned} \right\} \Rightarrow \tag{4.44}$$

$$\Rightarrow \mathbf{A} = \mathbf{A}_1^{-1}\mathbf{A}_2 = \begin{bmatrix} 1 - \dfrac{\Delta t^2}{2}\omega_h^2 & \Delta t[1 - \Delta t \xi_h \omega_h] \\[4mm] \dfrac{\omega_h^2 \Delta t(-2 + \omega_h^2 \Delta t^2 - 2\Delta t)}{4[1 + \Delta t \xi_h \omega_h]} & \dfrac{(-\omega_h^2 \Delta t^2 + \omega_h^3 \Delta t^3 \xi_h + 2 - 2\omega_h \Delta t \ \xi_h)}{2[1 + \Delta t \xi_h \omega_h]} \end{bmatrix}$$

Once the spectral radius is obtained, it can be concluded that the stability solution is only achieved provided that there is a time increment as a function of the maximum angular frequency and of the damping term,

$$(\omega_h)^{max} \Delta t \le 2(\sqrt{1 + \xi^2} - \xi) \tag{4.45}$$

This condition involves a high computer cost when dealing with a finite element problem solution. For instance, in a null viscosity problem $\xi = 0$, the structural maximum angular frequency for each mode $"h"$ must be obtained first and then the time step must satisfy $(\omega_h)^{max} \Delta t \le 2$ (see the same result in equation (4.41)). This can be carried out by simply obtaining the frequencies for each finite element and then getting the maximum frequency that will always meet the following condition[9],

---

[9] Cesari F. (1982). *Metodi di calcolo nella dinamica delle strutture*. Ed. Pitagora, Bologna.

$$(\omega_h)^{\max} \le \max_{\forall e}(\omega^e)^{\max} \le \frac{2(\sqrt{1+\xi^2}-\xi)}{\Delta t} \qquad (4.46)$$

Time equilibrium expression for Newmark with: $\beta = 0$ and $\gamma = \frac{1}{2}$ *(central differences)*

Substituting equations (4.42) into the dynamic equilibrium differential equation in time $(t + \Delta t)$, written in terms of forces, the linearized form in time of this equation is obtained. Thus,

$$m_h \ddot{\varphi}_h^{t+\Delta t} + d_h \dot{\varphi}_h^{t+\Delta t} + k_h \varphi_h^{t+\Delta t} - [f_h^{\text{ext}}]^{t+\Delta t} = 0$$

$$m_h \ddot{\varphi}_h^{t+\Delta t} + d_h \left[\dot{\varphi}_h^t + \ddot{\varphi}_h^t \frac{\Delta t}{2} + \ddot{\varphi}_h^{t+\Delta t} \frac{\Delta t}{2} \right] + k_h \left[\varphi_h^t + \dot{\varphi}_h^t \Delta t + \frac{1}{2} \ddot{\varphi}_h^t \Delta t^2 \right] - [f_h^{\text{ext}}]^{t+\Delta t} = 0 \qquad (4.47)$$

By grouping terms, the following linearized equation in time for the "h" degree-of-freedom can be written,

$$\left[m_h + d_h \frac{\Delta t}{2} \right] \ddot{\varphi}_h^{t+\Delta t} + \underbrace{\left\{ d_h \left[\dot{\varphi}_h^t + \ddot{\varphi}_h^t \frac{\Delta t}{2} \right] + k_h \left[\varphi_h^t + \dot{\varphi}_h^t \Delta t + \frac{1}{2} \ddot{\varphi}_h^t \Delta t^2 \right] - [f_h^{\text{ext}}]^{t+\Delta t} \right\}}_{\hat{f}_h^{t+\Delta t}} = 0 \qquad (4.48)$$

$$\left[m_h + d_h \frac{\Delta t}{2} \right] \ddot{\varphi}_h^{t+\Delta t} + \hat{f}_h^{t+\Delta t} = 0$$

The linear-elastic dynamic equilibrium system of equations could also be obtained from equation (4.1) without the diagonalization of the system (equation (4.7)). This is written as,

$$\left[\mathbb{M} + \mathbb{D}\frac{\Delta t}{2} \right] \ddot{\mathbf{U}}^{t+\Delta t} + \underbrace{\mathbb{D}\left[\dot{\mathbf{U}}^t + \ddot{\mathbf{U}}^t \frac{\Delta t}{2} \right] + \mathbb{K}\left[\mathbf{U}^t + \dot{\mathbf{U}}^t \Delta t + \ddot{\mathbf{U}}^t \frac{\Delta t^2}{2} \right] - \left[\mathbf{f}^{\text{ext}} \right]^{t+\Delta t}}_{\hat{\mathbf{f}}^{t+\Delta t}} = 0 \qquad (4.49)$$

4.6 Solution stability of second-order nonlinear symmetric systems

4.6.1 Stability of the linearized equation

Once the concept of stability solution of linear systems has been studied, the following nonlinear ordinary differential equations system, whose nonlinearity lies in the evolution of the internal forces, is considered

$$\boxed{\Delta \mathbf{f} = \mathbb{M}\,\ddot{\mathbf{U}}(t + \Delta t) + \mathbf{f}^{\text{int}}(\dot{\mathbf{U}}, \mathbf{U}, t + \Delta t) - \mathbf{f}^{\text{ext}}(t + \Delta t) = \mathbf{0}} \qquad (4.50)$$

Its linearization can be written, as observed in (2.77), by using a Taylor approximation truncated in the second term,

$$0={}^{i+1}\left[\Delta\mathbf{f}\right]_{\Omega}^{t+\Delta t}\cong{}^{i}\left[\Delta\mathbf{f}\right]_{\Omega}^{t+\Delta t}+{}^{i}\left(\frac{\partial\Delta\mathbf{f}}{\partial\mathbf{U}}\right)_{\Omega}^{t+\Delta t}\cdot{}^{i+1}\left[\Delta\mathbf{U}\right]_{\Omega}^{t+\Delta t}=$$

$$={}^{i}\left[\Delta\mathbf{f}\right]_{\Omega}^{t+\Delta t}+\underbrace{{}^{i}\left[\mathbb{M}\frac{\partial\ddot{\mathbf{U}}}{\partial\mathbf{U}}+\mathbb{K}^{T}+\mathbb{D}^{T}\frac{\partial\dot{\mathbf{U}}}{\partial\mathbf{U}}-\frac{\partial\mathbf{f}^{\text{ext}}}{\partial\mathbf{U}}\right]_{\Omega}^{t+\Delta t}}_{{}^{i}\mathbb{J}_{\Omega}^{t+\Delta t}}\cdot{}^{i+1}\left[\Delta\mathbf{U}\right]_{\Omega}^{t+\Delta t} \qquad (4.51)$$

It should be noted that in this equation (i) represents the iteration counter and (t) is the time. The solution of the equation is obtained by inverting the Jacobian operator ${}^{i}\mathbb{J}_{\Omega}^{t+\Delta t}$. Thus,

$$\boxed{{}^{i+1}\left[\Delta\mathbf{U}\right]_{\Omega}^{t+\Delta t}=-\left[{}^{i}\mathbb{J}_{\Omega}^{t+\Delta t}\right]^{-1}\cdot{}^{i}\left[\Delta\mathbf{f}\right]_{\Omega}^{t+\Delta t}} \qquad (4.52)$$

Various time algorithms can be used for the solution of this last equation; however, none of them can guarantee the solution's stability. In this formulation it is assumed that $\mathbb{M}$ is constant, $\mathbb{M}$, $\mathbb{K}$ and $\mathbb{D}$ are symmetric, $\mathbb{M}$ and $\mathbb{K}$ are positive defined, and $\mathbb{D}$ is positive semi-defined.

The concept of "stability of the linearized equation" is undoubtely the most widely known method to prove stability of nonlinear problems. In this case the stability of a linear algorithm, studied in previous sections, can only guarantee the necessary condition although not sufficient for stability.

4.6.2 Energy conservation algorithms

In linear behavior, the stability concept involves the energy conservation law imposing a limit to the unlimited growth of the solution (equation (4.20)). On the other hand, in a nonlinear problem the equilibrium linearization in time does not necessarily involve energy conservation. Normally, a pathological growth of energy occurs, as observed in some cases for unconditionally stable algorithms during the linearization. An example of this occurs when using the trapezoidal rule or Newmark's algorithm with $\beta=\frac{1}{4}$ and $\gamma=\frac{1}{2}$. This reasoning highlights whether the stability notions previously proposed for linear problems are the right ones.

This section presents a time solution algorithm, different from the one above mentioned, whose objective is to guarantee permanently the energy conservation condition and consequently enforcing the stability condition of the solution.

These algorithms of energy conservation are based on a modification of the trapezoidal rule (Newmark with $\beta=\frac{1}{4}$ and $\gamma=\frac{1}{2}$) used in nonlinear problems (see chapter 3). In particular, it should be observed that in the absence of external forces, the system conserves the energy both in nonlinear and elasto-dynamic problems. This involves that stability contidions are automatically met. The technique used to modify the trapezoidal rule defines an additional energy conservation through the method of Lagrange multipliers.

The trapezoidal rule algorithm is written in the Euler-Lagrange form linked to a natural functional (see Appendix-2). Also, it is assumed that damping is zero, which means that $\mathbf{f}^{\text{int}}(\dot{\mathbf{U}},\mathbf{U})\equiv\mathbf{f}_{\sigma}^{\text{int}}(\mathbf{U})$, where $\mathbf{f}_{\sigma}^{\text{int}}(\mathbf{U})$ can be interpreted as the gradient of a strain energy function,

$$\mathbf{f}_\sigma^{\mathrm{int}}(\mathbf{U}) = \frac{\partial}{\partial\mathbf{U}}\left[\int_t P_{\mathrm{int}}\,dt\right] = \frac{\partial}{\partial\mathbf{U}}\underbrace{\left[\int_V \rho\,\omega\,dV + \int_t \dot{K}\,dt\right]}_{E(\mathbf{U},\dot{\mathbf{U}})} \Rightarrow \mathbb{K}^T = \frac{\partial}{\partial\mathbf{U}}\mathbf{f}_\sigma^{\mathrm{int}}(\mathbf{U}) = \frac{\partial}{\partial\mathbf{U}}\left[\int_V \rho\,\omega\,dV\right] \quad (4.53)$$

where P_{int} is the introduced power, ω is the internal energy density, $\dot{K}$ is the kinetics power (see chapter B.2) and $E(\mathbf{U},\dot{\mathbf{U}}) = \int_V \rho\,\omega\,dV + \frac{1}{2}\dot{\mathbf{U}}^T : \mathbb{M} : \dot{\mathbf{U}}$ is the total energy system.

It is assumed that $\int_V \rho\omega\,dV \geq 0$ and $\mathbb{K}$ is symmetric, and the solution of the dynamic problem satisfies the *fundamental identity in energy*, which states that,

$$E(\mathbf{U},\dot{\mathbf{U}}) = E(\mathbf{U}_0,\dot{\mathbf{U}}_0) + \underbrace{\int_0^t \dot{U}(\tau) : f^{\mathrm{ext}}(\tau)\,d\tau}_{\substack{\text{Impulse of the}\\ \text{accumulated force } 0\leq\tau\leq t}} \quad (4.54)$$

In this last equation we can see that for a null external force the power is conserved and constant,

$$\text{If}\qquad \mathbf{f}^{\mathrm{ext}}(\tau) = \mathbf{0}\quad \Rightarrow\quad E(\mathbf{U},\dot{\mathbf{U}}) = E(\mathbf{U}_0,\dot{\mathbf{U}}_0) \quad (4.55)$$

In this case the trapezoidal rule can be written as,

$$\Delta\mathbf{f} = \mathbb{M}\,\ddot{\mathbf{U}}^{t+\Delta t} + \left[\mathbf{f}_\sigma^{\mathrm{int}}(\mathbf{U})\right]^{t+\Delta t} - \left[\mathbf{f}^{\mathrm{ext}}\right]^{t+\Delta t} = \mathbf{0} \quad (4.56)$$

which, linearized, leads to equation (4.35). In equation (4.56), according to the trapezoidal rule, the velocity and the displacement are expressed as,

$$\begin{cases} \mathbf{U}^{t+\Delta t} = \mathbf{U}^t + \dfrac{\Delta t}{2}\left[\dot{\mathbf{U}}^t + \dot{\mathbf{U}}^{t+\Delta t}\right] \\[2mm] \dot{\mathbf{U}}^{t+\Delta t} = \dot{\mathbf{U}}^t + \dfrac{\Delta t}{2}\left[\ddot{\mathbf{U}}^t + \ddot{\mathbf{U}}^{t+\Delta t}\right] \end{cases} \quad (4.57)$$

whose initial values are,

$$\begin{cases} \mathbf{U}(t=0) = \mathbf{U}_0 \\[1mm] \dot{\mathbf{U}}(t=0) = \dot{\mathbf{U}}_0 \\[1mm] \ddot{\mathbf{U}}(t=0) = \ddot{\mathbf{U}}_0 = \mathbb{M}^{-1}\left[\mathbf{f}^{\mathrm{ext}} - \mathbf{f}_\sigma^{\mathrm{int}}(\mathbf{U}_0)\right] \end{cases} \quad (4.58)$$

The nonlinear algebraic problem in $\mathbf{U}^{t+\Delta t}$ is obtained by susbtituting equations (4.57) into (4.56).

It is convenient to deal with the nonlinear algebraic problem linearized in time by using Newmark's procedure, as an Euler-Lagrange functional. This form also coincides with Hamilton's functional[7] expressed in $(t + \Delta t)$, as

$$\pi(\mathbf{U}^{t+\Delta t}) = E(\mathbf{U}^{t+\Delta t},\dot{\mathbf{U}}^{t+\Delta t}) + \left[\mathbf{U}^{t+\Delta t}\right]^T \cdot\left\{\hat{\mathbf{f}}^{t+\Delta t} - \left[\mathbf{f}^{\mathrm{int}}\right]^{t+\Delta t}\right\} \quad (4.59)$$

where $\hat{\mathbf{f}}^{t+\Delta t}$ is the linearized external force for the Newmark algorithm as shown in equation (4.35). Substituting $E(\mathbf{U}^{t+\Delta t},\dot{\mathbf{U}}^{t+\Delta t})$ and $\hat{\mathbf{f}}^{t+\Delta t}$ into equation (4.59) we obtain the functional,

$$\pi(\mathbf{U}^{t+\Delta t}) = \frac{2}{\Delta t^2}\left[\mathbf{U}^{t+\Delta t}\right]^T \cdot \mathbb{M}\,\mathbf{U}^{t+\Delta t} + \left[\int_V \rho\omega\, dV\right]^{t+\Delta t} +$$

$$-\left[\mathbf{U}^{t+\Delta t}\right]^T \cdot \left\{\left[\mathbf{f}^{\text{ext}}\right]^{t+\Delta t} + \mathbb{M}\cdot\left[\ddot{\mathbf{U}}^t + \frac{4}{\Delta t^2}\left(\Delta\mathbf{U}^t + \Delta t\,\dot{\mathbf{U}}^t\right)\right]\right\}$$

(4.60)

Its first variation, which makes stationary the functional as

$$\delta\pi(\mathbf{U}^{t+\Delta t}) = \delta\left[\mathbf{U}^{t+\Delta t}\right]^T \cdot \frac{\partial\pi}{\partial\mathbf{U}}\bigg|_{t+\Delta t} = \delta\left[\mathbf{U}^{t+\Delta t}\right]^T \left\{\frac{4}{\Delta t^2}\mathbb{M}\cdot\mathbf{U}^{t+\Delta t} + \mathbf{f}_\sigma^{\text{int}}(\mathbf{U}^{t+\Delta t}) -\right.$$

$$\left. -\left[\mathbf{f}^{\text{ext}}\right]^{t+\Delta t} - \mathbb{M}\cdot\left[\ddot{\mathbf{U}}^t + \frac{4}{\Delta t^2}\left(\Delta\mathbf{U}^t + \Delta t\,\dot{\mathbf{U}}^t\right)\right]\right\} = 0$$

(4.61)

This last expression is zero for any variation $\delta\left[\mathbf{U}^{t+\Delta t}\right] \neq 0$.

Modified trapezoidal rule.

The trapezoidal rule is modified to satisfy the following energy conservation equation during a time step Δt,

$$E^{t+\Delta t} - E^t = \frac{1}{2}\,(\Delta\mathbf{U}^{t+\Delta t})^T \cdot \left[\Delta\mathbf{f}^{\text{ext}}\right]^{t+\Delta t}$$

(4.62)

where $E^{t+\Delta t} = E(\mathbf{U}^{t+\Delta t}, \dot{\mathbf{U}}^{t+\Delta t})$. As observed in the above equation, the energy is conserved and constant for $\mathbf{f}^{\text{ext}} = cte$.

The formulation for the desired algorithm, following the "energy conservation condition" at $(t + \Delta t)$, is carried out by imposing the following functional obtained from equation (4.62), which will be the constraint equation,

$$\mathcal{L}(\mathbf{U}^{t+\Delta t}) = E(\mathbf{U}^{t+\Delta t}, \tfrac{2}{\Delta t}(\Delta\mathbf{U}^{t+\Delta t} - \tfrac{\Delta t}{2}\dot{\mathbf{U}}^t)) - E^t - \frac{1}{2}\,(\Delta\mathbf{U}^{t+\Delta t})^T \cdot \left[\Delta\mathbf{f}^{\text{ext}}\right]^{t+\Delta t} = 0$$

(4.63)

This condition $\mathcal{L}(\mathbf{U}^{t+\Delta t}) = 0$, combined with equations (4.57), implies that the energy conservation condition $E^{t+\Delta t} - E^t = \frac{1}{2}\,(\Delta\mathbf{U}^{t+\Delta t})^T \cdot \left[\Delta\mathbf{f}^{\text{ext}}\right]^{t+\Delta t}$ is satisfied.

Now, considering the Euler-Lagrange functional $\pi(\mathbf{U}^{t+\Delta t})$ that defines the nonlinear algebraic problem through the trapezoidal rule and the constraint equation $\mathcal{L}(\mathbf{U}^{t+\Delta t}) = 0$; then we formulate an *Euler-Lagrange functional with restrictions in energy conservation* by using the Lagrange multiplier m_λ. This is,

$$\overline{\pi}(\mathbf{U}^{t+\Delta t}) = \underbrace{\pi(\mathbf{U}^{t+\Delta t})}_{\text{functional}} + m_\lambda \underbrace{\mathcal{L}(\mathbf{U}^{t+\Delta t})}_{\text{constraint}}$$

(4.64)

By making zero the first variation, and obtaining the displacement field that makes this functional stationary,

$$\delta\overline{\pi}(\mathbf{U}^{t+\Delta t}) = \delta\left[\mathbf{U}^{t+\Delta t}\right]^T \cdot \left\{\frac{\partial\pi}{\delta\mathbf{U}}\bigg|^{t+\Delta t} + m_\lambda\frac{\partial\mathcal{L}}{\delta\mathbf{U}}\bigg|^{t+\Delta t}\right\} + \delta m_\lambda\,\mathcal{L}(\mathbf{U}^{t+\Delta t}) = 0$$

(4.65)

By substituting the respective functionals in this equation and by making zero the expression for each non-null variation of $\delta\left[\mathbf{U}^{t+\Delta t}\right]^T$ and δm_λ, the following algebraic system representing the movement equation with the energy conservation condition is obtained,

$$
\begin{aligned}
0 = (1+m_\lambda)&\left\{\frac{4}{\Delta t^2}\mathbb{M}\cdot\mathbf{U}^{t+\Delta t} + \mathbf{f}^{int}_\sigma(\mathbf{U}^{t+\Delta t})\right\} - \left(1-\frac{m_\lambda}{2}\right)\left[\mathbf{f}^{ext}\right]^{t+\Delta t} - \frac{m_\lambda}{2}\left[\mathbf{f}^{ext}\right]^t - \\
& - \mathbb{M}\cdot\left[\dot{\mathbf{U}}^t + \frac{4}{\Delta t^2}\left((1+m_\lambda)\mathbf{U}^t + \Delta t\left(1+\frac{m_\lambda}{2}\right)\dot{\mathbf{U}}^t\right)\right] = 0
\end{aligned}
$$

(4.66)

Through an iterative algorithm to solve simultaneously equations (4.63) and (4.66) the Lagrange multiplier m_λ can be obtained, which controls the force introduced to the system and the desired displacement, velocity and acceleration fields.

APPENDIX - 1

Computer program to obtain the spectral radius of a matrix.

Then a program, written in FORTRAN, giving the spectral radius of a matrix is presented. If this program is used to study the stability concept in the solution of Newmark's method presented in section 4.5.1, the following relationship between the spectral radius $\rho(\mathbf{A})$ and the ratio of the time increment and period $\Delta t / T$ is set.

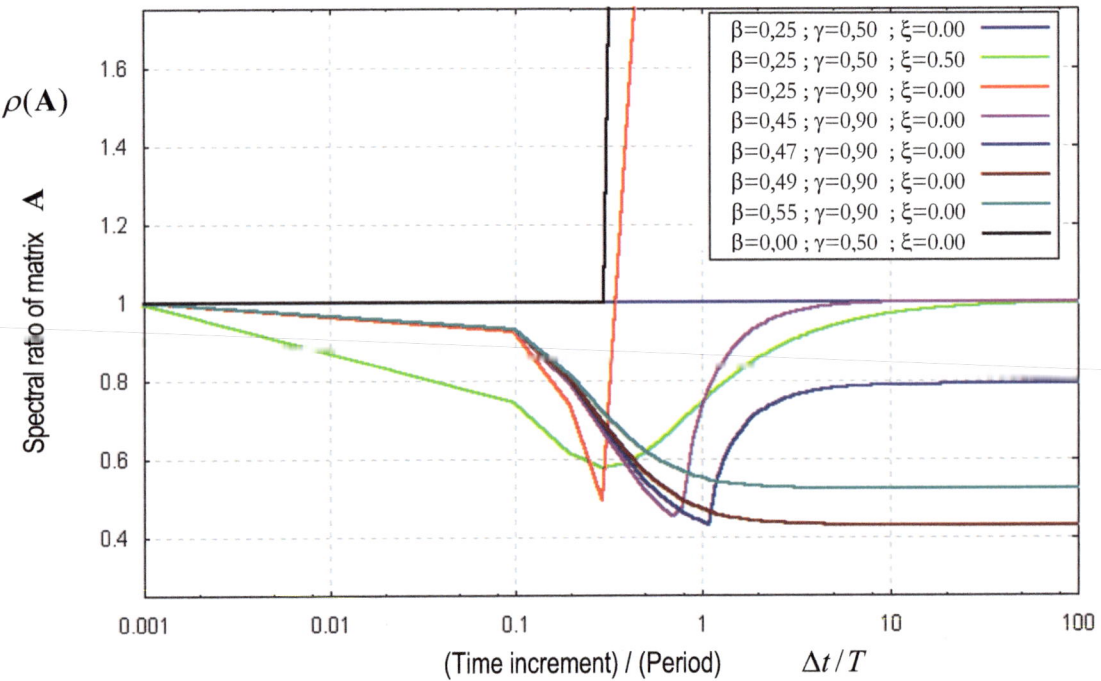

Figure 4.3 – Study of the solution stability for the Newmark method.

The last figure shows how the solution stability of the Newmark solution is guaranteed for $\beta = 0.25$ $\gamma = 0.5$, while the solution stability will be conditioned to the relation $\Delta t / T$ for any other case. Also, there exist some domain zones where the spectral radius is $\rho(\mathbf{A}) \neq 1$.

```
      PROGRAM Stability
C-----------------------------------------------------------------------
C
C      THIS PROGRAM SOLVES THE STABILITY PROBLEM BY THE SPECTRAL RATIO
METHOD
C
C            INTEGRATION METHOD:NEWMARK
C
C         Data Entry                             file.dts
          Spectral Ratio Results output      file.res
C         Vect. And Val results.own file.sal
C
C                      DATA ENTRY MANUAL
C
C      Ng              :  Number of degrees of freedom of the system
C      Dtini           :  Initial time Increment
C      Dtfin           :  Final time Increment
C      T               :  Period
C      B               :  Beta Newmark
C      G               :  Gama Newmark
C      Psi             :  Damping
C      Iread           :  1 (computes for Newmark´s matrix)
C                         2 (computes for centered dif. matrix )
c-----------------------
C      if Iread = 2, then:
C
C COEFFICIENT TO SET THE MATRIX
C
C   COEFmatrix(1 : 1)
C   COEFmatrix(1 : 2)
C   COEFmatrix(1 : 3)
C            .
C            .
C   COEFmatrix(Ng:Ng)
C
C-----------------------------------------------------------------------
      IMPLICIT INTEGER*4 (I - N)
      IMPLICIT REAL*8(A-H,O-Z)
      complex (8) RAIZ,vl1,vl2
      PARAMETER(NP=2)

      DIMENSION AUX1(NP,NP),AUX2(NP,NP),AUX3(NP,NP),
     .          COEFmatrix(NP,NP),A1(NP,NP),A2(NP,NP),
     .          A1inv(NP,NP)
      DIMENSION DVAL(NP),DVEC(NP,NP),RES(NP)
      DIMENSION IEIG(NP)
      CHARACTER ARCH1*20,ARCH2*20,ARCH3*20,CN*20
C
      PI=DACOS(-1.0D00)
C
C*** MATRICES RESET
C
      DO Ig=1,NP
       DO Jg=1,NP
          AUX1(Ig,Jg)=0.0
          AUX2(Ig,Jg)=0.0
          AUX3(Ig,Jg)=0.0
        IF(Jg.EQ.Ig)AUX2(Ig,Jg)=1.0
         ENDDO
      ENDDO
```

```
C*** FILE NAME READ

      WRITE(6,*)' LOAD FILE NAME, NO-ROOT >>>>'
      READ(5,111)CN
 111  FORMAT(1A20)
C
C*** ELIMINATION OF WHITES IN THE FILE NAME
C
      DO 444 IN1=1,20
 444  IF (CN(IN1:IN1).EQ.' ') GOTO 445
 445  IN1=IN1-1
C
      IN2=IN1+4
C
      ARCH1=CN(1:IN1)//'.dts'
      ARCH2=CN(1:IN1)//'.res'
      ARCH3=CN(1:IN1)//'.sal'
C
      OPEN(UNIT=1,FILE=ARCH1(1:IN2),STATUS='OLD') ! All data
      OPEN(UNIT=2,FILE=ARCH2(1:IN2),STATUS='NEW') ! Spectral radius
      OPEN(UNIT=3,FILE=ARCH3(1:IN2),STATUS='NEW') !
C*** DATA READ
C    Reading of all the problem data
C
      READ(1,*)Ng,Dtini,Dtfin,T,B,G,Psi,Iread
      W=2.0*Pi/T
      Dinc=(Dtfin-Dtini)/1000.0
      Dt=Dtini
C
      IF(Ng.GT.NP) THEN
        WRITE(6,*) ' ORDER OF MATRIX  > MAXIMUM SIZED'
        STOP
      ENDIF
C
C*** PRINT THE RATIO SPECTRAL FILE HEADER
      WRITE(2,910)B,G,Psi
 910  FORMAT('# Curve Dt/T vs.Spectral Radius.( Newmark: Beta:',1F5.3,
     .                               1x,' ,Gamma:',1F5.3,
     .                               1x,' ,Psi:',1F5.3,')'   )

 1111 CONTINUE
      IF(Iread.EQ.1)THEN       !NEWMARK
        Dt2=Dt*Dt
        W2=W*W

C     matrix A1

      A1(1,1)=1.0+(Dt2*B*W2)
      A1(1,2)=2.0*Dt2*B*Psi*W
      A1(2,1)=Dt*G*W2
      A1(2,2)=1.0+(2.0*Dt*G*Psi*W)

C     matrix A2

      A2(1,1)=1.0-((0.5*Dt2)*(1.0-2*B)*W2)
      A2(1,2)=Dt*(1.0-(Dt*(1.0-2.0*B)*Psi*W))
      A2(2,1)=-Dt*(1.0-G)*W2
      A2(2,2)=1.0-(2.0*Dt*(1.0-G)*Psi*W)
```

```
C       Matrix Inversion  A1

        CALL INGAUS(A1,A1inv,NP,NG)

C       NEWMARK SPECTRAL MATRIX CALCULATION

        DO Ig=1,Ng
         DO Jg=1,Ng
           COEFmatrix(Ig,Jg)=0.0
           DO Kg=1,Ng
           COEFmatrix(Ig,Jg)=COEFmatrix(Ig,Jg)+
       .                     A1inv(Ig,Kg)*A2(Kg,Jg)
           ENDDO
         ENDDO
        ENDDO
       ELSE  ! Central Differences
        Dt2=Dt*Dt
        W2=W*W

C       matrix A1

        x=1.0/((1.0/(W2*Dt2))+(Psi/(W*Dt)))
        y=Psi*x/(W*Dt)

        A1(1,1)=0.5*Dt*(1.0-x*0.5-y)
        A1(1,2)=(1.0-x*0.5-y)
        A1(2,1)=0.5*Dt2
        A1(2,2)=Dt


C       NEWMARK SPECTRAL MATRIX CALCULATION

        DO Ig=1,Ng
         DO Jg=1,Ng
           COEFmatrix(Ig,Jg)=A1(Ig,Jg)
         ENDDO
        ENDDO
       ENDIF


c     WRITE THE MATRIX TO BE ANALYZED

        WRITE(3,903)Dt
 903    FORMAT(' Spectral Matrix to be analized, for',1F12.5)
        DO 3 Ig=1,Ng
         WRITE(3,999)(COEFmatrix(Ig,Jg),Jg=1,Ng)
       DO 3 Jg=1,Ng
         AUX1(Ig,Jg)=COEFmatrix(Ig,Jg)
   3  CONTINUE

C*** EIGENVALUES, EIGENVECTORS AND SPECTRAL RATIO CALCULATION


      A1coef=0.5*(AUX1(1,1)+AUX1(2,2))
      A2coef=AUX1(1,1)*AUX1(2,2)-AUX1(1,2)*AUX1(2,1)
      DIS=(A1coef*A1coef)-A2coef
CC
      RAIZ=CDSQRT(DCMPLX(DIS))
      vl1=(A1coef+RAIZ)
    vl2=(A1coef-RAIZ)
C
      Pent1=DREAL(vl1)
```

```
         Pima1=DIMAG(vl1)
         vmod1=dsqrt(Pent1*Pent1+Pima1*Pima1)
C
         Pent2=DREAL(vl2)
         Pima2=DIMAG(vl2)
         vmod2=dsqrt(Pent2*Pent2+Pima2*Pima2)
C
      RADIO=vmod1
      IF(vmod2.ge.radio)RADIO=vmod2

C
C*** ANSWER PRINT OUT
C
c      WRITE(3,850)1,Pent1,Pima1
c       WRITE(3,850)2,Pent2,Pima2
       WRITE(2,100)(Dt/T),radio
C
       Dt=Dt+Dinc
       IF(Dt.LT.Dtfin)GOTO 1111
C
       STOP
 100   FORMAT(3(1X,F12.5))
 850   FORMAT(/,5x,I5,2x,' Eigenvalue - w=',1x,1E12.5,'i',1E12.5)
 999   FORMAT( 30(1X,1E12.5))
       END
C------------------------------------------------------------

       SUBROUTINE INGAUS(A,B,NP,N)
C***********************************************************************
C    SUBROUTINE FOR MATRICES INVERSION BY GAUSSIAN ELIMINATION
C    PARAMETERS:
C
C                 A ====>   MATRIX TO BE INVERTED
C                 B ====>   INVERTED MATRIX
C                 N ====>   MATRIX ORDER
C***********************************************************************
       IMPLICIT INTEGER*4 (I - N)
       IMPLICIT REAL*8(A-H,O-Z)

       DIMENSION A(NP,NP),B(NP,NP)
C
       DATA TOLSL /1.0E-30/

       DO I=1,N
         IF(DABS(A(I,I)).LT.TOLSL)A(I,I)=1.0E-15
       ENDDO

       DO I=1,N
         DO J=1,N
            B(I,J)=0.0
            IF(I.EQ.J)THEN
               B(I,J)=1.0D0
            ENDIF
         ENDDO
       ENDDO
       DO I=1,N
         P=A(I,I)
         IF(ABS(P).LT.TOLSL) THEN
          PRINT '(A72)',
     .            ' ERROR IN THE MATRICES INVERSION ROUTINE'
         STOP
```

```
            ENDIF
            TEMP=1.0D0/P
            DO J=1,N
                B(I,J)=B(I,J)*TEMP
                A(I,J)=A(I,J)*TEMP
            ENDDO
            DO K=1,N
                IF(K.NE.I) THEN
                    P=A(K,I)
                    DO J=1,N
                        A(K,J)=A(K,J)-P*A(I,J)
                        B(K,J)=B(K,J)-P*B(I,J)
                    ENDDO
                ENDIF
            ENDDO
        ENDDO
        RETURN
C
        END
C----------------------------------------------------------------------
```

APPENDIX - 2

Differential equation linked to a functional – Natural functional of Euler-Lagrange.

Given a functional depending on a function $f = f(x)$, on its derivative $f_x = df/dx$, on the own variable x and whose mathematical expression has the following form,

$$\pi = \int_{x=a}^{x=b} I(f, f_x, x)\, dx \tag{4.67}$$

and making stationary its first variation,

$$\delta\pi = 0 = \int_{x=a}^{x=b} \left[\frac{\partial I}{\partial f}\,\delta f + \frac{\partial I}{\partial f_x}\,\delta f_x \right] dx \tag{4.68}$$

after integrating by parts it is obtained $\displaystyle\int_{x=a}^{x=b} \frac{\partial I}{\partial f_x}\,\delta f_x\, dx = \frac{\partial I}{\partial f_x}\,\delta f_x \Big|_{x=a}^{x=b} - \int_{x=a}^{x=b} \frac{d}{dx}\left(\frac{\partial I}{\partial f_x}\right) \delta f\, dx$,

the following expression for the first variation,

$$\delta\pi = 0 = \int_{x=a}^{x=b} \left[\frac{\partial I}{\partial f} - \frac{d}{dx}\left(\frac{\partial I}{\partial f_x}\right) \right] \delta f\, dx + \frac{\partial I}{\partial f_x}\,\delta f_x \Big|_{x=a}^{x=b} \tag{4.69}$$

As $\delta f \neq 0$, then the following relations are obtained,

a) The Euler - Lagrange equation $\quad \dfrac{\partial I}{\partial f} - \dfrac{d}{dx}\left(\dfrac{\partial I}{\partial f_x}\right) = 0 \quad, \forall\ a \leq x \leq b,$

$$\tag{4.70}$$

b) The boundary condition $\qquad \dfrac{\partial I}{\partial f_x}\,\delta f_x \Big|_{x=a}^{x=b} = 0 \quad$ in : $x = a \quad$ and $\quad x = b$

5 Time-independent Models[*]

5.1 Introduction

Some basic concepts of the *theory of elasticity* and its mechanical variables are reviewed in this chapter. Particularly, a brief summary of the *classic plasticity theory* is presented as well as its modification in order to make it more general. Moreover, a brief presentation of the *continuous damage theory* will be offered. All this will be carried out within the kinematic system with small displacements that hypothetically introduce small deformations. Basic knowledge of the mechanics of continuous media[1,2,3] is recommended. Answers will be found to analyze the subject in depth. It is important, however, to set the criteria, hypothesis and notations and also to remember the most important concepts of the subject addressed in this work.

5.2 Elastic behavior

A point in a solid subjected to elastic behavior with small deformations must satisfy the following basic conditions to maintain its stability and compliance with the laws of mechanics,

1 – **Equilibrium conditions**

$$\frac{\partial \sigma_{ij}}{\partial x_j} = -\rho b_i \tag{5.1}$$

2 – **Equilibrium conditions at the boundary**

$$\boldsymbol{\sigma} \cdot \mathbf{n} = \mathbf{t} \quad ; \quad \sigma_{ij} n_j = t_i \tag{5.2}$$

3 – **Condition of infinitesimal deformation**

$$\boldsymbol{\varepsilon} = \nabla^S \mathbf{u} = \frac{1}{2}\left(\nabla_0 \mathbf{u} + \nabla_0^T \mathbf{u}\right) \quad ; \quad \varepsilon_{ij} = \nabla_j^S u_i = \frac{1}{2}\left(\frac{\partial u_i}{\partial x_j} + \frac{\partial u_j}{\partial x_i}\right) \tag{5.3}$$

---

[*] Note: this chapter is a summary of chapters 6, 9 and 10 of the book: S. Oller (2001). *Fractura mecánica – Un enfoque global*. CIMNE – Ediciones UPC. Barcelona.
[1] Malvern, L. (1969). *Introduction to the mechanics of continuous medium*. Prentice Hall, Englewood Cliffs, NJ.
[2] Lubliner, J. (1990). *Plasticity theory*. MacMillan, New York.
[3] Maugin, G. A. (1992). *The thermomechanics of plasticity and fracture*. Cambridge University Press.

such that in these three last equations t_i is the surface load applied on the boundary S; σ_{ij} is Cauchy's stresses tensor; n_j is the vector normal to the boundary S surrounding the solid[1,2,3]; b_i is the body loads per unit mass; $\rho = \partial M/\partial V$ is the mass density ; M is the mass and V is the volume; u_i is the displacement field and ε_{ij} is the infinitesimal deformation tensor.

The equations defining the classic elastic constitutive models are well established and are as follows:

"**Cauchy's elastic model**" is a classic among elastic models. In its formulation the problem variable is established through a linear tensor function with tensor arguments, as shown below,

$$\sigma_{ij} = f_{ij}(\varepsilon_{kl}) \quad ; \quad \varepsilon_{ij} = f_{ij}^{-1}(\sigma_{kl}) \tag{5.4}$$

So the stress is obtained from a model with a free variable that is the deformation in models that are based on free stress variables. These relations are invertible and reversible; therefore there is no energy dissipation in elasticity.

"**Green's classic model**", also known as hyper-elastic, where the problem variable depends on a potential density $\Psi = \Psi(\varepsilon_{ij})$, or on its complement $\overline{\Psi} = \overline{\Psi}(\sigma_{ij})$, which must be preset,

$$\sigma_{ij} = \frac{\partial \Psi(\varepsilon_{ij})}{\partial \varepsilon_{ij}} \quad ; \quad \varepsilon_{ij} = \frac{\partial \overline{\Psi}(\sigma_{ij})}{\partial \sigma_{ij}} \tag{5.5}$$

As with Cauchy's model, the stress comes from a potential based on the free deformation variable, or in inverse form, the deformation is obtained from a potential based on the free stress variable. Also, these relations are invertible and reversible, therefore there is no energy dissipation. This model contains the previous one and is the more general and comprehensive form to define the elastic behavior of a point in a solid.

"**Hypo-elastic models**" are based on a definition that usually comes from experimental observation. These models are not suitable when dealing with a nonlinear elastic material, as they may break the basic laws of thermodynamics. This is due to their arbitrary definition. Usually, they can be written as,

$$\dot{\sigma}_{ij} = f_{ij}(\dot{\varepsilon}_{ij};\sigma_{mn}) \quad ; \quad \dot{\varepsilon}_{ij} = g_{ij}(\varepsilon_{ij};\dot{\sigma}_{mn}) \tag{5.6}$$

All the models before mentioned must satisfy the concept of thermodynamic reversibility and independence between stresses and trajectory.

The generalized Hooke's law is a well-known linear relation that meets the three definitions above mentioned,

$$\sigma_{ij} = \mathbb{C}_{ijkl}\varepsilon_{kl} \tag{5.7}$$

where a mechanical operator called constitutive tensor $\mathbb{C}_{ijkl}$ appears. This is a fourth-order tensor that has 81 components in its more general expression but for isotropic and linear elasticity it is reduced to two independent components. These components are Young's modulus E and Poisson's coefficient ν. In the case of isotropy and using the symmetry of the fourth-order constitutive tensor[2] it can be written as a second-order tensor (square matrix) of 36 components.

An alternative way to write the linear elastic relation (5.7) is by Lamé's constants[1]. Here the stress is decomposed into two parts, one related to the effects of the volumetric changes and the second to the effects of deviations or distortions. This is,

$$\sigma_{ij} = \lambda \varepsilon_{kk}\delta_{ij} + \underbrace{2\mu(\varepsilon_{ij} - \tfrac{1}{3}\varepsilon_{kk}\delta_{ij})}_{e_{ij}} \quad \Rightarrow \quad \varepsilon_{ij} = \frac{1}{2\mu}\sigma_{ij} - \frac{\lambda}{2\mu(3\lambda + 2\mu)}\sigma_{kk}\delta_{ij} \tag{5.8}$$

where $\lambda = E/3(1 - 2\nu)$ and $\mu = E/2(1 + \nu)$ are Lamé's constants. This elastic constitutive law can also be written as,

$$\sigma_{ij} = 2G\varepsilon_{ij} + \frac{3\nu\kappa}{1+\nu}\varepsilon_{kk}\delta_{ij} \quad \Rightarrow \quad \varepsilon_{ij} = \frac{1}{2G}\sigma_{ij} - \frac{\nu}{3\kappa(1-2\mu)}\sigma_{kk}\delta_{ij} \tag{5.9}$$

where $\kappa = \dfrac{3\lambda + 2\mu}{3} = \dfrac{p}{\varepsilon_v}$ is the bulk modulus which identifies the volumetric deformation $\varepsilon_v = \mathrm{tr}(\varepsilon_{ij}) = \varepsilon_{kk}\delta_{ij} = 3\varepsilon_{oct}$ and the pressure or octahedral stress $p = \tfrac{1}{3}\mathrm{tr}(\sigma_{ij}) = \tfrac{1}{3}\sigma_{kk} = \sigma_{oct}$. Also, in equation (5.9) $G = \mu = \dfrac{\tau_{ij}}{\gamma_{ij}}$, where $\gamma_{ij} = 2\varepsilon_{ij}$ and $\tau_{ij} = \sigma_{ij}$ can be identified.

The same constitutive law of the above equation can be expressed as follows,

$$\sigma_{ij} = \frac{E}{1+\nu}\varepsilon_{ij} + \frac{E\nu}{(1+\nu)(1-2\nu)}\varepsilon_{kk}\delta_{ij} \quad \Rightarrow \quad \varepsilon_{ij} = \frac{1+\nu}{E}\sigma_{ij} - \frac{\nu}{E}\sigma_{kk}\delta_{ij} \tag{5.10}$$

where $E = \dfrac{\sigma_{ij}}{\varepsilon_{ij}}$, $\nu = -\dfrac{\varepsilon_{22}}{\varepsilon_{11}}$ represent the elasticity modulus and Poisson's coefficient.

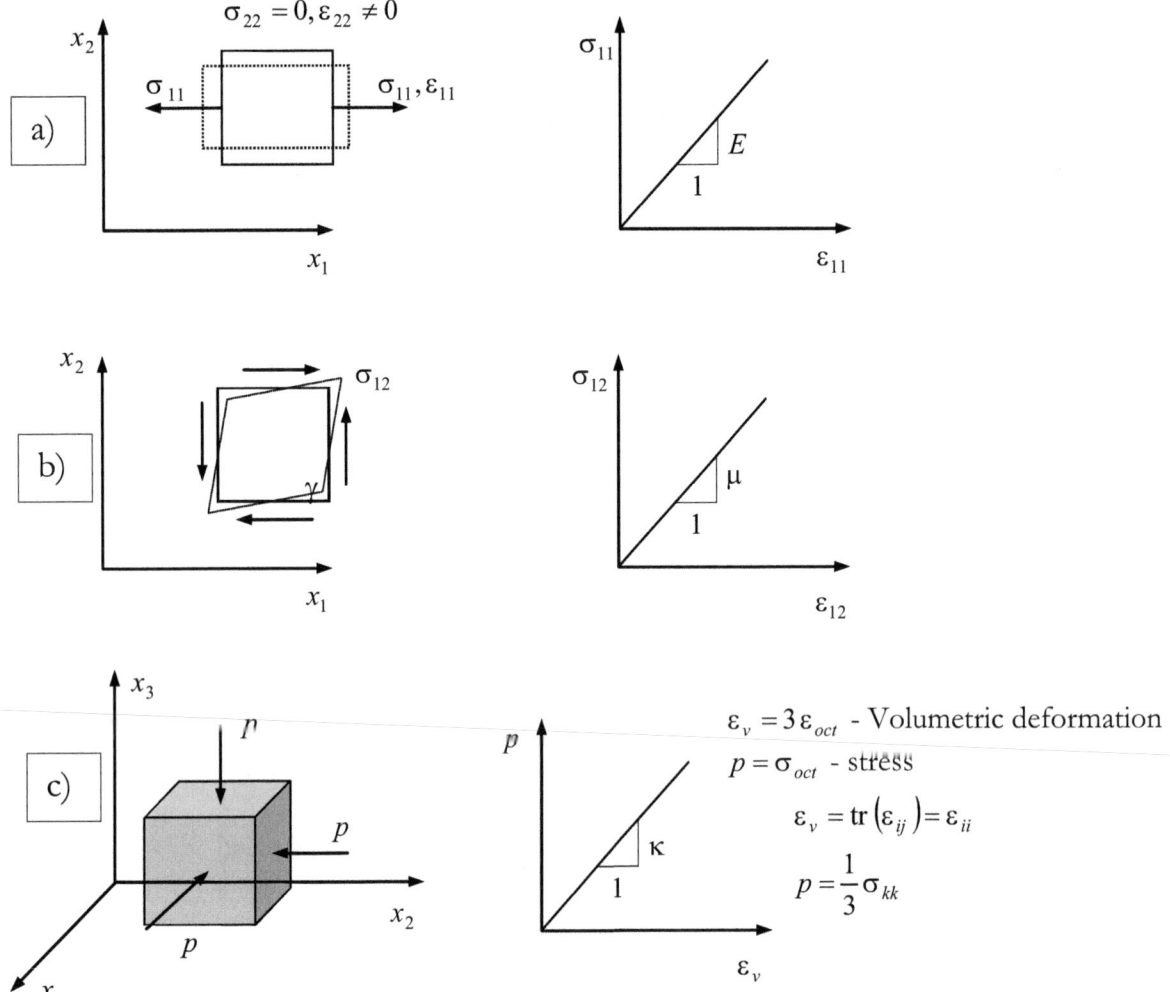

Figure 5.1 – Elastic behavior: a) Simple traction, b) Shear stress, c) Hydrostatic state.

5.2.1 Invariants of the tensors

The tensor invariant is a scalar entity that does not change its size when changing the coordinate system. Let two mechanical entities be assumed, the stresses tensor and the deformations tensor,

$$\sigma_{ij} = \mathbf{\sigma} = \begin{bmatrix} \sigma_{11} & \sigma_{12} & \sigma_{13} \\ \sigma_{21} & \sigma_{22} & \sigma_{23} \\ \sigma_{31} & \sigma_{32} & \sigma_{33} \end{bmatrix} \quad ; \quad \varepsilon_{ij} = \mathbf{\varepsilon} = \begin{bmatrix} \varepsilon_{11} & \varepsilon_{12} & \varepsilon_{13} \\ \varepsilon_{21} & \varepsilon_{22} & \varepsilon_{23} \\ \varepsilon_{31} & \varepsilon_{32} & \varepsilon_{33} \end{bmatrix} \tag{5.11}$$

As is well known, this tensor form is only a conventional representation of the stress and deformation state of a point in a solid through its components in three orthogonal planes. Nevertheless, it may be possible to see how the stress or deformation state changes in a point depending on the plane considered. But this is not real as it changes only the form in which such stress state is represented. Nevertheless, there is a reliable way to be freed from this un-objectiveness, which is through the so-called invariants of the stress and deformation tensors. In the following expressions the most common expressions for these invariants are presented.

With the stress tensor and the deformation tensor, their invariants can be obtained and are expressed as follows:

Stress and deformation tensor invariants:

$$
\begin{aligned}
I_1 &= \sigma_{ii} = \sigma_{11} + \sigma_{22} + \sigma_{33} \\
I_2 &= \frac{1}{2}\left(\sigma_{ij}\sigma_{ij} - \sigma_{kk}^2\right) \\
I_3 &= \det[\sigma_{ij}] = |\sigma_{ij}|
\end{aligned}
\qquad
\begin{aligned}
I_1' &= \varepsilon_{ii} = \varepsilon_{11} + \varepsilon_{22} + \varepsilon_{33} \\
I_2' &= \frac{1}{2}\left(\varepsilon_{ij}\varepsilon_{ij} - \varepsilon_{kk}^2\right) \\
I_3' &= \det[\varepsilon_{ij}] = |\varepsilon_{ij}|
\end{aligned}
\qquad (5.12)
$$

Stress and deformation deviatoric tensor invariants:

$$
\begin{aligned}
J_1 &= 0 \\
J_2 &= \frac{1}{2}s_{ij}s_{ij} \\
J_3 &= \frac{1}{3}s_{ij}s_{jk}s_{ki}
\end{aligned}
\qquad
\begin{aligned}
J_1' &= \varepsilon_{ii} = 0 \\
J_2' &= \frac{1}{2}e_{ij}e_{ij} \\
J_3' &= \frac{1}{3}e_{ij}e_{jk}e_{ki}
\end{aligned}
\qquad (5.13)
$$

where the stress deviatoric tensor is represented as the following expression, $s_{ij} = \sigma_{ij} - \phi\delta_{ij} = \sigma_{ij} - \dfrac{\sigma_{kk}}{3}\delta_{ij}$, and the deformation deviatoric tensor is expressed as, $e_{ij} = \varepsilon_{ij} - \dfrac{\varepsilon_v}{3}\delta_{ij} = \varepsilon_{ij} - \dfrac{\varepsilon_{kk}}{3}\delta_{ij}$. There are also mathematical forms to express these invariants and some of these will be studied in other sections of the book.

5.3 Nonlinear elasticity

5.3.1 Introduction

Below is the representation of the constitutive nonlinear reversible elastic behavior of an ideal material. Therefore, it is convenient to start with a hyper-elastic formulation that depends on the energy density and it can be defined as a function of the deformation field (a problem-free variable) as follows,

$$
\Psi = \int_{t=0}^{t} \sigma_{ij}\dot{\varepsilon}_{ij}\, dt \qquad (5.14)
$$

or by its complementary form, complementary energy density as a function of the stress field (a problem-free variable) as,

$$
\overline{\Psi} = \int_{t=0}^{t} \varepsilon_{ij}\dot{\sigma}_{ij}\, dt \qquad (5.15)
$$

Also, a nonlinear elastic constitutive law could be defined based on the hypo-elastic concepts, in other words, generalizing the linear equations.

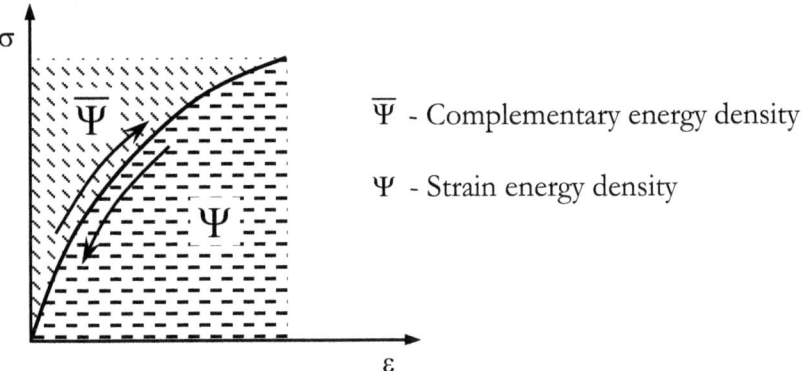

Figure 5.2 – Complementary and strain energy.

5.3.2 Nonlinear hyper-elastic model

Starting with the general definition of potentials from the invariants,

$$\Psi = \Psi\left(I_1^{'}; I_2^{'}; I_3^{'}\right) \quad \Rightarrow \quad \sigma_{ij} = \frac{\partial \Psi}{\partial \varepsilon_{ij}} \qquad \text{strain - based}$$

$$\overline{\Psi} = \overline{\Psi}\left(I_1; I_2; I_3\right) \quad \Rightarrow \quad \varepsilon_{ij} = \frac{\partial \overline{\Psi}}{\partial \sigma_{ij}} \qquad \text{stress - based}$$

(5.16)

5.3.2.1 Stress-based hyper-elastic model

The following potential is chosen based on the stress $\overline{\Psi}\left(I_1; J_2\right) = aJ_2 + bI_1 J_2$, and it is assumed that the first invariant of the stress tensor and the second invariant of the stress deviatoric have a special influence on the material. Thus, the following constitutive law is obtained,

$$\varepsilon_{ij} = \frac{\partial \overline{\Psi}}{\partial \sigma_{ij}} = \frac{\partial \overline{\Psi}}{\partial I_1}\frac{\partial I_1}{\partial \sigma_{ij}} + \frac{\partial \overline{\Psi}}{\partial J_2}\frac{\partial J_2}{\partial \sigma_{ij}} = \frac{\partial \overline{\Psi}}{\partial I_1}\delta_{ij} + \frac{\partial \overline{\Psi}}{\partial J_2}\delta_{ij}$$

$$\varepsilon_{ij} = bJ_2 \delta_{ij} + \left(a + bI_1\right)\delta_{ij}$$

(5.17)

Assuming it is an uniaxial traction problem, once the first invariant stress tensor and the second deviatoric invariant are substituted by the model previously defined, the following law is reduced as,

$$I_1 = \sigma \quad ; \quad J_2 = \frac{1}{3}\sigma^2$$

$$s_{11} = \frac{2}{3}\sigma \quad ; \quad s_{22} = s_{33} = -\frac{\sigma}{3} \quad ; \quad s_{12} = s_{13} = s_{23} = 0$$

(5.18)

$$\varepsilon_{11} = \varepsilon = \frac{b}{3}\sigma^2 + \left(a + b\sigma\right)\frac{2}{3}\sigma = \sigma^2 b + \frac{2a}{3}\sigma$$

(5.19)

such that it is necessary to parameterize the model (obtaining parameters a and b) and for this a lab test must be carried out as shown in Figure 5.3.

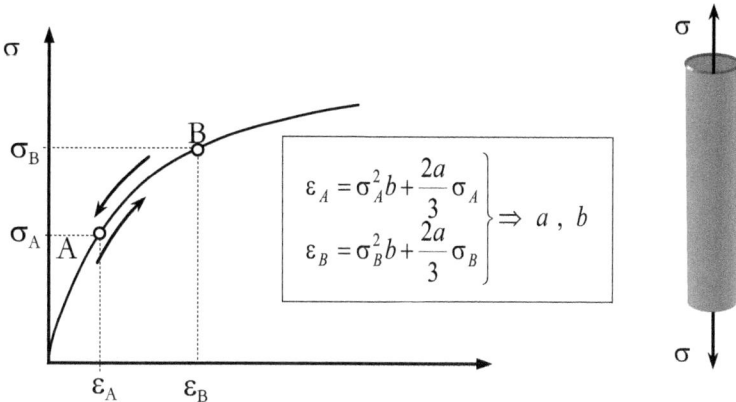

Figure 5.3 – Nonlinear elastic traction model.

5.3.2.2 Stability postulates

The stability postulates or Drucker's[1] postulates must be considered in all these formulations as they indirectly guarantee the compliance with the second law of thermodynamics. This postulate has originally been deducted for plasticity problems but can be applied to nonlinear elasticity problems. Plasticity problems will be studied further.

Let a solid of volume V and external surface S be considered, under external load of surface $\mathbf{t}$ and loads of volume $\mathbf{b}$, that produce a displacement state $\mathbf{u}$, deformation $\boldsymbol{\varepsilon}$ and a stress $\boldsymbol{\sigma}$, in each point of the solid. Let us now consider an arbitrary change in the magnitude of such loads $\dot{\mathbf{t}}$ and $\dot{\mathbf{b}}$, that produces an increment in the displacement states $\dot{\mathbf{u}}$, deformation $\dot{\boldsymbol{\varepsilon}}$ and stress $\dot{\boldsymbol{\sigma}}$. Now, it is said that the material behavior will be stable if the two following postulates are satisfied (Drucker's postulates).

Stability requirements:
1. The magnitude change work in the external agents must always be positive.

$$\int_V \dot{\boldsymbol{\sigma}} : \dot{\boldsymbol{\varepsilon}} \; dV = \int_S \dot{\mathbf{t}} \cdot \dot{\mathbf{u}} \; dS + \int_V \dot{\mathbf{b}} \cdot \dot{\mathbf{u}} \; dV > 0 \tag{5.20}$$

2. A cyclic magnitude change work experimented by the external agents cannot be negative.

$$\oint_V \dot{\boldsymbol{\sigma}} : \dot{\boldsymbol{\varepsilon}} \; dV = \oint_S \dot{\mathbf{t}} \cdot \dot{\mathbf{u}} \; dS + \oint_V \dot{\mathbf{b}} \cdot \dot{\mathbf{u}} \; dV \geq 0 , \text{ Cyclic Stability} \tag{5.21}$$

Free-energy existence:

These stability criteria that are applied to elastic materials, where all the deformations are recoverable, is a necessary and sufficient condition to guarantee the existence of a free-energy $\Psi(\boldsymbol{\varepsilon})$ and its complementary $\overline{\Psi}(\boldsymbol{\sigma})$ and, accordingly, the existence of two constitutive laws: the deformation free variable $\boldsymbol{\sigma}(\boldsymbol{\varepsilon})$ and the stress free variable $\boldsymbol{\varepsilon}(\boldsymbol{\sigma})$. Thus,

$$\Psi(\boldsymbol{\varepsilon}) = \int_{\boldsymbol{\varepsilon}} \boldsymbol{\sigma} : d\boldsymbol{\varepsilon} \quad \Rightarrow \quad \boldsymbol{\sigma}(\boldsymbol{\varepsilon}) = \frac{\partial \Psi(\boldsymbol{\varepsilon})}{\partial \boldsymbol{\varepsilon}}$$

$$\Psi(\boldsymbol{\sigma}) = \int_{\boldsymbol{\sigma}} \boldsymbol{\varepsilon} : d\boldsymbol{\sigma} \quad \Rightarrow \quad \boldsymbol{\varepsilon}(\boldsymbol{\sigma}) = \frac{\partial \Psi(\boldsymbol{\sigma})}{\partial \boldsymbol{\sigma}}$$

(5.22)

Necessary and sufficient stability condition:

Every hyper-elastic or Green-type constitutive relation meets the stability conditions before mentioned, provided that the energy potential is defined positive. To show this, let us consider a constitutive relation of the type $\boldsymbol{\sigma}(\boldsymbol{\varepsilon}) = \partial \Psi(\boldsymbol{\varepsilon})/\partial \boldsymbol{\varepsilon}$, such that any variation in the external agents can produce the following change in the stress,

$$\dot{\boldsymbol{\sigma}}(\boldsymbol{\varepsilon}) = \frac{\partial \boldsymbol{\sigma}(\boldsymbol{\varepsilon})}{\partial \boldsymbol{\varepsilon}} : \dot{\boldsymbol{\varepsilon}} = \frac{\partial^2 \Psi(\boldsymbol{\varepsilon})}{\partial \boldsymbol{\varepsilon} \otimes \partial \boldsymbol{\varepsilon}} : \dot{\boldsymbol{\varepsilon}} \tag{5.23}$$

The necessary and sufficient condition to satisfy the load stability criterion for the whole volume, (see equation 5.20), and cyclic load, (see equation 5.21), is that all and every point of this solid carry out an specific positive second-order work,

$$\dot{\boldsymbol{\sigma}} : \dot{\boldsymbol{\varepsilon}} > 0 \tag{5.24}$$

Substituting equation (5.23) into equation (5.24), the following quadratic form is obtained,

$$\dot{\boldsymbol{\sigma}}(\boldsymbol{\varepsilon}) : \dot{\boldsymbol{\varepsilon}} = \frac{\partial \boldsymbol{\sigma}(\boldsymbol{\varepsilon})}{\partial \boldsymbol{\varepsilon}} : \dot{\boldsymbol{\varepsilon}} : \dot{\boldsymbol{\varepsilon}} = \frac{\partial^2 \Psi(\boldsymbol{\varepsilon})}{\partial \boldsymbol{\varepsilon} \otimes \partial \boldsymbol{\varepsilon}} : \dot{\boldsymbol{\varepsilon}} : \dot{\boldsymbol{\varepsilon}} > 0 \tag{5.25}$$

such that the compliance of the stability condition is guaranteed provided the Hessian is defined positive. Then,

$$H_{ijkl} = \mathbf{H} \quad ; \quad \det(\mathbf{H}) = \det\left|\frac{\partial^2 \Psi(\boldsymbol{\varepsilon})}{\partial \boldsymbol{\varepsilon} \otimes \partial \boldsymbol{\varepsilon}}\right| > 0 \tag{5.26}$$

With an alternative form, the stability condition in stress-based models is guaranteed if the complementary Hessian is also defined positive,

$$\dot{\boldsymbol{\varepsilon}}(\boldsymbol{\sigma}) : \dot{\boldsymbol{\sigma}} = \frac{\partial \boldsymbol{\varepsilon}(\boldsymbol{\sigma})}{\partial \boldsymbol{\sigma}} : \dot{\boldsymbol{\sigma}} : \dot{\boldsymbol{\sigma}} = \frac{\partial^2 \overline{\Psi}(\boldsymbol{\sigma})}{\partial \boldsymbol{\sigma} \otimes \partial \boldsymbol{\sigma}} : \dot{\boldsymbol{\sigma}} : \dot{\boldsymbol{\sigma}} > 0 \quad \Rightarrow \quad \det(\overline{\mathbf{H}}) = \det\left|\frac{\partial^2 \overline{\Psi}(\boldsymbol{\sigma})}{\partial \boldsymbol{\sigma} \otimes \partial \boldsymbol{\sigma}}\right| > 0 \tag{5.27}$$

The last equations can guarantee the existence of the free energy and its positive condition. This guarantees constitutive equations have a unique inverse.

Convexity condition:

The potential convexity is guaranteed in compliance with the stability criteria expressed by equations (5.26) and (5.27). It is understood that a convex function, as any tangent to the curve $\Psi(\boldsymbol{\varepsilon}) = cte$ or $\overline{\Psi}(\boldsymbol{\sigma}) = cte$, cuts it in another point of such curve.

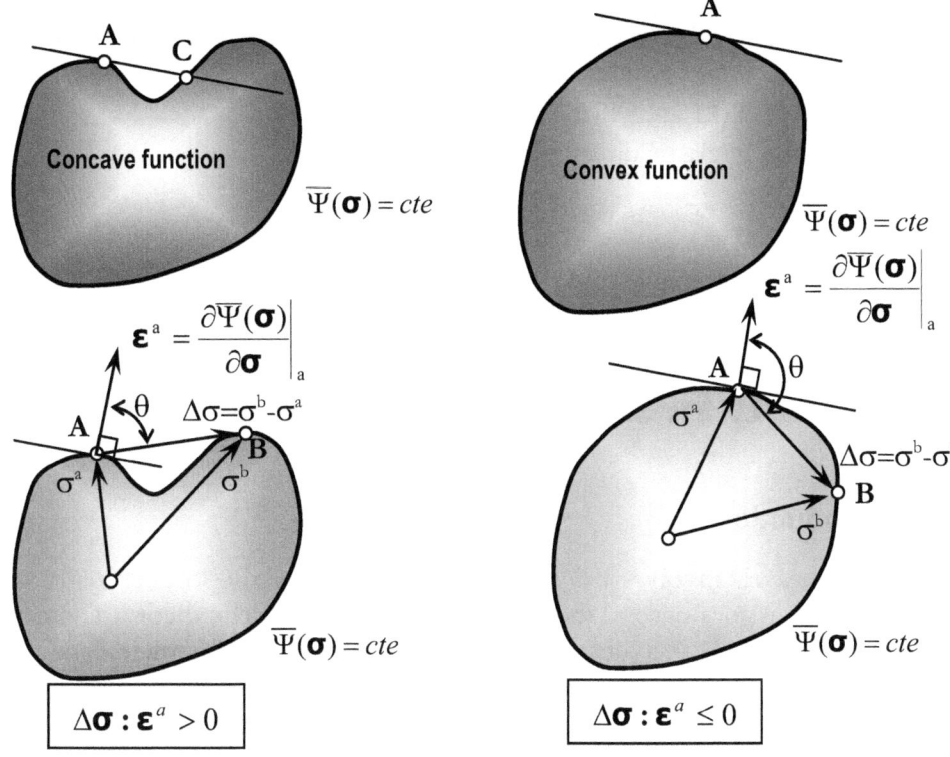

Figure 5.4 – Concave and convex functions.

Mathematically speaking a potential function is convex if it always complies with the following relation in any of the two stress states, $\Delta\boldsymbol{\sigma}:\boldsymbol{\varepsilon}^a = (\boldsymbol{\sigma}^b - \boldsymbol{\sigma}^a):\boldsymbol{\varepsilon}^a \leq 0$.

It is said that a potential function is concave when there are at least two stress states $\boldsymbol{\sigma}^a$ and $\boldsymbol{\sigma}^b$ that comply with the following relation, $\Delta\boldsymbol{\sigma}:\boldsymbol{\varepsilon}^a = (\boldsymbol{\sigma}^b - \boldsymbol{\sigma}^a):\boldsymbol{\varepsilon}^a > 0$.

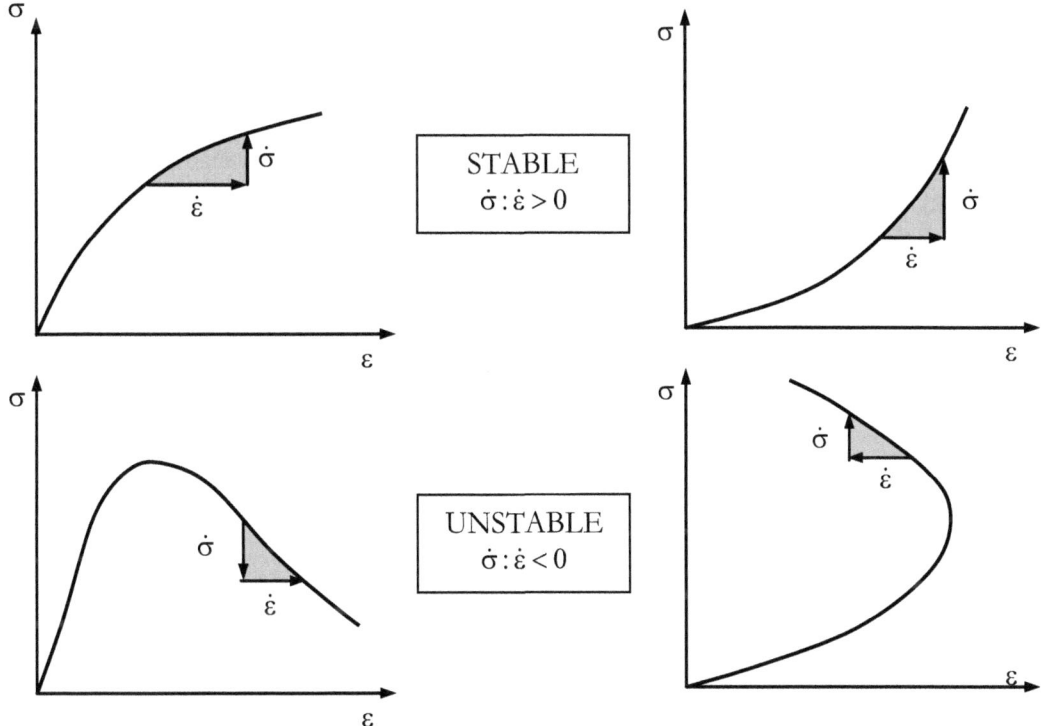

Figure 5.5 – Drucker's stability postulates.

Basically, to comply with Drucker's postulate it is convenient to consider the following recommendations,

a. The potentials $\Psi(\boldsymbol{\varepsilon})$ and $\overline{\Psi}(\boldsymbol{\sigma})$ must be positive defined

b. There must be an elastic direct relation and its reverse.

$$\sigma_{ij} = F(\varepsilon_{ij}) \quad \Leftrightarrow \quad \varepsilon_{ij} = f(\sigma_{ij}) \tag{5.28}$$

c. Potentials $\Psi(\boldsymbol{\varepsilon})$ and $\overline{\Psi}(\boldsymbol{\sigma})$ must be convex functions.

5.4 Plasticity in small deformations

5.4.1 Introduction

The plasticity theory involves the mechanical behavior of the loaded solids within an application range that goes beyond the one defined by the elasticity theory. There are other theories that comply with this idea and require more precision for their definition. This mathematical theory was first formulated in 1872 to represent the distortion phenomenon affecting the metal lattice[4] when subjected to deformations. Its mathematical structure is currently used to represent the macroscopic behavior of a point in a solid without developing a metal plasticity phenomenon at microscopic level in the strict sense[1,2,3]. The plasticity in small deformations assumes deformations in one point $\boldsymbol{\varepsilon} = \boldsymbol{\varepsilon}^e + \boldsymbol{\varepsilon}^p$, is decomposed in one elastic part $\boldsymbol{\varepsilon}^e$ and another plastic irreversible one, $\boldsymbol{\varepsilon}^p$. The latter is responsible for a non-conservative energetic behavior that depends on the previous path. The plasticity theory formulation is based on the mechanics of continuous solids and represents the macroscopic physical behavior of an ideal solid with the following features:

- An initial linear or nonlinear elastic period (section OA' of Figure 5.6).

- A behavior called elasto-plastic that follows the initial period (section A'E of Figure 5.6), where the stress field does not increase proportionally to the deformation field and in which deformations result from an addition of a recoverable part $\boldsymbol{\varepsilon}^e$ (elastic section) and another irrecoverable part $\boldsymbol{\varepsilon}^p$ (plastic section). This inelastic part of the deformation is observed at the beginning of the discharging process which is always assumed as elastic.

The point A' showing the separation between the two mechanical states is known as the yield limit for metallic materials and as a discontinuous limit for frictional materials and is defined through a function in the stress space called plastic yield function or discontinuity function respectively.

---

[4] Tresca, H. E. (1872). Mémorie sur l'coulement des corps solides. Acad. Sci. Paris, 20, 75-135.

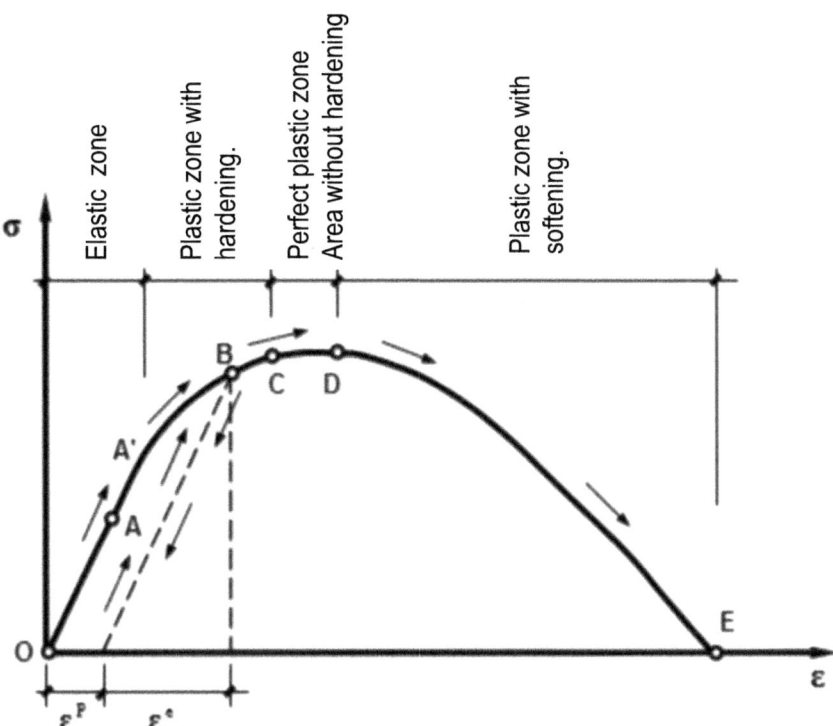

Figure 5.6 – Schematic uniaxial behavior of an ideal elasto-plastic material.

In Figure 5.6 the schematic uniaxial behavior of a point of an ideal elasto-plastic material is shown. When the loading process starts, an elastic linear area is observed which goes up to point A called proportionality limit. Next a nonlinear elastic process continues until point A´. This is called elasticity limit from where the elasto-plastic behavior starts and is characterized by a sustained decrease of the tangent stiffness module due to the action of the irreversible inelastic mechanisms. If a loading process starts during the elasto-plastic behavior of the point, only the elastic part $\boldsymbol{\varepsilon}^{e}$ is recovered from the total of strain $\boldsymbol{\varepsilon}$, and the plastic strain $\boldsymbol{\varepsilon}^{p}$ is the non-recoverable part. Within the elasto-plastic behavior, three regions can be distinguished. (See Figure 5.6):

- One where a stress increase is found, section (A'-C), called elasto-plastic with hardening.
- Another one where the point being analyzed does not experiment any stress change, section (C-D), called perfect elasto-plastic area or null hardening.
- Finally, an area where the stress decreases while the deformation increases, section (D-E), elasto-plastic period with softening.

This ideal elasto-plastic material can be used to represent in a fairly good manner the macroscopic behavior of different real materials (metallic and non-metallic) through a simple modification of the limits defined before.

Of these concepts above, there are two important aspects to deal with in the plasticity mathematical theory:

- The yield or discontinuous criterion $\mathbb{F}(\boldsymbol{\sigma};\mathbf{q})=0$, that helps to set the beginning of the inelastic behavior during the loading process and the subsequent evolution of the boundaries of the elastic domain within the stress space (see Figure 5.7).

- The behavior beyond the elastic limit, known as elasto-plastic, is defined from the formulation of: (i) one deformation decomposition into one elastic part and another plastic one $\boldsymbol{\varepsilon}=\boldsymbol{\varepsilon}^{e}+\boldsymbol{\varepsilon}^{p}$; (ii) a rule of plastic flow $\mathbf{g}(\boldsymbol{\sigma})$; (iii) and some internal variables $\mathbf{q}(\boldsymbol{\sigma},\mathbf{q})$, which depend on the evolution of the elasto-plastic process. It will be later noted that the plastic deformation $\boldsymbol{\varepsilon}^{p}$ can be handled as an internal variable, so the plastic flow rule can be assumed as the explicit evolution rule of this internal variable.

These two basic concepts will be studied in detail in the following sections.

5.4.2 Discontinuity behavior or plastic yield criterion

The *discontinuity or yield criterion* is a scalar function of tensor arguments that defines the elastic domain. This criterion is normally represented by a function that will be called hereafter as plastic yield function (see Figure 5.7) and has the following mathematical form,

$$\mathbb{F}(\boldsymbol{\sigma};\mathbf{q})=0 \tag{5.29}$$

where $\boldsymbol{\sigma}$ is Cauchy's stress tensor and $\mathbf{q}$ is the set of internal variables grouped as "back stress". This function sets the limit of the nonlinear behavior. Any stress state out of the domain bounded by this surface is inadmissible. Therefore, the elasto-plastic solution process consists of forcing the stress state to be at the border or inside the elasto-plastic function (see Figure 5.7). For uniaxial processes, this function is perfectly established as it is a scalar that is compared to another scalar that represents the threshold between an elastic and a plastic behavior. For multiaxial behaviors, the yield function behaves as a translator of uniaxial and multiaxial states. Once the uniaxial stress equivalent to the multiaxial state is obtained, it is compared to a scalar with the threshold obtained in the lab for uniaxial problems.

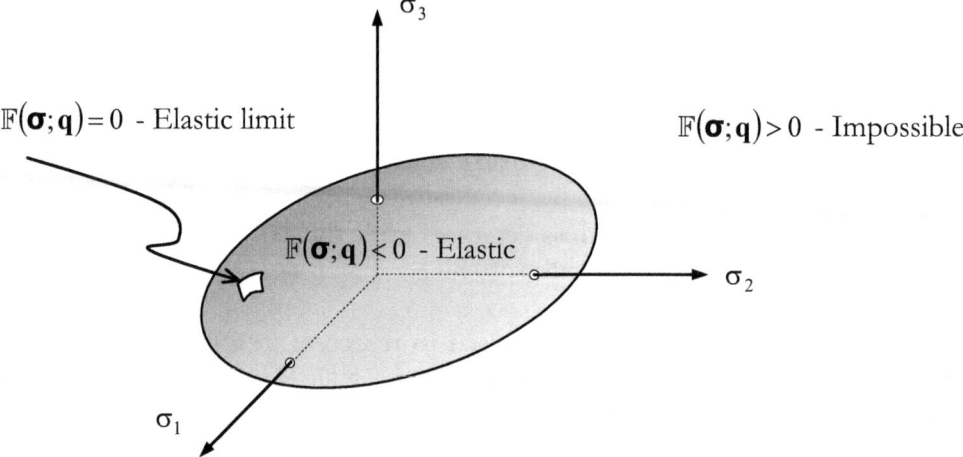

Figure 5.7 – Elastic domain.

The evolution law of internal variables $\mathbf{q}$ can be written in a general form depending on the state of the free variable, stress in this case and current magnitude of all the internal variables as follows,

$$\dot{\mathbf{q}} = \dot{\lambda} \; \mathbf{H}(\boldsymbol{\sigma}; \mathbf{q}) \tag{5.30}$$

where λ is a non-negative scalar called plastic consistency factor and $\mathbf{H}(\boldsymbol{\sigma}; \mathbf{q})$ is a tensor function for the evolution description of each internal variable.

The plasticity theory only admits two mechanical behavior states for each point of an ideal solid: "the elastic state" or "the plastic-elastic state". The situation of any point at a given time t of the quasi-static loading process is defined based on the plastic consistency condition also known as Prager's consistency condition that will be detailed further. Thus,

- The deformation process of a point is elastic if the following condition is satisfied,

$$\mathbb{F}(\boldsymbol{\sigma}; \mathbf{q}) < 0 \quad \text{or if} \quad \dot{\mathbb{F}}(\boldsymbol{\sigma}; \mathbf{q}) = \frac{\partial \mathbb{F}}{\partial \boldsymbol{\sigma}} : \dot{\boldsymbol{\sigma}} + \frac{\partial \mathbb{F}}{\partial \mathbf{q}} : \dot{\mathbf{q}} < 0 \quad \text{(Unload)} \tag{5.31}$$

-The elasto-plastic deformation process of a point occurs if,

$$\mathbb{F}(\boldsymbol{\sigma}; \mathbf{q}) = 0 \quad \text{and} \quad \dot{\mathbb{F}}(\boldsymbol{\sigma}; \mathbf{q}) = \frac{\partial \mathbb{F}}{\partial \boldsymbol{\sigma}} : \dot{\boldsymbol{\sigma}} + \frac{\partial \mathbb{F}}{\partial \mathbf{q}} : \dot{\mathbf{q}} \geq 0 \quad \text{(Load)} \tag{5.32}$$

These functions are symmetrical for isotropic materials in the stress space and in the latter the stress state is defined through their invariants. Experimental research carried out on non-porous solids such as metallic materials prove that the incidence of the hydrostatic pressure on the plastic deformation is negligible and the latter depends mainly on the deviatoric stress. This guarantees that the volumetric deformation will always be elastic, as is the case of incompressible solids. For this particular case of metallic and isotropic materials the plastic yield criterion (equation (5.29)) is reduced as:

Isotropic metallic material $\quad \mathbb{F}(\boldsymbol{\sigma}; \mathbf{q}) = \mathbb{F}(J_2, J_3; \mathbf{q}) = 0 \tag{5.33}$

where J_2 and J_3 are the second and third invariants of the stress deviator tensor respectively.

For frictional materials, it must be remembered that frictional forces among particles increase with pressure between their faces. This effect can be observed in the importance of the spherical or hydrostatic stress (first invariant of the stress tensor I_1), which must be considered for the model formulation through a discontinuity limit criterion. This is:

Frictional Isotropic Material $\quad \mathbb{F}(\boldsymbol{\sigma}; \mathbf{q}) = \mathbb{F}(I_1, J_2, J_3; \mathbf{q}) = 0 \tag{5.34}$

The general representation of the yield or discontinuity functions expressed in equations (5.33) and (5.34) is carried out through a surface in the stress space where the main directions of this space set the reference axes. This space is called High-Westergaard[1,2] stress space (see Figures 5.8 and 5.9). An alternative way of describing these functions is through their decomposition into planes (see Figure 5.9). This is:

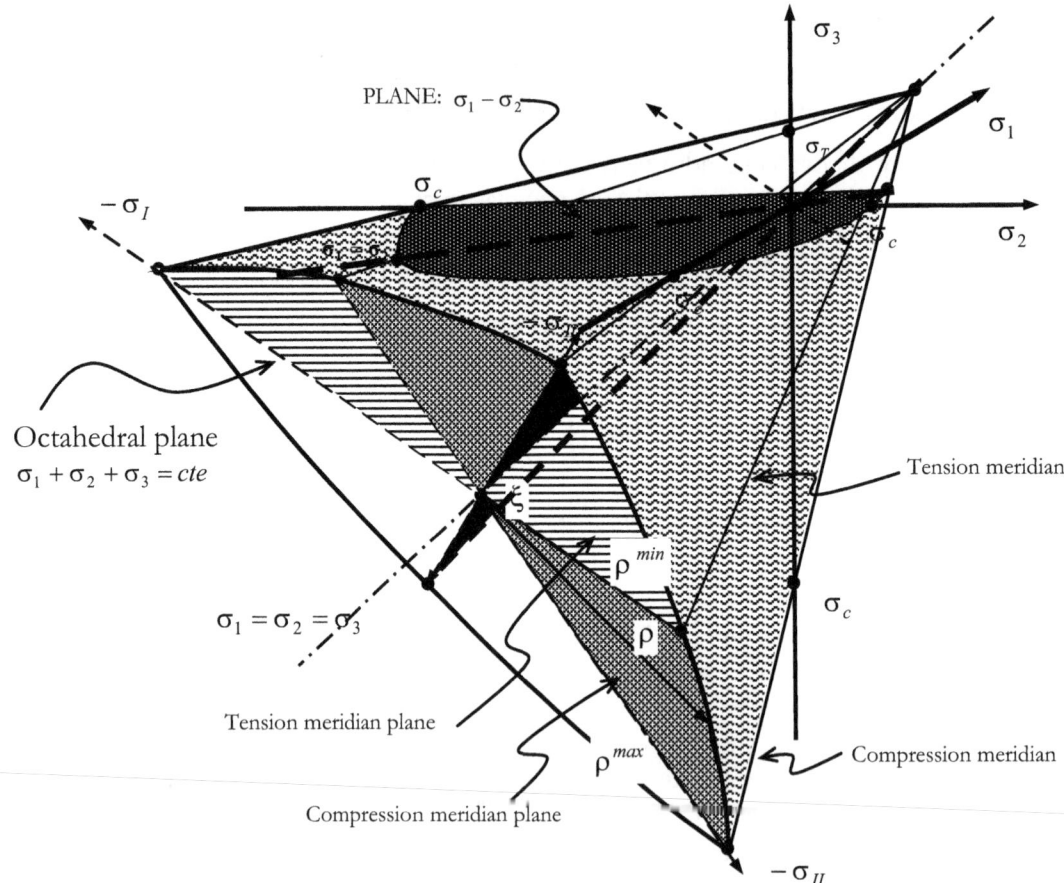

Figure 5.8 – Representation of a generic criterion of plastic yield in the main stress space.

Octahedral planes: The diagonal space of the stresses is cut orthogonally by these planes (a line defined by $\sigma_1 = \sigma_2 = \sigma_3$) and therefore they form a similar angle with the three main stress axes $(\sigma_1, \sigma_2, \sigma_3)$ which define the compressive octant or total traction. In these planes, the first invariant of the stress tensor is constant $I_1 = cte$. Through this invariant, *its position* can be obtained from the origin of the stress space $\xi = \sqrt{3}\ \sigma_{oct} = \sqrt{3}(I_1/3) = I_1/\sqrt{3}$. *Its shape* depends on two other invariants: the octahedral radius $\rho = \sqrt{3}\ \tau_{oct} = \sqrt{2J_2}$, and Lode angle similarity $\theta = \arcsen\left[\left(3\sqrt{3}\ J_3\right)/\left(2J_2^{3/2}\right)\right]$. The octahedral plane passing through the origin of the "diagonal space" $\xi = 0$ is called plane Π. The intersection of these planes and the yield surface defines curves in the main stress space called *yield functions under octahedral planes*.

Meridian planes of maximum compression: these planes are orthogonal to the octahedral planes and are defined by the line describing the diagonal space $(\sigma_1 = \sigma_2 = \sigma_3)$, and for each of the lines describing the octahedral radius ρ, for $\theta = 1\pi/6, 5\pi/6, 9\pi/6$. These planes cut the main stress axes σ_i, in points with the same uniaxial compression stress σ_C. The intersection of these planes with the yield surface defines curves in the stress space called yield functions under compression meridians.

Main Planes: these are the planes defined by the intersection of two of the three main stress directions. The intersection of these planes with the yield surface defines curves in the stress space called yield functions according to the main stress planes.

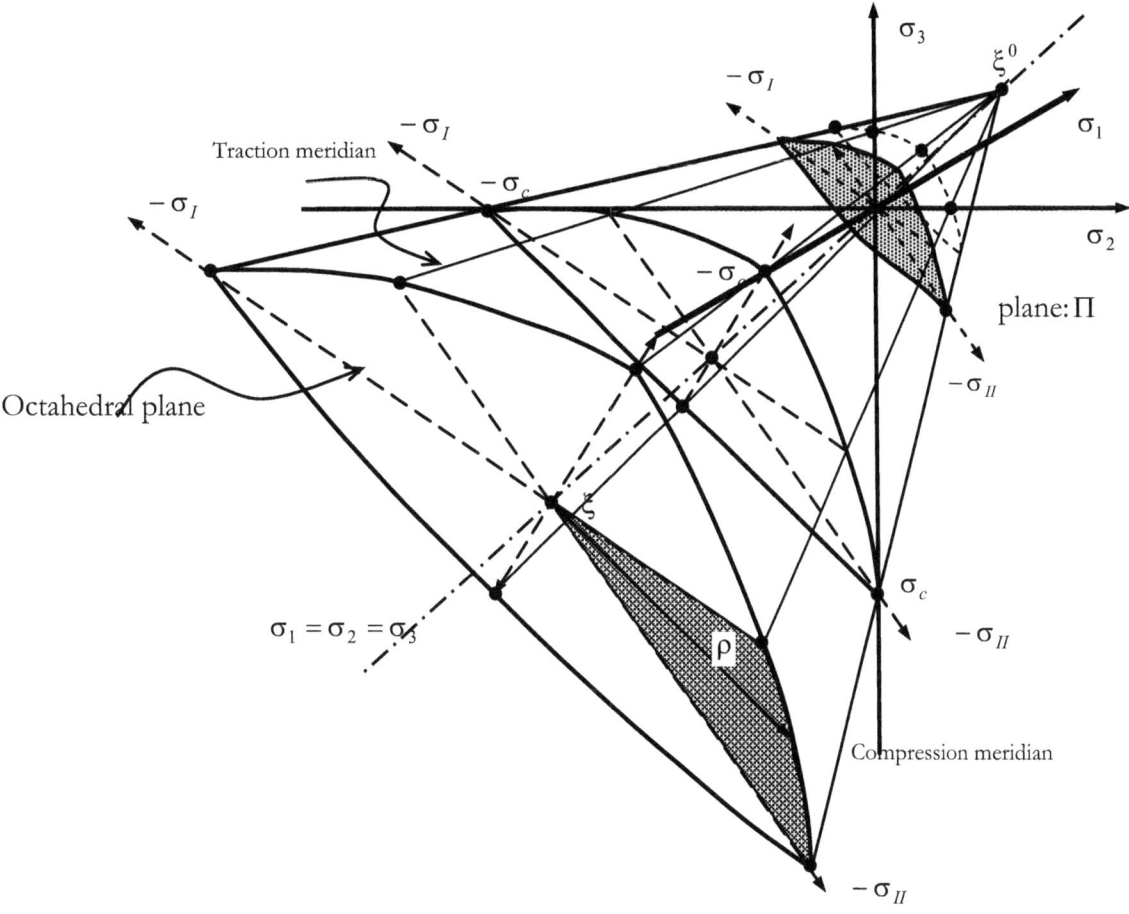

Figure 5.9 – Representation of a generic criterion of plastic yield decomposed into octahedral, meridian and main planes.

5.5 Elasto-plastic behavior

There is no single theory for the representation of the elasto-plastic behavior of materials. There are various approximations to the problem depending on their objective. The classic approximations to the elasto-plastic problem will be presented.

5.5.1 The Levy-Mises theory

One way of modeling the elasto-plastic behavior of a point in a solid is through the Levy-Mises theory. This admits, as a first hypothesis, that the temporal increase of the total deformation is equal to the temporal increase of the plastic deformation during the elasto-plastic process. This assumption implies that the elastic deformation is closer to zero and also that Young's modulus gets bigger during this period [1,2]. Then:

$$\dot{\boldsymbol{\varepsilon}} = \dot{\boldsymbol{\varepsilon}}^{P} \quad \Rightarrow \quad \dot{\boldsymbol{\varepsilon}}^{e} \cong \boldsymbol{0} \quad \text{or,} \quad E \to \infty \tag{5.35}$$

This theory also assumes, as a second hypothesis, that an ideal solid to be modeled is plastically incompressible $\varepsilon_v^P = 0$; as a result of this and from the previous hypothesis then the temporal increase of the deviator tensor of the plastic deformation is equal to the temporal increase of the total plastic deformation tensor. Thus,

$$\dot{\boldsymbol{\varepsilon}}^{\mathrm{P}} = \dot{\varepsilon}^{\mathrm{P}}_{\mathrm{oct}}\mathbf{1} + \dot{\mathbf{e}}^{\mathrm{P}} \quad \Rightarrow \quad \dot{\boldsymbol{\varepsilon}}^{\mathrm{P}} \equiv \dot{\mathbf{e}}^{\mathrm{P}} \quad \text{or,} \quad \dot{\boldsymbol{\varepsilon}} \equiv \dot{\mathbf{e}} \tag{5.36}$$

where the octahedral deformation is null and is defined as $3\varepsilon^{P}_{oct} = \varepsilon^{P}_{v} = 0$ and the unit vector is equal to $\mathbf{1} = \{1,1,1,0,0,0\}^{T}$. From these two hypotheses, it follows that the material behaves as a rigid plastic not influenced by volume changes due to hydrostatic pressure. This matches very well with the behavior of metallic materials.

The Levy-Mises theory states that the principal directions of plastic deformation should coincide with the stress ones which leads to the third hypothesis defining the so-called flow rule,

$$\dot{\mathbf{e}} \equiv \dot{\boldsymbol{\varepsilon}} = \dot{\lambda}\,\mathbf{s} \tag{5.37}$$

where $\mathbf{s} = \boldsymbol{\sigma} - \mathbf{I}\cdot\mathrm{tr}(\boldsymbol{\sigma})/3$ is the stress deviator tensor.

5.5.2 The Prandtl-Reus theory

It is a generalization of the Levy-Mises theory. The main difference with the latter is that the total deformation results from the contribution of an elastic part and a plastic one,

$$\dot{\boldsymbol{\varepsilon}} = \dot{\boldsymbol{\varepsilon}}^{e} + \dot{\boldsymbol{\varepsilon}}^{P} \tag{5.38}$$

where the temporal increase of the elastic deformation $\dot{\boldsymbol{\varepsilon}}^{e}$ follows the elasticity theory laws and the temporal increase of the plastic deformation tensor $\dot{\boldsymbol{\varepsilon}}^{P}$ will be obtained as a scale of the stress deviator tensor $\mathbf{s}$, and therefore the volumetric part of the plastic deformation tensor will be null. This hypothesis is called the Prandtl-Reus flow rule:

$$\dot{\boldsymbol{\varepsilon}}^{P} = \dot{\lambda}\,\mathbf{s} \quad \Rightarrow \quad \dot{\varepsilon}^{P}_{v} = 0 \tag{5.39}$$

The plastic consistency factor λ is obtained in this case from the stress space and the principal deformations.

$$\dot{\boldsymbol{\varepsilon}}^{P}_{i} = \dot{\lambda}\mathbf{s}_{i} \longrightarrow (\dot{\varepsilon}^{P}_{i} - \dot{\varepsilon}^{P}_{j}) = \dot{\lambda}(\mathbf{s}_{i} - \mathbf{s}_{j}) \qquad \{i, j \in 1,2,3\} \tag{5.40}$$

By squaring both members and adding component by component, the following is obtained

$$\left(\dot{\varepsilon}^{P}_{1} - \dot{\varepsilon}^{P}_{2}\right)^{2} + \left(\dot{\varepsilon}^{P}_{2} - \dot{\varepsilon}^{P}_{3}\right)^{2} + \left(\dot{\varepsilon}^{P}_{1} - \dot{\varepsilon}^{P}_{3}\right)^{2} = \dot{\lambda}^{2}\left[\left(\mathbf{s}_{2} - \mathbf{s}_{1}\right)^{2} + \left(\mathbf{s}_{2} - \mathbf{s}_{3}\right)^{2} + \left(\mathbf{s}_{1} - \mathbf{s}_{3}\right)^{2}\right] \tag{5.41}$$

but keeping in mind that $(s_{i} - s_{j}) = (\sigma_{i} - \sigma_{j})$, the plastic consistency factor results from the previous equation,

$$6\dot{J}^{\prime P}_{2} = \dot{\lambda}6J_{2} \quad \Rightarrow \quad \dot{\lambda} = \sqrt{\frac{\dot{J}^{\prime P}_{2}}{J_{2}}} \tag{5.42}$$

where $\dot{J}^{\prime P}_{2} = \frac{1}{2}\left(\dot{\mathbf{e}}^{P} : \dot{\mathbf{e}}^{P}\right)$ is the second invariant of the temporal increase of the deviator tensor of the plastic deformations and $\dot{\mathbf{e}}^{P}$ y $J_{2} = \frac{1}{2}(\mathbf{s} : \mathbf{s})$ is the second invariant of the stress deviator tensor $\mathbf{s}$. Substituting (5.42) into (5.39), the plastic deformation tensor is obtained for the Prandtl-Reus theory,

$$\dot{\boldsymbol{\varepsilon}}^P = \sqrt{\frac{\dot{J}_2^{'P}}{J_2}}\,\mathbf{s} = \sqrt{\dot{J}_2^{'P}}\,\frac{\mathbf{s}}{\sqrt{J_2}} \qquad (5.43)$$

5.6 The classic plasticity theory

When the stress state of a point in an ideal solid reaches the initial discontinuity $\mathbb{F}(\boldsymbol{\sigma};\mathbf{q})=0$ and satisfies, in turn, the plastic consistency condition $\dot{\mathbb{F}}(\boldsymbol{\sigma};\mathbf{q})=0$, it is assumed that this point is in an elasto-plastic state. The classic plasticity theory in small deformations[2] admits as valid the Prandtl-Reus hypothesis about the total deformation decomposition,

$$\boldsymbol{\varepsilon} = \mathbb{C}^{-1}:\boldsymbol{\sigma}+\boldsymbol{\varepsilon}^P = \boldsymbol{\varepsilon}^e + \boldsymbol{\varepsilon}^P \qquad (5.44)$$

where the plastic deformation $\boldsymbol{\varepsilon}^P$ represents the fundamental internal variable of the elasto-plastic problem and its definition is as follows:

$$\dot{\boldsymbol{\varepsilon}}^P = \dot{\lambda}\,\mathbf{H}_{\varepsilon^P} = \dot{\lambda}\,\frac{\partial \mathbb{G}(\boldsymbol{\sigma},\mathbf{q})}{\partial \boldsymbol{\sigma}} = \dot{\lambda}\,\mathbf{g} \qquad (5.45)$$

This expression is also known as the normality rule – normal to the potential plastic surface $\mathbb{G}(\boldsymbol{\sigma},\mathbf{q})=cte.$ –, a generalization of equation (5.39), and λ is a non-negative scalar called plastic consistency parameter which is determined by the Prager consistency condition and which provides the magnitude of the temporal increase of the plastic $\dot{\boldsymbol{\varepsilon}}^P$ and will be presented further. The potential plastic function is determined by experimental studies and it defines the temporal increase direction of the plastic potential.

A plastic flow occurs when the plastic yield surface is hypothetically adopted as the potential plastic surface $\mathbb{G}(\boldsymbol{\sigma},\mathbf{q}) \equiv \mathbb{F}(\boldsymbol{\sigma},\mathbf{q})$. Here equation (5.45) is reduced to,

$$\dot{\boldsymbol{\varepsilon}}^P = \dot{\lambda}\,\frac{\partial \mathbb{F}(\boldsymbol{\sigma},\mathbf{q})}{\partial \boldsymbol{\sigma}} = \dot{\lambda}\,\mathbf{f} \qquad (5.46)$$

and it is said that the flow rule is associated with the plastic yield surface. Otherwise, it is said that the flow is not associated.

If the Von Mises function is adopted as a potential function, then Prandtl-Reus flow is obtained (equation (5.39)),

$$\text{if } \mathbb{G} = \left[\mathbb{G}\right]^{\text{von Mises}} \quad \Rightarrow \quad \mathbf{g} \equiv \mathbf{s} \qquad (5.47)$$

5.6.1 Plastic unit or specific work

The total work developed in a unit volume of an ideal elasto-plastic solid, in a quasi-static process during a pseudo-time increment $(t \rightarrow t+dt)$, is called rate or temporal increment of unit work or specific work,

$$\dot{\omega} = \boldsymbol{\sigma} : \dot{\boldsymbol{\varepsilon}} = \boldsymbol{\sigma} : \left(\dot{\boldsymbol{\varepsilon}}^e + \dot{\boldsymbol{\varepsilon}}^P \right) = \boldsymbol{\sigma} : \dot{\boldsymbol{\varepsilon}}^e + \boldsymbol{\sigma} : \dot{\boldsymbol{\varepsilon}}^P = \dot{\omega}^e + \dot{\omega}^P \qquad (5.48)$$

This way of writing the energy temporal variation is known as *coupled elasticity* and is only valid for elasto-plastic problems whose elastic deformations are infinitesimal and therefore the assumption of additivity deformations is accepted, (equation (5.44)).

Focusing on the plastic part of the energy, the expression defining the plastic work for Prandtl-Reus is developed as follows,

$$\dot{\omega}^P = \boldsymbol{\sigma} : \dot{\boldsymbol{\varepsilon}}^P = \left(\frac{1}{3} I_1 \mathbf{1} + \mathbf{s} \right) : \dot{\boldsymbol{\varepsilon}}^P = \sigma_{oct} \underbrace{\dot{\varepsilon}_v^P}_{0} + \tau_{oct} \dot{\gamma}_{oct}^P = \underbrace{\dot{\omega}_k^P}_{\text{volumetric part}} + \underbrace{\dot{\omega}_G^P}_{\text{desviatoric part}}$$

$$\dot{\omega}^P = \boldsymbol{\sigma} : \dot{\boldsymbol{\varepsilon}}^P = \dot{\omega}_G^P = \tau_{oct} \dot{\gamma}_{oct}^P = \sqrt{\frac{2}{3} J_2} \ \sqrt{\frac{8}{3} \dot{J}_2^{\prime P}} = \dot{\lambda} (\mathbf{s} : \mathbf{s}) \qquad (5.49)$$

where $\sigma_{oct} = p = I_1 / 3$ is the normal octahedral stress or pressure, $\tau_{oct} = \sqrt{2 J_2 / 3}$ is the tangent octahedral stress or deviation, $\varepsilon_{oct}^P = \varepsilon_v^P / 3 = I_1^{\prime P} / 3$ is the normal octahedral deformation, and $\gamma_{oct}^P = \sqrt{8 J_2^{\prime P} / 3}$ is the octahedral deviation.

Substituting the Prandtl-Reus flow rule (equation (5.43)) then:

$$\dot{\omega}^P = \lambda (\mathbf{s} : \mathbf{s}) = \frac{1}{\dot{\lambda}} \left(\dot{\boldsymbol{\varepsilon}}^P : \dot{\boldsymbol{\varepsilon}}^P \right) = \sqrt{\frac{J_2}{\dot{J}_2^{\prime P}}} \left(\dot{\mathbf{e}}^P : \dot{\mathbf{e}}^P \right) \qquad (5.50)$$

where:

$$\dot{J}_2^{\prime P} = \frac{1}{2} \left(\dot{\mathbf{e}}^P : \dot{\mathbf{e}}^P \right) \qquad \dot{e}_{ij}^P = \dot{\varepsilon}_{ij}^P - \frac{1}{3} \underbrace{\dot{\varepsilon}_v^P \delta_{ij}}_{0} \qquad (5.51)$$

Such that in metals the second invariant of the deformation deviatoric tensor can be witten as $\dot{J}_2^{\prime P} = \frac{1}{2} \left(\dot{\varepsilon}_{ij}^P \ \dot{\varepsilon}_{ij}^P \right)$, so the plastic work, equation (5.50), is then

$$\dot{\omega}^P = \sqrt{2 J_2} \cdot \frac{\left(\dot{\boldsymbol{\varepsilon}}^P : \dot{\boldsymbol{\varepsilon}}^P \right)}{\sqrt{\dot{\boldsymbol{\varepsilon}}^P : \dot{\boldsymbol{\varepsilon}}^P}} = \sqrt{2 J_2} \ \sqrt{\dot{\boldsymbol{\varepsilon}}^P : \dot{\boldsymbol{\varepsilon}}^P} \qquad (5.52)$$

The potential function $\mathbb{G}$, which substituted into (5.45) yields a plastic flow equivalent to Prandtl-Reus, is the von Mises function, $\mathbb{G} = [\mathbb{G}]^{\text{von Mises}}$,

$$\mathbf{g} = \mathbf{s} = \frac{\partial [\mathbb{G}]^{\text{von Mises}}}{\partial \boldsymbol{\sigma}} \qquad (5.53)$$

where,

$$[\mathbb{G}]^{\text{von Mises}} = J_2 - \mathcal{K}^2 = 0 \xrightarrow{\text{also}} [\mathbb{G}]^{\text{von Mises}} = \sqrt{3 J_2} - \sqrt{3} \mathcal{K} = 0 \qquad (5.54)$$

being $\overline{\sigma} = \sqrt{3 J_2}$ the *effective stress* or equivalent von Mises uniaxial stress which, substituted into (5.52), provides the expression for the rate of the plastic work as follows,

$$\dot{\omega}^P = \sqrt{\frac{2}{3}}\overline{\sigma} \cdot \sqrt{\dot{\boldsymbol{\varepsilon}}^P : \dot{\boldsymbol{\varepsilon}}^P} = \overline{\sigma}\,\dot{\overline{\varepsilon}}^P \tag{5.55}$$

Through this expression *the effective plastic deformation* can be written in a general form as:

$$\dot{\overline{\varepsilon}}^P = \sqrt{\gamma\,\dot{\boldsymbol{\varepsilon}}^P : \dot{\boldsymbol{\varepsilon}}^P} \tag{5.56}$$

such that for the Prandtl-Reus (o von Mises) plasticity cases $\gamma \equiv 2/3$. For the remaining cases the magnitude must be found.

In general, the plastic work is used as the hardening variable or the effective plastic deformation can be obtained from this work,

$$\overline{\varepsilon}^P = \int_0^t \dot{\overline{\varepsilon}}^P dt = \int_0^t \frac{\dot{\omega}^P}{\overline{\sigma}} dt = \int_0^t \sqrt{\frac{2}{3}\left(\dot{\boldsymbol{\varepsilon}}^P : \dot{\boldsymbol{\varepsilon}}^P\right)}\, dt \tag{5.57}$$

The previous equation can be simplified for a radial loading problem, in other words, when all the stress tensor components maintain their proportion throughout the loading $\dfrac{\sigma_{11}}{\sigma_{11}^0} = \dfrac{\sigma_{22}}{\sigma_{22}^0} = \dots$. This is,

$$\overline{\varepsilon}^P = \sqrt{\frac{2}{3}\left(\boldsymbol{\varepsilon}^P : \boldsymbol{\varepsilon}^P\right)} \tag{5.58}$$

5.6.2 Plastic loading surface. Plastic hardening variable

Figure 5.6 describes the schematic uniaxial behavior of an ideal elasto-plastic solid. Four areas of different behaviors can be identified. One of these areas follows strictly the elasticity theory laws and the other three are governed by the plasticity theory. The limit between the elastic and the plastic areas is set through the *yield surface or discontinuity surface,* and from such limit this surface can move in the stress space, follow the evolution of the plastic process and transform itself into the so called *plastic loading surface.* This function representing the plastic loading surface is simply the discontinuity or yield limit function updated (5.29) for each of the internal variable value **q** at every moment of the pseudo time *t* of the elasto-plastic process. The phenomenon governing this *yield surface* change of position in the stress space is known as *plastic hardening.* This hardening behavior can be isotropic or kinematic and will be presented below.

A simple way of introducing the elasto-plastic behavior hardening is through the *plastic loading function* $\mathbb{F}(\boldsymbol{\sigma};\mathbf{q}) = 0$. This is defined as a scalar tensor and homogenous function of first degree in the stresses[*].

$$\mathbb{F}(\boldsymbol{\sigma};\mathbf{q}) = f(\boldsymbol{\sigma}) - \mathcal{K} = 0 \tag{5.59}$$

Thus, the stress function $f(\boldsymbol{\sigma})$ is set to translate from a stress tensor state into another equivalent scalar. This scalar is compared to the evolution of the plastic hardening $\mathcal{K}$, which is related to the evolution of the equivalent uniaxial stress $\overline{\sigma} = \mathcal{K}$.

---

[*] NOTE: $f(\boldsymbol{\sigma})$ is a homogenous function of n degree in stresses, provided that, $f(\alpha,\boldsymbol{\sigma}) \equiv \alpha^n f(\boldsymbol{\sigma})$.

5.6.2.1 Isotropic hardening

It is said that there is an isotropic hardening when there is a homothetic movement of the plastic loading surface. This movement can be:

Positive: when the homothetic movement of the plastic loading surface is an expansion movement (see Figure 5.10) it is said that it is an isotropic hardening elasto-plastic process.

Null: when there is no evolution of the plastic loading surface during the elasto-plastic process, it is said that this is an isotropic perfectly elasto-plastic process.

Negative: when there is a contraction homothetic movement on the plastic loading surface, it is said that this is an isotropic softening elasto-plastic process.

The isotropic hardening, the homothetic movement of the plastic loading function, is controlled by the evolution of the *plastic hardening function* $\mathcal{K}$, which is generally defined as an internal variable $\mathbf{q}$. The evolution of this internal variable depends on the mechanical process.

This internal variable evolution depends on its own mechanical process and the latter is conditioned by an evolution rule whose formulation must adapt itself to the solid behavior (See equation (5.30)).

In the classic plasticity, *the internal variable of the plastic hardening* is usually expressed as a *plastic hardening function* $\mathcal{K}(\kappa^P)$, which, in turn, depends on the *internal variable of plastic hardening* κ^P. Then

$$\mathcal{K}(\kappa^P) = f\left(\kappa^P\right) \quad \text{with:} \begin{cases} \kappa^P = \overline{\varepsilon}^P \\ \kappa^P = \omega^P \end{cases} \tag{5.60}$$

Defining the hardening function as an internal variable of the plastic process a more general formulation arises, so there are greater possibilities to represent the behavior of a large variety of solids.

$$\dot{\kappa}^P = \dot{\lambda}\, \mathrm{H}_\kappa(\boldsymbol{\sigma};\mathbf{q}) = \dot{\lambda}\left[\mathbf{h}_\kappa(\boldsymbol{\sigma};\mathbf{q}):\frac{\partial \mathbb{G}(\boldsymbol{\sigma};\kappa^P)}{\partial \boldsymbol{\sigma}}\right]$$

$$\dot{\mathcal{K}} = \dot{\lambda}\, \mathrm{H}_\mathcal{K}(\boldsymbol{\sigma};\mathbf{q}) = h_\mathcal{K}(\boldsymbol{\sigma};\mathbf{q})\, \dot{\kappa}^P \tag{5.61}$$

where the tensor function $\mathbf{h}_\kappa(\boldsymbol{\sigma};\mathbf{q})$ and the scalar function $h_\mathcal{K}(\boldsymbol{\sigma};\mathbf{q})$ depend on the updated tensor state and the internal variables. In the simplest case of the plasticity theory the following relations are identified,

$$\mathbf{h}_\kappa \equiv \boldsymbol{\sigma} \;\Rightarrow\; \dot{\kappa}^P = \mathbf{h}_\kappa : \dot{\boldsymbol{\varepsilon}}^P \equiv \boldsymbol{\sigma}:\dot{\boldsymbol{\varepsilon}}^P = \dot{\omega}^P = \overline{\sigma}\,\dot{\overline{\varepsilon}}^P$$

from where, $\tag{5.62}$

$$\dot{\mathcal{K}} = h_\mathcal{K}\,\dot{\kappa}^P = \frac{\partial \mathcal{K}(\kappa^P)}{\partial \kappa^P}\,\dot{\kappa}^P$$

such that this latter $\mathcal{K}(\kappa^P) = f(\kappa^P)$ is a function as expressed in equation (5.60).

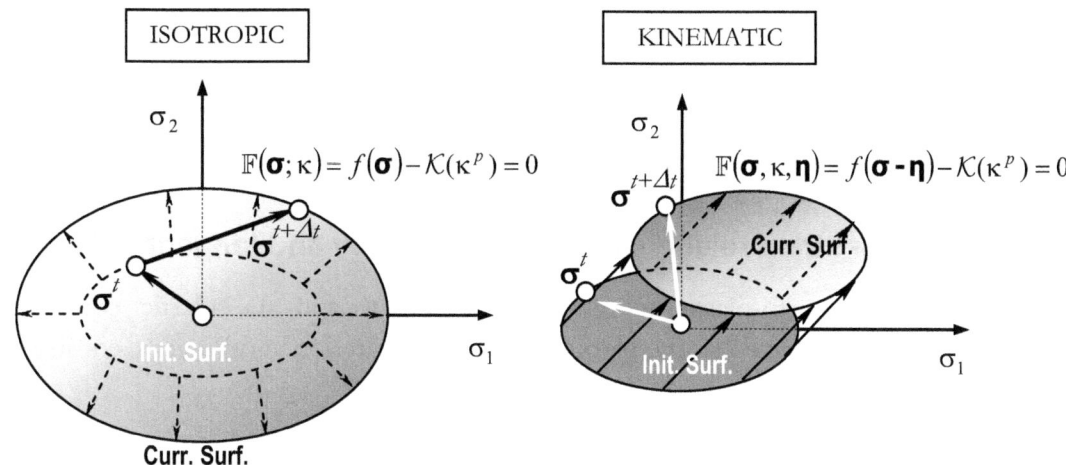

Figure 5.10 – Plastic loading surface. Isotropic and kinematic movements.

5.6.2.2 Kinematic hardening

The kinematic hardening, a translational motion of the plastic loading surface, remains controlled by the *kinematic plastic hardening internal variable* $\boldsymbol{\eta}$, which defines the stress space origin. The continuous change of position of this coordinate origin during the elasto-plastic process induces a translational movement of the yield surface that can or cannot be combined with its own expanding or contracting isotropic movement. In the most general case, the plastic loading function can be written as:

$$\mathbb{F}(\boldsymbol{\sigma};\mathbf{q}) = f(\boldsymbol{\sigma} - \boldsymbol{\eta}) - \mathcal{K} = 0 \tag{5.63}$$

where the plastic hardening can be defined according to Prager and Melan[1] as $\dot{\boldsymbol{\eta}} = \boldsymbol{\beta}\dot{\kappa}^p = c_k\,\dot{\boldsymbol{\varepsilon}}^P$, with $\boldsymbol{\beta} = \sqrt{c_k}\,\dot{\boldsymbol{\varepsilon}}^P / \dot{\bar{\varepsilon}}^P$. The expression c_k depends on the type of function of the plastic potential being used. In the simplest case, there is the following form for the von Mises potential function,

$$c_k = \frac{2}{3}h_k \tag{5.64}$$

where h_k is a material property to be determined by tests. This same property can be used to define the following most general expression for c_k that best fits the behavior of metals with the kinematic hardening[5],

$$c_k = \left[\frac{1}{\dot{\varepsilon}^p_{rs}\,\dot{\varepsilon}^p_{rs}} \cdot \sqrt{\frac{\sigma_{ij}\,\varepsilon^p_{ij}}{f(\sigma_{kl} - \eta_{kl})}}\right] \cdot h_k \tag{5.65}$$

The origin of the plastic loading function can also be set to represent Bauschinger's effect, by following an expression to decompose the kinematic behavior as follows[2,6],

---

[5] Chaboche J. L. (1983). On the constitutive equations of materials under monotonic or cyclic loading. *Rech. Aérosp. 1983-5*. France

$$\dot{\eta}_{ij} = c_k \, \dot{\varepsilon}_{ij}^p - a_k \, \eta_{ij} \, \dot{\bar{\varepsilon}}^p \tag{5.66}$$

where the effective plastic deformation $\dot{\bar{\varepsilon}}^p$ is obtained by equation (5.56) and c_k and a_k are two parameters to be determined.

5.6.3 Stress- strain relation. Plastic consistency and tangent stiffness

The constitutive elasto-plastic tangent law $\dot{\boldsymbol{\sigma}} = \mathbb{C}_T : \dot{\boldsymbol{\varepsilon}}$ and the plastic consistency parameter λ can be obtained by the plastic yield general criterion and Prager's[1] consistency condition. Thus,

$$\left.\begin{array}{l} \mathbb{F}(\boldsymbol{\sigma};\mathbf{q}) = f(\boldsymbol{\sigma} - \boldsymbol{\eta}) - \mathcal{K} = 0 \\[2mm] \dot{\mathbb{F}} = \dfrac{\partial \mathbb{F}}{\partial \boldsymbol{\sigma}} : \dot{\boldsymbol{\sigma}} + \dfrac{\partial \mathbb{F}}{\partial \boldsymbol{\eta}} : \dot{\boldsymbol{\eta}} + \underbrace{\dfrac{\partial \mathbb{F}}{\partial \mathcal{K}}}_{-1} \dot{\mathcal{K}} = 0 \end{array}\right\} \Rightarrow \dfrac{\partial \mathbb{F}}{\partial \boldsymbol{\sigma}} : \dot{\boldsymbol{\sigma}} + \dfrac{\partial \mathbb{F}}{\partial \boldsymbol{\eta}} : \dot{\boldsymbol{\eta}} - \dot{\mathcal{K}} = 0 \tag{5.67}$$

Substituting equation (5.61) and $\dot{\boldsymbol{\eta}} = \boldsymbol{\beta}\dot{\kappa}^p = c_k \dot{\boldsymbol{\varepsilon}}^p$ into equation (5.67), then

$$\frac{\partial \mathbb{F}}{\partial \boldsymbol{\sigma}} : \mathbb{C} : \left(\dot{\boldsymbol{\varepsilon}} - \dot{\boldsymbol{\varepsilon}}^p\right) + c_k \, \frac{\partial \mathbb{F}}{\partial \boldsymbol{\eta}} : \dot{\boldsymbol{\varepsilon}}^p - h_\mathcal{K}\left(\mathbf{h}_\kappa : \dot{\boldsymbol{\varepsilon}}^p\right) = 0$$

$$\left[\frac{\partial \mathbb{F}}{\partial \boldsymbol{\sigma}} : \mathbb{C} : \dot{\boldsymbol{\varepsilon}}\right] - \lambda\left[\frac{\partial \mathbb{F}}{\partial \boldsymbol{\sigma}} : \mathbb{C} : \frac{\partial \mathbb{G}}{\partial \boldsymbol{\sigma}} - c_k \frac{\partial \mathbb{F}}{\partial \boldsymbol{\eta}} : \frac{\partial \mathbb{G}}{\partial \boldsymbol{\sigma}} + h_\mathcal{K}\,\mathbf{h}_\kappa : \frac{\partial \mathbb{G}}{\partial \boldsymbol{\sigma}}\right] = 0 \tag{5.68}$$

From this last expression the plastic consistency factor λ can be obtained, which is considered as the factor evaluating the distance between an inadmissible tensor state outside the domain and the plastic loading surface. Thus,

$$\lambda = \frac{\dfrac{\partial \mathbb{F}}{\partial \boldsymbol{\sigma}} : \mathbb{C} : \dot{\boldsymbol{\varepsilon}}}{\underbrace{\left[-c_k \dfrac{\partial \mathbb{F}}{\partial \boldsymbol{\eta}} : \dfrac{\partial \mathbb{G}}{\partial \boldsymbol{\sigma}} + h_\mathcal{K}\,\mathbf{h}_\kappa : \dfrac{\partial \mathbb{G}}{\partial \boldsymbol{\sigma}}\right] + \left(\dfrac{\partial \mathbb{F}}{\partial \boldsymbol{\sigma}} : \mathbb{C} : \dfrac{\partial \mathbb{G}}{\partial \boldsymbol{\sigma}}\right)}_{A}} \quad \text{being} \quad \lambda \geq 0 \tag{5.69}$$

where A is the plastic hardening parameter. In a simple case of the classic theory of plasticity without kinematic hardening $c_k = 0$, this parameter becomes the slope of the equivalent uniaxial stress curve $\bar{\sigma}(\bar{\varepsilon}^p) = \mathcal{K}(\bar{\varepsilon}^p)$ vs. $\bar{\varepsilon}^p$. To prove this, a hardening function is considered $\mathcal{K}(\bar{\varepsilon}^p) = f(\bar{\varepsilon}^p)$ and its slope is defined as,

$$A = \frac{d\mathcal{K}(\bar{\varepsilon}^p)}{d\bar{\varepsilon}^p} = \frac{d\mathcal{K}(\bar{\varepsilon}^p)}{d\kappa^p}\frac{d\kappa^p}{d\bar{\varepsilon}^p} \tag{5.70}$$

From the latter and equation (5.62), the equation denominator of equation (5.69) is verified

---

[6] Ohno N. and Wang J. (1993). Kinematic hardening rules with critical state of dynamic recovery, Part I: Formulation and basic features for ratcheting behavior. *International Journal of Plasticity*. Vol. 9 pp. 375-390.

$$A = -\frac{\partial \mathbb{F}}{\partial \kappa^p}\overline{\sigma} = h_\mathcal{K}\,\mathbf{h}_\kappa : \frac{\partial \mathbb{G}}{\partial \boldsymbol{\sigma}} \tag{5.71}$$

Substituting equation (5.69) into the tangent constitutive equation,

$$\dot{\boldsymbol{\sigma}} = \mathbb{C} : \left(\dot{\boldsymbol{\varepsilon}} - \dot{\boldsymbol{\varepsilon}}^P\right) \tag{5.72}$$

Then the following tangent elasto-plastic law is obtained,

$$\dot{\boldsymbol{\sigma}} = \left\{ \mathbb{C} - \frac{\left[\mathbb{C} : \dfrac{\partial \mathbb{G}}{\partial \boldsymbol{\sigma}}\right] \otimes \left[\dfrac{\partial \mathbb{F}}{\partial \boldsymbol{\eta}} : \mathbb{C}\right]}{\underbrace{\left[-c_k\,\dfrac{\partial \mathbb{F}}{\partial \boldsymbol{\eta}} : \dfrac{\partial \mathbb{G}}{\partial \boldsymbol{\sigma}} + h_\mathcal{K}\mathbf{h}_\kappa : \dfrac{\partial \mathbb{G}}{\partial \boldsymbol{\sigma}}\right]}_{A} + \left(\dfrac{\partial \mathbb{F}}{\partial \boldsymbol{\sigma}} : \mathbb{C} : \dfrac{\partial \mathbb{G}}{\partial \boldsymbol{\sigma}}\right)} \right\} : \dot{\boldsymbol{\varepsilon}} \quad \Rightarrow \quad \dot{\boldsymbol{\sigma}} = \mathbb{C}_T : \dot{\boldsymbol{\varepsilon}} \tag{5.73}$$

where $\mathbb{C}_T$ is the continuous tangent constitutive tensor. In Table 5.1 Euler's implicit algorithm is shown, which can be used to integrate the constitutive equation in an efficient manner.

5.7 Drucker's stability postulate and maximum plastic dissipation

Euler's second postulate defines the local stability of the behavior of a point in a solid subjected to a stress-strain state (see sections 5.3.2.2). In the nonlinear problem, this postulate is related to the maximum plastic dissipation axiom.

Consider a point in a solid subjected to a stress state $\boldsymbol{\sigma} = \boldsymbol{\sigma}\left(\boldsymbol{\varepsilon}; \boldsymbol{\varepsilon}^P; \mathbf{q}\right)$ and deformation $\boldsymbol{\varepsilon}$, such that their magnitudes were in the previous moment $\boldsymbol{\sigma}^* = \boldsymbol{\sigma}\left(\boldsymbol{\varepsilon}^*; \boldsymbol{\varepsilon}^P; \mathbf{q}\right)$ and $\boldsymbol{\varepsilon}^*$. It is said that the behavior has been stable if the following inequality is satisfied,

$$\boldsymbol{\sigma} : \dot{\boldsymbol{\varepsilon}}^P \geq \boldsymbol{\sigma}^* : \dot{\boldsymbol{\varepsilon}}^P \quad \rightarrow \quad \left(\boldsymbol{\sigma} - \boldsymbol{\sigma}^*\right) : \dot{\boldsymbol{\varepsilon}}^P \geq 0 \tag{5.74}$$

where it can be observed that the subsequent stress state $\boldsymbol{\sigma}$ is required to be greater than the previous one $\boldsymbol{\sigma}^*$.

And with the following approximation,

$$\boldsymbol{\varepsilon} - \boldsymbol{\varepsilon}^* \approx \dot{\boldsymbol{\varepsilon}}\,dt \quad \Rightarrow \quad \boldsymbol{\sigma} - \boldsymbol{\sigma}^* = \mathbb{C} : \dot{\boldsymbol{\varepsilon}}\,dt \tag{5.75}$$

and substituting the latter into equation (5.76), then the following particular form of Druckers second postulate is obtained

$$\dot{\boldsymbol{\varepsilon}} : \mathbb{C} : \dot{\boldsymbol{\varepsilon}}^P \geq 0 \tag{5.76}$$

This same conclusion is reached through the local form of maximum plastic dissipation (M.D.P.)[2], that is written as,

$$\dot{\boldsymbol{\varepsilon}}:\left(\frac{\partial\Xi}{\partial\boldsymbol{\varepsilon}^e}\right)\geq 0 \tag{5.77}$$

where the plastic dissipation Ξ, for problems without rigidity degradation, is then written,

$$\Xi = \boldsymbol{\sigma}:\dot{\boldsymbol{\varepsilon}}^P - \dot{\Psi} \geq 0 \tag{5.78}$$

Substituting the latter into the M.D.P. expression, then

$$\dot{\boldsymbol{\varepsilon}}:\mathbb{C}:\dot{\boldsymbol{\varepsilon}}^P \geq 0 \tag{5.79}$$

From the latter and equation (5.76) it follows that Drucker's stability postulate totally coincides with the maximum plastic dissipation axiom.

---

1. Computation of the predictive stress at current time "$t + \Delta t$", equilibrium iteration "i", convergence counter of the constitutive model "$k = 1$"

$$_{k-1}^{i}[\boldsymbol{\sigma}]^{t+\Delta t} = \mathbb{C}:\left([\boldsymbol{\varepsilon}]^{t+\Delta t} - {}^{i-1}[\boldsymbol{\varepsilon}^{P}]^{t+\Delta t}\right)$$

$$_{k-1}^{i}[\mathbf{q}]^{t+\Delta t} = {}^{i-1}[\mathbf{q}]^{t+\Delta t}$$

2. Verification of the plastic yield condition:

   a. If: $\mathbb{F}\left(_{k-1}^{i}[\boldsymbol{\sigma}]^{t+\Delta t};{}_{k-1}^{i}[\mathbf{q}]^{t+\Delta t}\right) < 0$, then $\left\{\begin{array}{c}{}^{i}[\boldsymbol{\sigma}]^{t+\Delta t} = {}_{k-1}^{i}[\boldsymbol{\sigma}]^{t+\Delta t} \\ {}^{i}[\mathbf{q}]^{t+\Delta t} = {}_{k-1}^{i}[\mathbf{q}]^{t+\Delta t}\end{array}\right\}$ and go to EXIT

   b. If: $\mathbb{F}\left(_{k-1}^{i}[\boldsymbol{\sigma}]^{t+\Delta t};{}_{k-1}^{i}[\mathbf{q}]^{t+\Delta t}\right) \geq 0$, then the integration of the constitutive equ. starts

3. Integration of the equation itself,

$$\Delta\lambda = \frac{\mathbb{F}\left(_{k-1}^{i}[\boldsymbol{\sigma}]^{t+\Delta t};{}_{k-1}^{i}[\mathbf{q}]^{t+\Delta t}\right)}{\underbrace{_{k-1}^{i}\left[-c_k\frac{\partial\mathbb{F}}{\partial\boldsymbol{\eta}}:\frac{\partial\mathbb{G}}{\partial\boldsymbol{\sigma}} + h_{\mathcal{K}}\,\mathbf{h}_{\kappa}:\frac{\partial\mathbb{G}}{\partial\boldsymbol{\sigma}}\right]^{t+\Delta t} + {}_{k-1}^{i}\left(\frac{\partial\mathbb{F}}{\partial\boldsymbol{\sigma}}:\mathbb{C}:\frac{\partial\mathbb{G}}{\partial\boldsymbol{\sigma}}\right)^{t+\Delta t}}_{_{k-1}^{i}A^{t+\Delta t}}}$$

$$_{k}^{i}[\boldsymbol{\sigma}]^{t+\Delta t} = {}_{k-1}^{i}[\boldsymbol{\sigma}]^{t+\Delta t} - \Delta\lambda\,\mathbb{C}:{}_{k-1}^{i}\left(\frac{\partial\mathbb{G}}{\partial\boldsymbol{\sigma}}\right)^{t+\Delta t}$$

The internal variables and the tangent constitutive tensor are updated with the new stress

$$_{k}^{i}[\mathbf{q}]^{t+\Delta t} = {}_{k-1}^{i}[\mathbf{q}]^{t+\Delta t}$$

$$_{k}^{i}[\mathbb{C}_{T}]^{t+\Delta t} = {}_{k}^{i}\left[\mathbb{C} - \frac{\left[\mathbb{C}:\frac{\partial\mathbb{G}}{\partial\boldsymbol{\sigma}}\right]\otimes\left[\frac{\partial\mathbb{F}}{\partial\boldsymbol{\sigma}}:\mathbb{C}\right]}{-c_k\frac{\partial\mathbb{F}}{\partial\boldsymbol{\eta}}:\frac{\partial\mathbb{G}}{\partial\boldsymbol{\sigma}} + h_{\mathcal{K}}\,\mathbf{h}_{\kappa}:\frac{\partial\mathbb{G}}{\partial\boldsymbol{\sigma}} + \left(\frac{\partial\mathbb{F}}{\partial\boldsymbol{\sigma}}:\mathbb{C}:\frac{\partial\mathbb{G}}{\partial\boldsymbol{\sigma}}\right)}\right]^{t+\Delta t}$$

4. It gets $k = k + 1$ and returns to point 2

Table 5.1 – Integration of the constitutive elasto-plastic equation by the Euler implicit algorithm.

5.8 Stability condition

Drucker's stability condition is also known as *local stability condition* and it only refers to the stability behavior of a point in a solid. When this condition is satisfied in all the points of the solid the stability of the whole set can be guaranteed; however, it is not necessary to verify all the points to guarantee the set's stability. This can be observed in softening materials, in which the local stability condition may not be satisfied in some points, but does not involve any stability loss of the global solid. The stability of the whole solid can be proved by a weaker condition known as *global stability condition*. A brief presentation of these concepts will be presented next.

5.8.1 Local stability

Drucker's second postulate (see equation (5.74)), entails a necessary and sufficient stability condition for plasticity softening problems and the associated rule flow, but it is only a sufficient condition for plasticity softening problems and/or not associated rule flow. Below it is proved that by requiring convexity in the yield functions, plastic potential and associated flow in materials softening, the fulfillment of Drucker's second postulate can be guaranteed.

$$\dot{\boldsymbol{\varepsilon}}:(\mathbb{C}):\dot{\boldsymbol{\varepsilon}}^p = \dot{\boldsymbol{\varepsilon}}:(\mathbb{C}):\left(\dot{\lambda}\frac{\partial \mathbb{G}}{\partial \boldsymbol{\sigma}}\right) \geq 0 \tag{5.80}$$

But the plastic consistency factor λ is a non-negative scalar. Thus, the previous inequality can also be written as,

$$\dot{\lambda} \geq 0 \quad , \quad \dot{\boldsymbol{\varepsilon}}:(\mathbb{C}):\frac{\partial \mathbb{G}}{\partial \boldsymbol{\sigma}} \geq 0 \tag{5.81}$$

Additionally, by analyzing the plastic consistency factor, it can be deduced that,

$$\dot{\lambda} = \frac{\dfrac{\partial \mathbb{F}}{\partial \boldsymbol{\sigma}}:\mathbb{C}:\dot{\boldsymbol{\varepsilon}}}{A + \dfrac{\partial \mathbb{G}}{\partial \boldsymbol{\sigma}}:\mathbb{C}:\dfrac{\partial \mathbb{F}}{\partial \boldsymbol{\sigma}}} \geq 0 \tag{5.82}$$

if the hardening is positive $A > 0$, and $\mathbb{G}$ and $\mathbb{F}$ are convex functions, so that the plastic consistency factor is negative, the following condition must be satisfied

$$\dot{\boldsymbol{\varepsilon}}:\mathbb{C}:\frac{\partial \mathbb{G}}{\partial \boldsymbol{\sigma}} \geq 0 \tag{5.83}$$

In order to guarantee that the inequalities (5.81) and (5.83) are satisfied at the same time, the associated plastic flow must occur. In other words,

$$\frac{\partial \mathbb{G}}{\partial \boldsymbol{\sigma}} \propto \frac{\partial \mathbb{F}}{\partial \boldsymbol{\sigma}} \quad , \quad \text{Associated plastic flow} \tag{5.84}$$

5.8.2 Global stability

As mentioned before, in softening materials, Drucker's postulate is transformed into a sufficient, but not necessary, stability condition

The necessary stability condition must be formulated at global level, in other words, in the whole solid. Let a *stable configuration* be considered $\Omega_* \subset \mathbb{R}^3$ where the total potential energy is $\Pi^* = P_d^* - P_{in}^*$, where the stress variables involved are $\boldsymbol{\sigma}^*$, deformation $\boldsymbol{\varepsilon}^*$, surface forces $\mathbf{t}^*$, volume $\mathbf{b}^*$ and displacement $\mathbf{u}^*$, and where the solid equilibrium is satisfied. By giving a virtual displacement $\delta\mathbf{u}$ to this stable configuration, a new configuration $\Omega_t \subset \mathbb{R}^3$ is obtained, and its variables are $\mathbf{u} = \mathbf{u}^* + \delta\mathbf{u} \rightarrow \boldsymbol{\varepsilon} = \boldsymbol{\varepsilon}^* + \delta\boldsymbol{\varepsilon} \quad \boldsymbol{\sigma} = \boldsymbol{\sigma}^* + \delta\boldsymbol{\sigma}; \ \mathbf{b} = \mathbf{b}^* \ ; \mathbf{t} = \mathbf{t}^*$, and the total energy will be $\Pi = P_d - P_{in} \cong \Pi^* + \delta\Pi + \frac{1}{2!}\delta^2\Pi$. It is said this new configuration will be in equilibrium if while applying the virtual displacement, the variation of the total potential energy developed is null. This is

$$\Pi = \Pi^* + \delta\Pi + \frac{1}{2!}\delta^2\Pi + \cdots \ \Rightarrow \ \Delta\Pi = \Pi - \Pi^* \cong \underset{0}{\underbrace{\delta\Pi}} + \frac{1}{2!}\delta^2\Pi + \cdots \tag{5.85}$$

where the first energy variation or stationary condition of the functional is $\delta\Pi = \int_V \boldsymbol{\sigma} : \delta\boldsymbol{\varepsilon}\, dV - \oint_S \mathbf{t} \cdot \delta\mathbf{u}\, dS - \int_V \rho\mathbf{b} \cdot \delta\mathbf{u}\, dV = 0$. As the virtual displacement $\delta\mathbf{u}$ and its derivative fields $\delta\boldsymbol{\varepsilon}$ are arbitrary, the equilibrium equation is obtained. The second variation of the total potential energy or stationary condition of the functional can be written as $\delta^2\Pi = \int_V \delta\boldsymbol{\sigma} : \delta\boldsymbol{\varepsilon}\, dV - \oint_S \delta\mathbf{t} \cdot \delta\mathbf{u}\, dS - \int_V \rho\,\delta\mathbf{b} \cdot \delta\mathbf{u}\, dV$. Although knowing the functional expression of the total potential energy [7] is not always possible, its corresponding variables can be known by using the principle of the virtual works, as established in the previous equations. By substituting the equilibrium condition into equation (5.85), the total increase of the virtual work is equal to the second variation of the functional and it is, in turn, the concave or convex stability condition of the functional,

$$\Delta\Pi \cong \underset{0}{\underbrace{\delta\Pi}} + \frac{1}{2!}\delta^2\Pi \quad \begin{cases} > 0 \ \Rightarrow \ \text{The original configuration is stable} \\ \qquad \text{for any virtual displacement.} \\ < 0 \ \Rightarrow \ \text{The original configuration is unstable} \\ \qquad \text{for this virtual displacement.} \end{cases} \tag{5.86}$$

By substituting the state previously defined ($\mathbf{u} = \mathbf{u}^* + \delta\mathbf{u} \rightarrow \boldsymbol{\varepsilon} = \boldsymbol{\varepsilon}^* + \delta\boldsymbol{\varepsilon} \quad \boldsymbol{\sigma} = \boldsymbol{\sigma}^* + \delta\boldsymbol{\sigma}$; $\mathbf{b} = \mathbf{b}^* \ ; \mathbf{t} = \mathbf{t}^*$) into the second variation of the functional, then

$$\Delta\Pi \cong \frac{1}{2}\int_V \delta\boldsymbol{\sigma} : \delta\boldsymbol{\varepsilon}\, dV \quad \Rightarrow \quad \begin{cases} > 0 \ \text{Stable original configuration} \\ < 0 \ \text{Unstable original configuration} \end{cases} \tag{5.87}$$

According to Bažant[8], by applying this condition to softening materials a limit size can be defined of volume V_p, of the area where an unstable plastic behavior can occur, according to Drucker's local stability postulate (softening area). Thus, in the rest of the solid where the behavior is elastic and its volume is $V_0 = V - V_p$, this second order work is positive such

---

[7] Washizu, K. (1974). *Variational methods in elasticity and plasticity.* Pergamon Press.
[8] Bažant, Z. (1986). Mechanics of distributed cracking. *Applied Mech. Rev. - Vol. 39, No. 5 pp. 675-705.*

that it compensates the negative work, giving a non-null second variation of the functional. This way a new stable configuration is achieved for a solid whose hardening behavior is found in one part of the domain V_0 and of softening in the other V_p. This is,

$$\Delta\Pi \cong \frac{1}{2}\int_V \delta\boldsymbol{\sigma} : \delta\boldsymbol{\varepsilon}\, dV = \left[\frac{1}{2}\int_{V_0} \delta\boldsymbol{\sigma} : \delta\boldsymbol{\varepsilon}\, dV + \frac{1}{2}\int_{V_p} \delta\boldsymbol{\sigma} : \delta\boldsymbol{\varepsilon}\, dV\right] \cong \left[\Delta\Pi_{V_0} + \Delta\Pi_{V_p}\right] > 0 \qquad (5.88)$$

From all of the above it can be concluded that in a point of a solid Drucker's stability condition (equation 5.83) may be violated without the global behavior being necessarily unstable.

5.9 Condition of unicity of solution

Given a point of a solid in equilibrium state (origin configuration), to which two virtual displacements are applied, $\delta\mathbf{u_1}$ and $\delta\mathbf{u_2}$, the potential energy difference between the two configurations are measured. It can be observed that this energy difference is equal to the one obtained by applying one single virtual displacement of magnitude $\Delta(\delta\mathbf{u}) = \delta\mathbf{u_2} - \delta\mathbf{u_1}$, producing a change of strain $\Delta(\delta\boldsymbol{\varepsilon})$ and stress $\Delta(\delta\boldsymbol{\sigma})$. From these stress and deformation fields a second-order virtual work is obtained equal to

$$\Delta(\delta^2\Pi) = \int_V \Delta(\delta\boldsymbol{\sigma}) : \Delta(\delta\boldsymbol{\varepsilon})\, dV \begin{cases} = 0 & \text{There is no unicity of the solution,} \\ \neq 0 & \text{There is unicity of the solution.} \end{cases} \qquad (5.89)$$

If this second variation is null $\Delta(\delta^2\Pi) = 0$ during the virtual displacement change $\Delta(\delta\mathbf{u})$, it means that the stress in both final configurations are the same $\delta\boldsymbol{\sigma_2} = \delta\boldsymbol{\sigma_2}$; in other words, $\Delta(\delta\boldsymbol{\sigma}) = \delta\boldsymbol{\sigma_2} - \delta\boldsymbol{\sigma_2} = 0$. Therefore, there are two admissible and independent of each other kinematic states $\delta\mathbf{u_2} \neq \delta\mathbf{u_1}$, but the stress increase is the same $\delta\boldsymbol{\sigma_2} = \delta\boldsymbol{\sigma_2}$, which implies that there is not a single solution and there is a bifurcation in the answer.

Given the same aforementioned kinematic configurations, if a non-null second variation is obtained $\Delta(\delta^2\Pi) \neq 0$, it is said the unicity of the solution is guaranteed.

5.10 Kuhn-Tucker. Loading-unloading condition

The loading-unloading condition and Prager's plastic consistency condition are satisfied simultaneously by Kuhn-Tucker's[3] three conditions, which is another way of presenting the axiom of maximum plastic dissipation M.P.D (5.7). This is,

$$\begin{cases} \dot{\lambda} \geq 0 \\ \mathbb{F}(\boldsymbol{\sigma}; \mathbf{q}) \leq 0 \\ \dot{\lambda}\, \mathbb{F}(\boldsymbol{\sigma}; \mathbf{q}) = 0 \end{cases} \qquad (5.90)$$

From these three conditions the following it can be deduced briefly,

$$
\begin{cases}
\mathbb{F} < 0 & \Rightarrow \dot{\lambda} = 0 \quad \text{elastic behavior or unloading,} \\
\mathbb{F} = 0 & \Rightarrow \begin{cases} \dot{\lambda} > 0 \quad \text{plastic behavior or loading,} \\ \dot{\lambda} = 0 \quad \text{neutral plastic load,} \end{cases} \\
\mathbb{F} > 0 & \Rightarrow \text{incompatible state.}
\end{cases}
\tag{5.91}
$$

5.11 Yield or plastic discontinuity classic criteria

A brief, almost enumerative, description of the yield or plastic discontinuity criteria is presented. The objective of this presentation is to highlight the most significant features of each of them.

A great number of criteria about yield and plastic discontinuity have been formulated during the last years to better represent the plastic behavior of ideal solids. There are other better suited criteria for the representation of the behavior of metal and other materials that work better for geomaterials. In general, the formulation and/or use of these criteria require considering the following basic behavior characteristics:

- **Metallic materials** have traction and compression strength of the same magnitude. The hydrostatic pressure, first invariant stress tensor I_1, has very little influence on the determination of the plastic yield state. The permanent volume changes are negligible (temporary increase of permanent volumetric deformation, null distance $\dot{\varepsilon}_v^p = 0$), which means that the shape and size of the yield surface cross section (octahedral plane) is maintained unaltered in low as much as in high stresses (not depending on the third invariant of the deviatoric stress tensor J_3), for example,: the cylindrical shape of the von Mises shape. The temporary increase of plastic deformation $\dot{\boldsymbol{\varepsilon}}^p$ depends on the stress deviatoric tensor $\mathbf{s}$ at each moment of the quasi-static loading process, using satisfactorily Prandtl-Reus flow rule, which involves using the general form of the flow rule (equation 5.46), with one of von Mises plastic potential functions.

- **Frictional materials** of the stony concrete type, soils, ceramics, etc., have less strength to traction than to compression. The hydrostatic pressure $p = I_1/3$ has more influence on the plastic yield condition for low and moderate stresses than on high hydrostatic stresses. The solid suffers unrecoverable volume changes showing dilatancy phenomena $\dot{\varepsilon}_v^p \neq 0$. The shape and dimension of the yield surface cross section (octahedral plane) are different for low and for high stresses, shifting from an almost triangular shape to a circular one respectively (for low hydrostatic pressures depend on the third invariant of the deviatoric stress tensor J_3 and gets freed from it at high pressures). The plastic deformation has a different direction from the direction given by the gradient of the yield surface, which makes it necessary to formulate a potential plastic surface different from the plastic yield (non-associated plasticity). In these materials and unlike metals, the yield criteria depends, among others, on three variables: internal cohesion among particles c, the internal friction angle among particles ϕ and the internal dilatancy ψ. These can be treated as internal variables of the same process itself, or also expressed as a dependent function in an explicit form of the evolution of the internal variables $\mathbf{q}$.

From this brief description the need to formulate different yield and plastic potential criteria to consider the requirements for each type of materials becomes obvious.

5.11.1 The Rankine criterion of maximum traction stress

This criterion was formulated by Rankine in 1876 and is based on one single parameter, the maximum uniaxial tension strength σ_T^{max}. Additionally, it is influenced by the first invariant of the stress tensor I_1 and by the second and third invariants of the deviatoric stress tensor J_2, J_3, respectively. This criterion helps to set in the limits in a simple way where the fracturing process starts in a point of a solid. This hypothesis leads to the assumption that fractures occur when the maximum main stress reaches the value of the uniaxial tension strength $\mathcal{K}(\kappa) = \sigma_T^{max}(\kappa)$. The different mathematical forms to express this criterion are the following:

- As a function of the principal stresses,

$$\mathbb{F}\left(\boldsymbol{\sigma}; \sigma_T^{max}\right) = \max[\sigma_i] - \sigma_T^{max}(\kappa) = 0 \tag{5.92}$$

- As a function of the invariants of the stress tensor and its deviatoric stress tensors,

$$\mathbb{F}\left(I_1; J_2; \theta; \sigma_T^{max}\right) = 2\sqrt{3J_2}\,\cos\left(\theta + \frac{\pi}{6}\right) + I_1 - 3\sigma_T^{max}(\kappa) = 0 \tag{5.93}$$

- As a function of the cylindrical coordinates,

$$\mathbb{F}\left(\rho; \xi; \theta; \sigma_T^{max}\right) = \sqrt{2}\,\rho\,\cos\left(\theta + \frac{\pi}{6}\right) + \xi - \sqrt{3}\,\sigma_T^{max}(\kappa) = 0 \tag{5.94}$$

where $\xi = \sqrt{3}\,\sigma_{oct} = \sqrt{3}\,(I_1/3) = I_1/\sqrt{3}$, the octahedral radio $\rho = \sqrt{3}\,\tau_{oct} = \sqrt{3}\sqrt{\dfrac{2J_2}{3}} = \sqrt{2J_2}$, and Lode's similarity angle $\theta = \text{arcsen}\left[\left(3\sqrt{3}\,J_3\right)/\left(2J_2^{3/2}\right)\right]$.

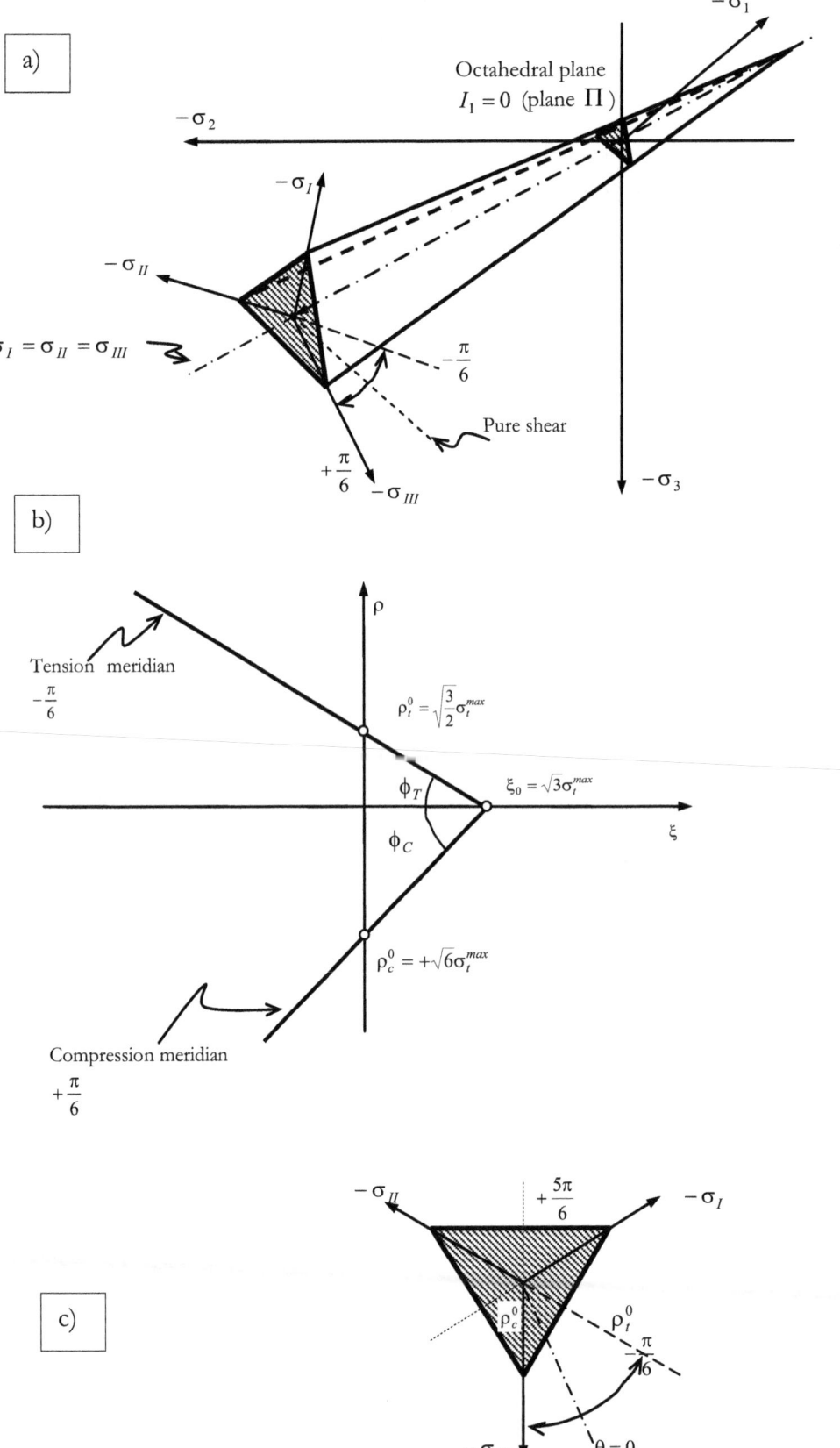

Figure 5.11 – Rankine yield surface: a) In the principal stress space, b) In the maximum tension and compression meridians, c) In the octahedral plane I_1=0 or Π plane.

5.11.2 The Tresca criterion of maximum shear stress

This criterion was formulated by Tresca in 1864. Similarly to Rankine criterion, it also depends on one single parameter which is the maximum tangent strength τ^{max}. Moreover, it considers the second and third invariants of the deviatoric stress tensor J_2, J_3, respectively, neglecting the influence of the first invariant of the stress tensor I_1. This is suitable for the metal's behavior representation, and its great limitation is due to the lack of continuity in its derivatives' definition describing the outer normal to the surface. According to this criterion, the plastic yield is reached when the value of the plastic hardening function $\mathcal{K}(\kappa) = \tau^{max}(\kappa)$, that has the meaning of a shear strength, reaches the maximum strength to the tangent stresses τ^{max}. The various mathematical forms to express this criterion are the following:

- As a function of the principal stresses,

$$\mathbb{F}\left(\boldsymbol{\sigma}; \tau^{max}\right) = \max\left[\frac{1}{2}\left|\sigma_i - \sigma_j\right|\right] - \tau^{max}(\kappa) = 0 \tag{5.95}$$

- As a function of the invariants of the deviatoric stresses tensor,

$$\mathbb{F}\left(J_2; \theta; \tau^{max}\right) = \sqrt{\frac{J_2}{3}}\cos\theta - \tau^{max}(\kappa) = 0$$

or multiplying by $2\sqrt{2}$ results as a function of the effective uniaxial stress, $\tag{5.96}$

$$\mathbb{F}\left(J_2; \theta; \overline{\sigma}\right) = 2\sqrt{J_2}\cos\theta - \overline{\sigma}(\kappa) = 0$$

- As a function of cylindrical coordinates,

$$\mathbb{F}\left(\rho; \theta; \overline{\sigma}\right) = \rho\cos\theta - \frac{\sqrt{2}}{2}\overline{\sigma}(\kappa) = 0 \tag{5.97}$$

being $\rho = \sqrt{3}\,\tau_{oct} = \sqrt{2J_2}$, and the angle of similarity of Lode $\theta = \arcsen\left[\left(3\sqrt{3}\,J_3\right)/\left(2\,J_2^{3/2}\right)\right]$.

The insensitivity of the octahedral plane to pressure keeps it constant or equal to the plane π. This octahedral plane represents a regular hexagon. From the interaction between the traction meridian plane $(\theta = -\pi/6)$, and the yield surface arises a null slope line, parallel to the one resulting from the interaction between the compression meridian plane $(\theta = +\pi/6)$, and the yield surface. Both meridian lines cut the octahedral shear axis $\rho_C^0 = \rho_T^0 = \pm\sqrt{2/3}\,\overline{\sigma}(\kappa)$. In the main plane $\sigma_1, \sigma_3, \sigma_2 = 0$, representing a deformed hexagon as per the stresses axis $\sigma_1 = \sigma_3$.

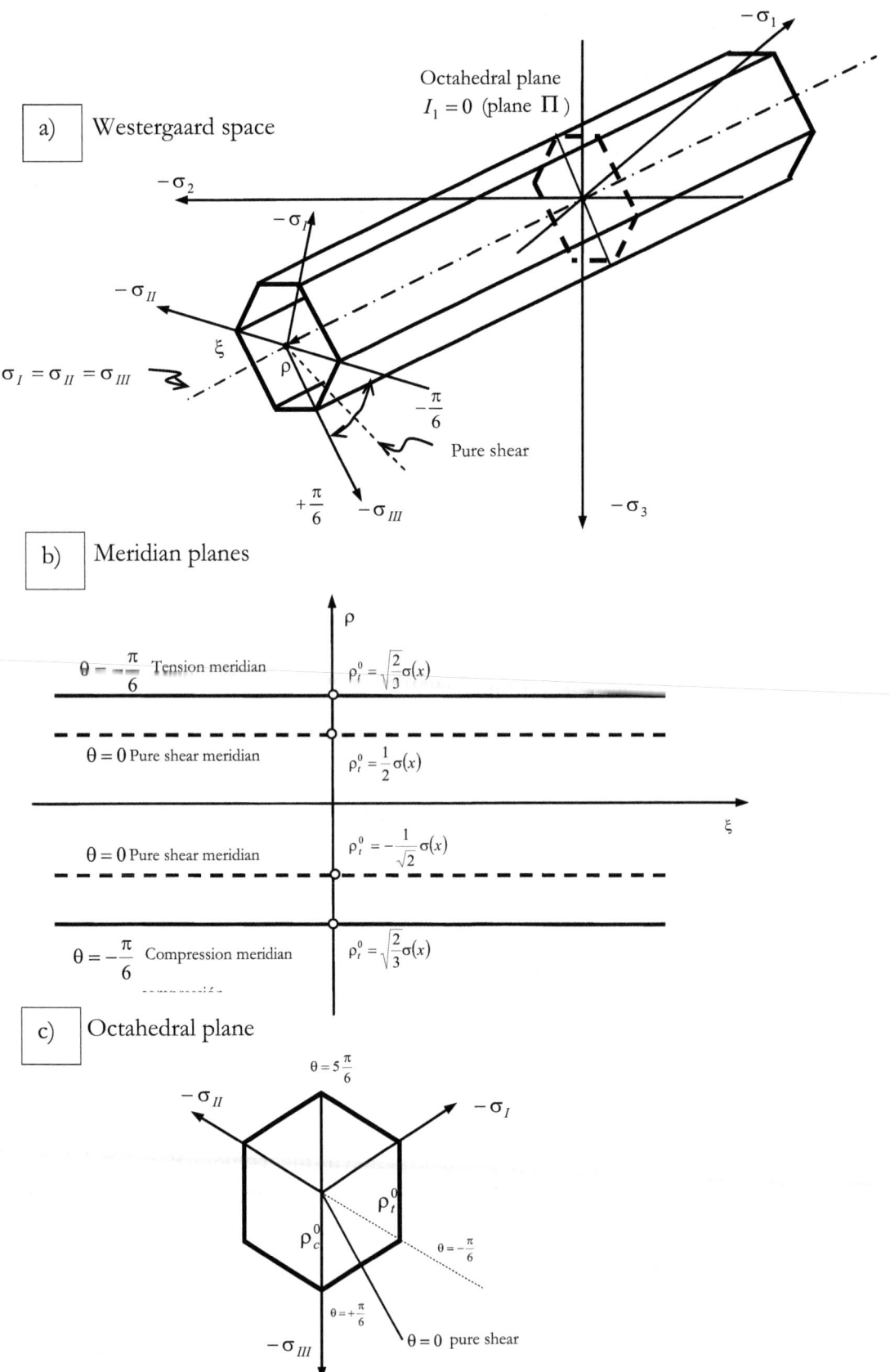

Figure 5.12 – Tresca yield surface: a) In the main stress space, b) As per maximum tension and compression meridians, c) As per the octahedral plane $I_1=0$ or Π plane.

5.11.3 The Von Mises criterion of octahedral shear stress

This criterion was formulated by von Mises in 1913, and like the two former, depends only on one single parameter, the maximum octahedral shear strength τ_{oct}^{max}. Moreover, it only considers the second invariant of the stress deviatoric tensor J_2, neglecting hence the influence of the first invariant of the stress tensor I_1 and the third invariant of the stress deviatoric tensor J_3. This is the most suitable criterion to represent the behavior of metallic materials. According to this criterion, the plastic yield is reached when the value of the plastic hardening function $\mathcal{K}(\kappa) = \tau_{oct}^{max}(\kappa)$, which means a shear strength, and reaches the maximum octahedral strength to the octahedral τ_{oct}^{max}. The different mathematical forms to express this criterion are the following,

As a function of the main stresses,

$$\mathbb{F}\left(\boldsymbol{\sigma}; \tau_{oct}^{max}\right) = \frac{1}{6}\left[\left(\sigma_1 - \sigma_2\right)^2 + \left(\sigma_2 - \sigma_3\right)^2 + \left(\sigma_3 - \sigma_1\right)^2\right] - \left[\tau_{oct}^{max}(\kappa)\right]^2 = 0 \tag{5.98}$$

As a function of the second invariant of the stress deviatoric tensor,

$$\mathbb{F}\left(J_2; \tau_{oct}^{max}\right) = J_2 - \left[\tau_{oct}^{max}(\kappa)\right]^2 = 0 \quad \Rightarrow \quad \mathbb{F}\left(J_2; \tau_{oct}^{max}\right) = \sqrt{J_2} - \tau_{oct}^{max}(\kappa) = 0$$

or as a function of the effective uniaxial stress $\overline{\sigma}(\kappa) = \sqrt{3}\,\tau_{oct}^{max}(\kappa)$, $\tag{5.99}$

$$\mathbb{F}\left(J_2; \overline{\sigma}\right) = \sqrt{3J_2} - \overline{\sigma}(\kappa) = 0$$

- As a function of the cylindrical coordinates,

$$\mathbb{F}\left(\rho; \overline{\sigma}\right) = \sqrt{\frac{3}{2}}\,\rho - \overline{\sigma}(\kappa) = 0 \tag{5.100}$$

being $\rho = \sqrt{3}\,\tau_{oct} = \sqrt{2J_2}$.

The insensibility to pressure makes the octahedral plane remain constant and equal to Π plane. This octahedral plane represents a circle. From the intersection of the traction meridian plane $(\theta = -\pi/6)$, with the yield surface a zero slope line arises, paralell to the one resulting from the intersection of the compression meridian plane $(\theta = +\pi/6)$, with the yield surface. Both meridian lines cut the stress axis of the octahedral shear $\rho_C^0 = \rho_T^0 = \pm\sqrt{2/3}\,\overline{\sigma}(\kappa)$ similarly to the Tresca criterion. In the main plane $\sigma_1, \sigma_3, \sigma_2 = 0$ represents an ellipse whose major axis coincides with the stress axis $\sigma_1 = \sigma_3$.

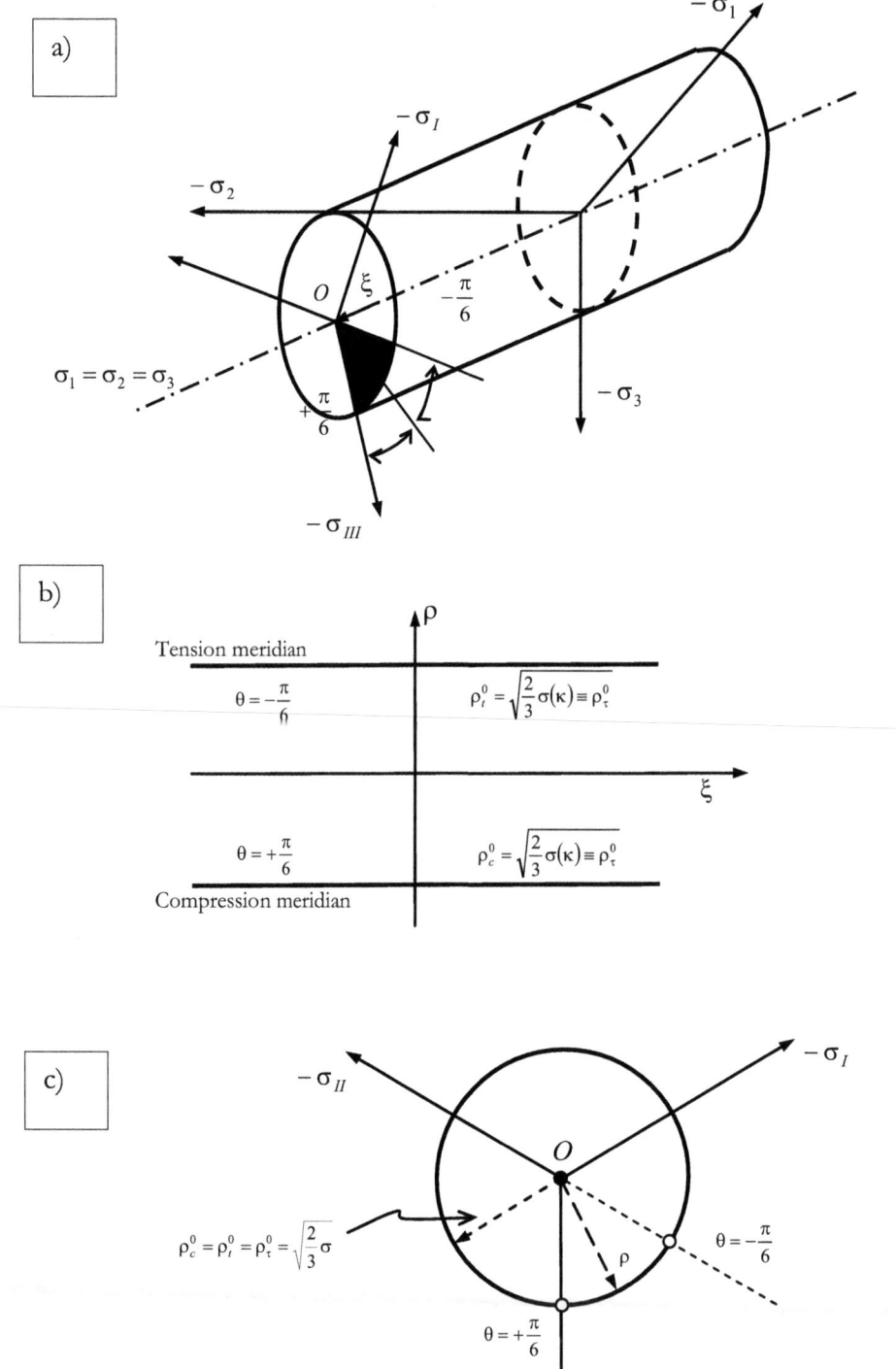

Figure 5.13 – Von Mises yield surface: a) In the principal stress space, b) In maximum tension and compression meridians c) In the octahedral plane I_1=0 or Π plane.

5.11.4 The Mohr-Coulomb criterion of octahedral shear stress

This criterion was first formulated by Coulomb in 1773 and later developed more thoroughly by Mohr in 1882. It is based on two parameters: the cohesion c and the internal friction angle ϕ among particles. It includes the first invariant of the stress tensor in its mathematical expression I_1 and the second and third invariants of the deviatoric stress tensor J_2, J_3 respectively. This criterion can be used to set the limits where the fracturing process starts in friction materials or in geomaterials. The strength at a point is increased by the friction among particles τ, and the latter depends on the normal stress σ_n and the cohesion c among them. Thus, the following simple form of the Mohr Coulomb criterion can be presented as (see Figure 5.14),

$$\mathbb{F}(\boldsymbol{\sigma};c;\phi)=\left|\tau\right|-c-\sigma_n\tan\phi=0 \tag{5.101}$$

In the extreme case that $\phi=0$, the Mohr-Coulomb criterion tends to the Tresca criterion, hence $\tau=c=\mathcal{K}$.

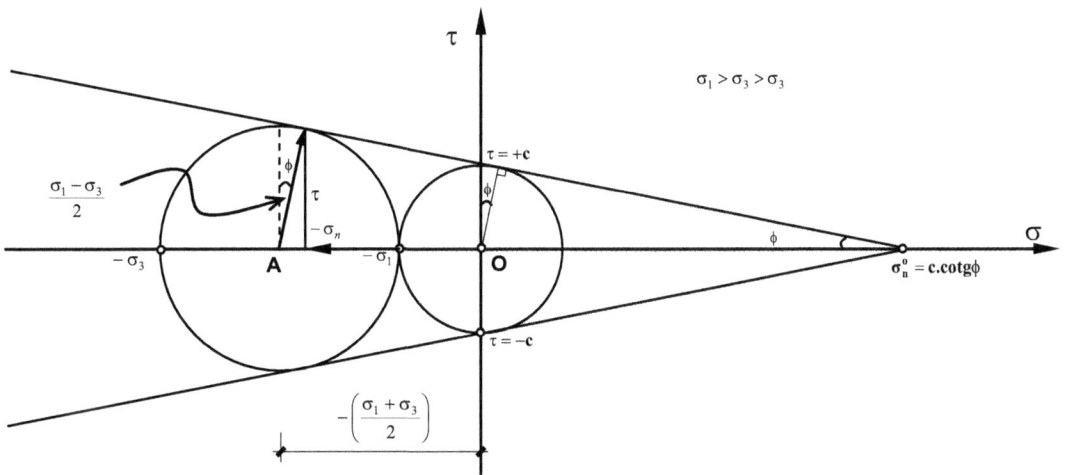

Figure 5.14 – Flat representation of the stress state in a point according to the Mohr-Coulomb criterion.

By observing Figure 5.14, equation (5.101) can be re-written as a function of the main stresses,

$$\mathbb{F}(\boldsymbol{\sigma};c;\phi)=\left[\left(\frac{\sigma_1-\sigma_3}{2}\right)\cos\phi\right]-c-\left[-\left(\frac{\sigma_1+\sigma_3}{2}\right)-\left(\frac{\sigma_1-\sigma_3}{2}\right)\text{sen}\,\phi\right]=0$$

$$\Rightarrow\ \mathbb{F}(\boldsymbol{\sigma};c;\phi)=\left(\frac{\sigma_1-\sigma_3}{2}\right)+\left(\frac{\sigma_1+\sigma_3}{2}\right)\text{sen}\,\phi-c\ \phi\cos=0 \tag{5.102}$$

where σ_1 and σ_3 represent the major and minor principal stresses respectively. Accordingly, it can be deduced that the Mohr-Coulomb criterion neglects the effect of the intermediate principal stress σ_2, which is a great limitation. Nevertheless, this problem can be solved by expressing its formulation as a function of the invariants of the stress tensor and its invariants,

$$\mathbb{F}(I_1; J_2; \theta; c; \phi) = \frac{I_1}{3} \operatorname{sen} \phi + \sqrt{J_2} \left(\cos \theta - \frac{\operatorname{sen} \theta \operatorname{sen} \phi}{\sqrt{3}} \right) - \sqrt{6} \, c(\kappa) \cos \phi = 0 \qquad (5.103)$$

In cylindrical coordinates it is expressed as,

$$\mathbb{F}(\rho; \xi; \theta; c; \phi) = \sqrt{2} \, \xi \operatorname{sen} \phi + \sqrt{3} \, \rho \left(\cos \theta - \frac{\operatorname{sen} \theta \operatorname{sen} \phi}{\sqrt{3}} \right) - \sqrt{6} \, c(\kappa) \cos \phi = 0 \qquad (5.104)$$

where $\xi = \sqrt{3} \, \sigma_{oct} = \sqrt{3} \, (I_1/3) = I_1/\sqrt{3}$, the octahedral radio $\rho = \sqrt{3} \, \tau_{oct} = \sqrt{2 J_2}$, and Lode similarity angle $\theta = \operatorname{arcsen} \left[\left(3\sqrt{3} \, J_3 \right) / \left(2 J_2^{3/2} \right) \right]$.

These functions describe in the principal stress space a deformed hexagonal-based pyramid, whose axis coincides with the isostatic pressure axis $\sigma_1 = \sigma_2 = \sigma_3$. The pressure increase makes the octahedral plane grow. This octahedral plane represents a deformed hexagon. By the interaction between the tension meridian plane $(\theta = -\pi/6)$ and the yield surface a slope line arises $(2\sqrt{2} \operatorname{sen}\phi) / (3 + \operatorname{sen}\phi)$, which cuts the octahedral tangent stress in $\rho_T^0 = (2c\sqrt{6} \cos\phi)/(3 + \operatorname{sen}\phi)$ and the pressure axis in $\xi^0 = \sqrt{3} \, c \cot g\phi$. By the interaction between the compression meridian plane $(\theta = +\pi/6)$ and the yield surface, a slope line arises $(2\sqrt{2} \operatorname{sen}\phi)/(3 - \operatorname{sen}\phi)$, greater than the tension meridian and which cuts the octahedral tangent stress in $\rho_C^0 = (2c\sqrt{6} \cos\phi)/(3 - \operatorname{sen}\phi)$. In the principal plane $\sigma_1, \sigma_3, \sigma_2 = 0$ represents a deformed hexagon whose major axis coincides with the stress axis $\sigma_1 = \sigma_3$.

From the functions describing the Mohr-Coulomb yield criterion, it is clear that their main characteristic is their capacity to distinguish the traction behavior from the compression behavior. Hence, according to this criterion the strength relation of tension and compression satisfies the following expression (see Figure 5.15),

$$R_{Mohr} = \frac{\left| \sigma_C^0 \right|}{\left| \sigma_T^0 \right|} = \tan^2 \left(\frac{\pi}{4} + \frac{\phi}{2} \right) \qquad (5.105)$$

This definition has an important limitation for the criterion adaptation into a particular material as this co-relation does not normally take place in real materials. In order to achieve a good co-relation between strengths and the internal friction angle it is necessary to modify the Mohr-Coulomb[9] criterion.

---

[9] Oller, S. (1991). *Modelización numérica de materiales friccionales*. CIMNE Nro. 3.

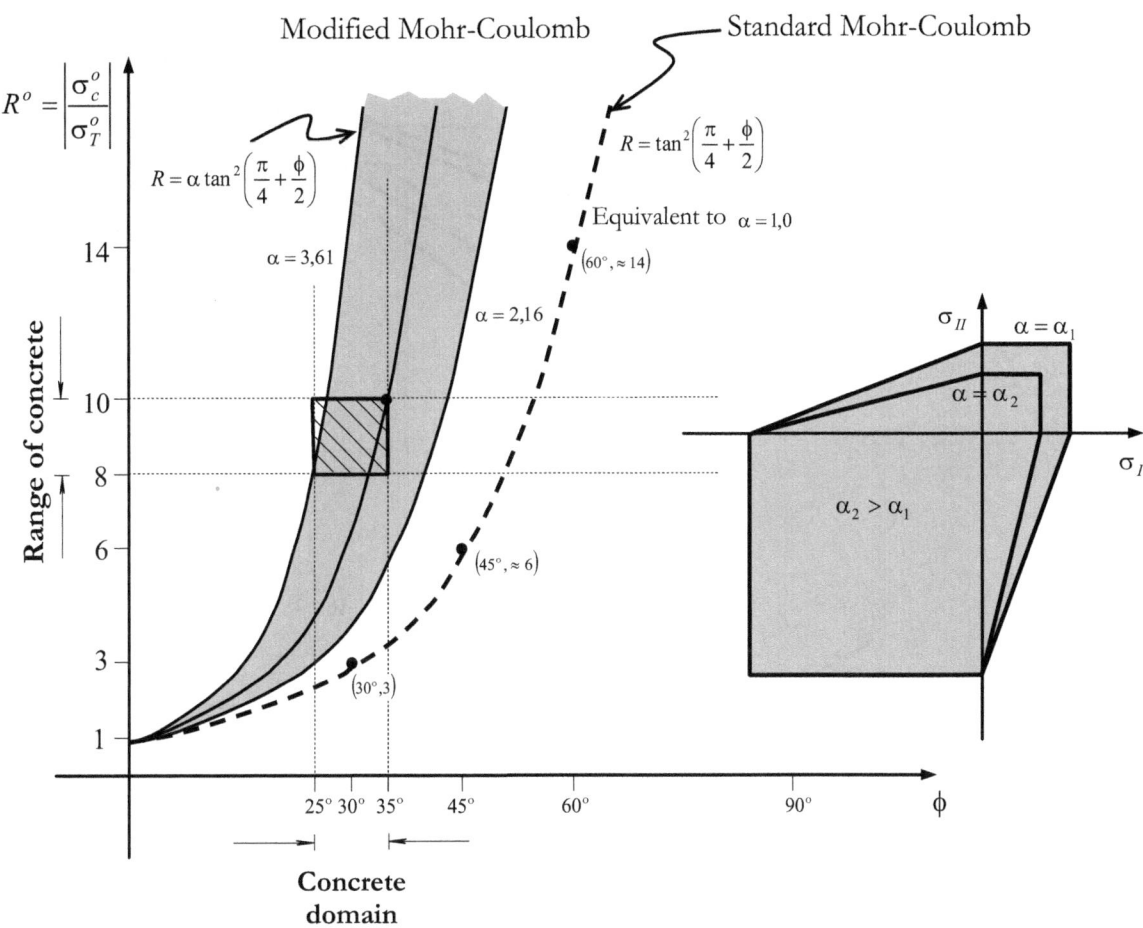

Figure 5.15 – Relation between ϕ and R_{Mohr}.

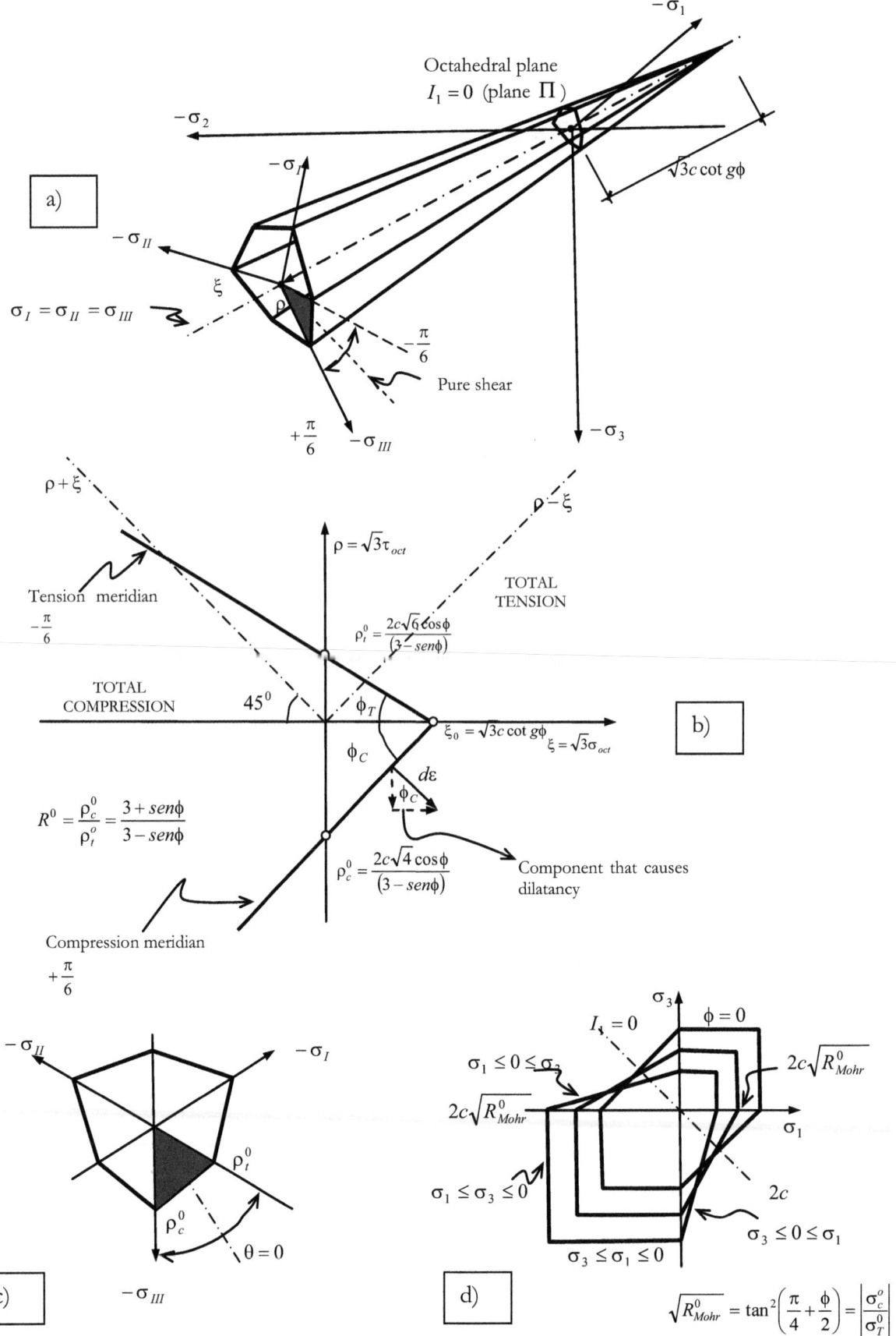

Figure 5.16 – Mohr-Coulomb yield surface: a) In the principal stress space, b) In the maximum tension and compression meridians, c) In the octahedral plane $I_1 = 0$ or Π plane.

5.11.5 The Drucker-Prager criterion

This criterion formulated by Drucker and Prager in 1952 is considered as a smoothed approximation to the Mohr-Coulomb criterion. However, the mathematical formulation arises from a generalization of the Von Mises criterion to include the influence of pressure, through the first invariant of the stress tensor I_1 and the internal friction angle ϕ. It also depends on the second invariant of the deviatoric stress tensor J_2, neglecting the influence of the third invariant of the deviatoric stress tensor J_3. This criterion depends on two parameters: the friction angle among particles ϕ and the cohesion c. The various forms of mathematically expressing this criterion are the following,

- As a function of the invariants of the stress tensor and its deviatoric tensor,

$$\mathbb{F}(I_1; J_2; c; \phi) = \overline{\alpha}(\phi) I_1 + \sqrt{J_2} - \overline{\mathcal{K}}(\kappa; \phi) = 0 \qquad (5.106)$$

- As a function of the cylindrical coordinates,

$$\mathbb{F}(\rho; \xi; c; \phi) = \overline{\alpha}(\phi) \sqrt{6}\ \xi + \rho - \sqrt{2}\ \overline{\mathcal{K}}(\kappa; \phi) = 0 \qquad (5.107)$$

being the variable pressure dependent $\xi = \sqrt{3}\ \sigma_{oct} = \sqrt{3}\,(I_1/3) = I_1/\sqrt{3}$, the octahedral radio $\rho = \sqrt{3}\ \tau_{oct} = \sqrt{2 J_2}$ and the hardening function that must be later adapted to the Mohr-Coulomb criterion, the results are $\overline{\mathcal{K}}(\kappa; \phi) = 6\ c(\kappa)\ \cos\phi/(3\sqrt{3} + \sqrt{3}\ \mathrm{sen}\ \phi)$ and $\overline{\alpha}(\phi) = 2\,\mathrm{sen}\,\phi/(3\sqrt{3} + \sqrt{3}\,\mathrm{sen}\,\phi)$. These two functions describe a cone inscribed in Mohr-Coulomb's pyramid. Both criteria coincide in the traction meridians. In case the cone confines Mohr-Coulomb's pyramid, the traction and compression meridians of both surfaces coincide and the following functions are obtained from them $\overline{\mathcal{K}}(\kappa; \phi) = 6\ c(\kappa)\ \cos\phi/(3\sqrt{3} - \sqrt{3}\,\mathrm{sen}\,\phi)$ and $\overline{\alpha}(\phi) = 2\,\mathrm{sen}\,\phi/(3\sqrt{3} - \sqrt{3}\,\mathrm{sen}\,\phi)$. Both particular cases describe two completely different behaviors.

The octahedral plane represents a circle, whose radio varies as a function of the pressure. From the interaction of the traction $(\theta = -\pi/6)$ and the yield surface a slope line $\sqrt{\overline{\alpha}}$ arises, that cuts the axis of the octahedral tangent in $\rho_T^0 = \sqrt{2}\ \overline{\mathcal{K}}$ and the axis of pressures in $\xi^0 = \overline{\mathcal{K}}/\sqrt{3}\ \overline{\alpha}$. As a result of the interaction between the compression meridian plane $(\theta = +\pi/6)$ with the yield surface, a line with a slope similar to the slope obtained for the traction meridian arises. In the main plane $\sigma_1, \sigma_3, \sigma_2 = 0$ represents an ellipse displaced from its center (pressure influence), whose major axis coincides with the stress axis $\sigma_1 = \sigma_3$.

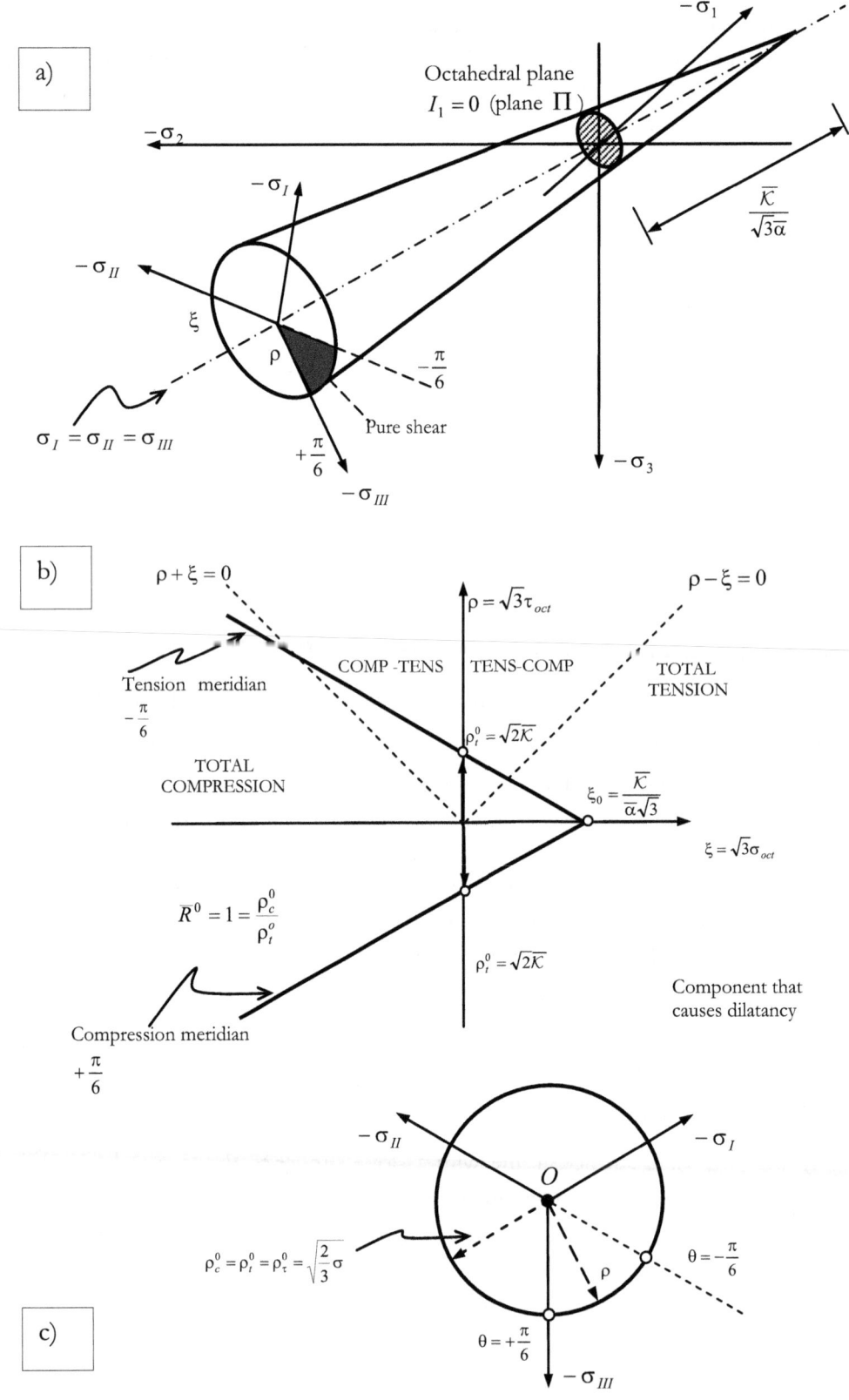

Figure 5.17 – Drucker-Prager yield surface: a) In the principal stress space, b) In maximum tension and compression meridians, c) In the octahedral plane $I_1=0$ or Π plane.

5.12 Geomaterials' plasticity

In this section a general constitutive model is presented, which is very suitable for the representation of ductile and brittle materials. It is common to find models for ductile and brittle materials such as metals, but it is not always possible to find models as efficient as the former to represent the behavior of fragile geomaterials. Accordingly, the model described below was initially formulated for fragile materials[10,11]; however, it can be used to represent the behavior of ductile materials after doing some particularizations in the parameters defining it (Figure 5.19). Among fragile materials, attention has been focused on the so-called "Frictional Materials" (see Figure 5.18). From the materials fitting in this classification all the ceramics, soils consisting of frictional components such as sand, among others, and more particularly about concrete, on which more emphasis will be made, can be named.

Frictional materials are the materials whose relation between their strength and pressure depends on the internal frictional angle. They exhibit dilatancy, in other words, a change of apparent volume, when they are subjected to tangent stresses.

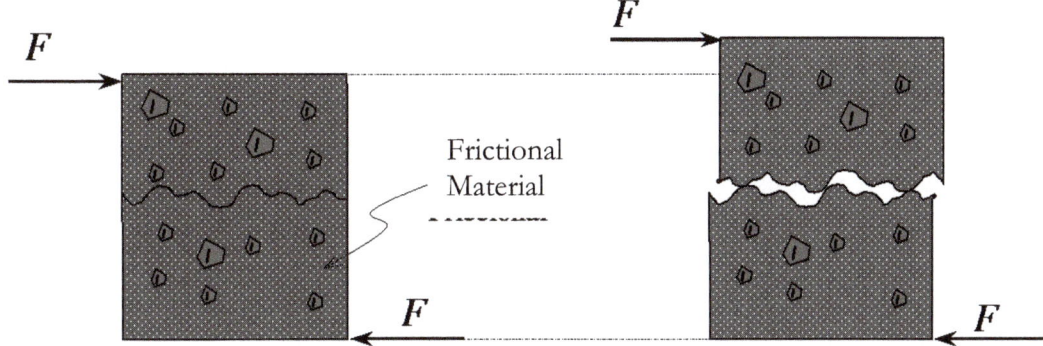

Figure 5.18 – Dilatancy phenomenon.

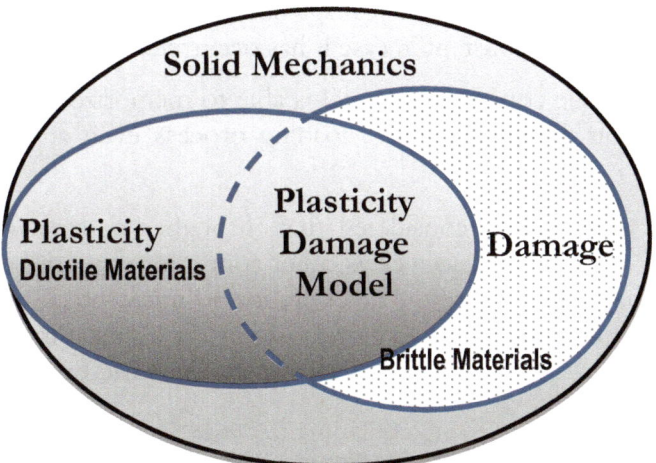

Figure 5.19 – Simple representation of the theories contributing to the "Plastic Damage Model" formulation.

---

[10] Oller, S. (1991). *Modelización de materiales friccionales.* CIMNE No. 3. Barcelona.
[11] Luccioni, B. (1993). *Formulación de un Modelo Constitutivo para Materiales Ortótropos*, Tesis Doctoral. Universidad Nacional de Tucumán. Argentina.

5.13 Basis of the "plastic-damage" model

The name *plastic damage model*[12] is due to the hypothesis about considering that the nonlinear inelastic behavior of a frictional-cohesive solid is a consequence of the formation and development of micro fractures. The chances of using *the mathematical theory of the plasticity* to represent the behavior of a fracturable-frictional material assumes that the non-recoverable strain, or micro-cracking deformation, can be accepted as it is understood in the classic plasticity theory, although the physical meaning of this plastic phenomenon is different.

The mathematical theory of the classical plasticity is based on an isotopic formulation for each point (see previous sections), which means that the plastic yield function or discontinuity threshold is subjected to a homothetic movement governed by the evolution of the plastic hardening variable κ^p.

Here the term *damage* is used as a synonym for *deterioration,* although its meaning will also be further associated with the *stiffness loss* phenomenon (stiffness degradation). The physical interpretation of this isotropic damage (deterioration) can be understood by the phenomenon of non-directional damage that each of the real solid points suffers when a fracture occurs. Thus, in a first analysis, the concept of the non-directional damage proves to be opposed to the macroscopic damage (fracture), as the latter is a directional phenomenon (see Figure 5.13).

By approaching the behavior to a fracture through a continuous formulation[13], it can be admitted as a hypothesis that the *directional macroscopic damage* (fracture) comes from a non-*directional microscopic* behavior of the points located at a certain area of the solid, called *damage zone.* Based on this, a fracture will be defined as the geometrical place of the points that have suffered non-directional damage (see Figure 5.20).

The damage concentration of a loaded solid within a zone whose dimensions are reduced regarding the total size of the solid is due to the strain location or concentration developed in such zone of the solid. Here, some points are subjected to a stress-strain behavior with softening, in other words, with stress loss or deformation growth (softening). On the other hand, the points that are out of the damage zone and that experiment an unloading process keep their damage level constant in case it has occurred.

The damage-plastic constitutive model is able to memorize the macro-directionality of a solid deterioration during the whole loading process even admitting non-radial loading conditions[*].

As a result of this, it is considered that through the *strain localization phenomenon* the plasticity theory can be used, as well as other continuous theories, as a base to formulate a constitutive model representing the deterioration by micro-cracks in the frictional materials. This gives a tool that represents to a great extent the damage mechanisms of many fragile materials (concrete), and which rigorously satisfy the basic principles of the mechanics of continuous media[14].

---

[12] Lubliner, J.; Oliver J.; Oller S. and Oñate E. (1989). A Plastic-Damage Model for Concrete. International Journal of Solids and Structures, Vol.25, No.3, pp. 299,326.

[13] Oller S. (2001). *Fracura mecánica – Un enfoque global.* CIMNE – Ediciones UPC. Barcelona

[*] **Note:** The point behavior of a solid is said to be "radial " or "proportional" if during the whole loading process the relationship $\sigma_{11}/\sigma_{11}^0 = \sigma_{12}/\sigma_{12}^0 = \cdots = \sigma_{33}/\sigma_{33}^0$ is satisfied, among the initial and current stress tensor components, respectively.

[14] Malvern, L. (1969). *Introduction to the mechanics of continuous medium.* Prentice Hall, Englewood Cliffs, NJ.

The formulation of the plastic damage model is based on the *mechanics of solids,* particularly in the *theory of plasticity* and the *continuous damage theory.* It uses *the finite element method* and various *numerical techniques* required to control and guarantee the problem solution (see Figure 5.21), as a framework for the structural problem solution.

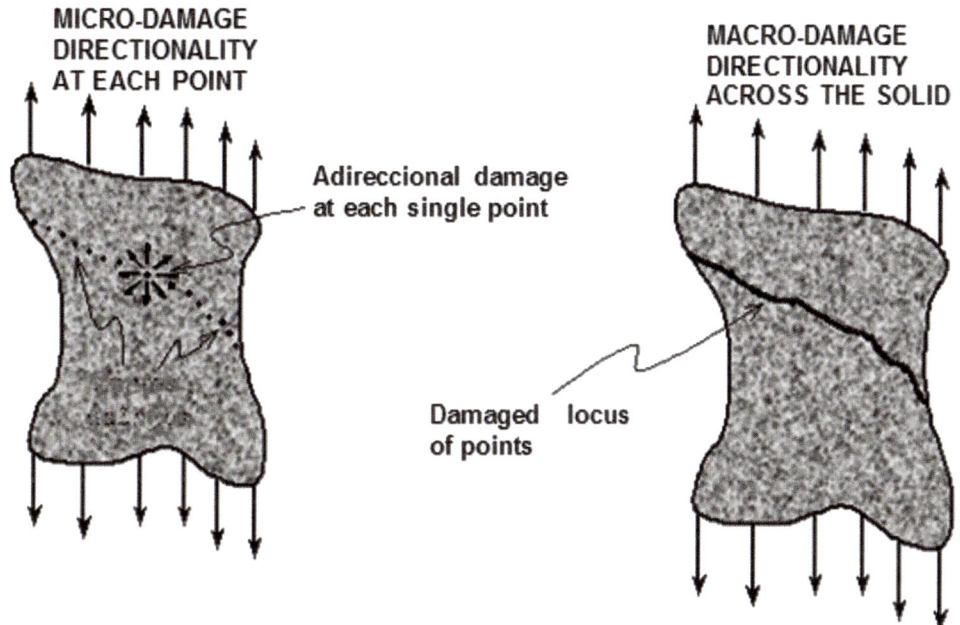

Figure 5.20 – Micro and macro damage directionality.

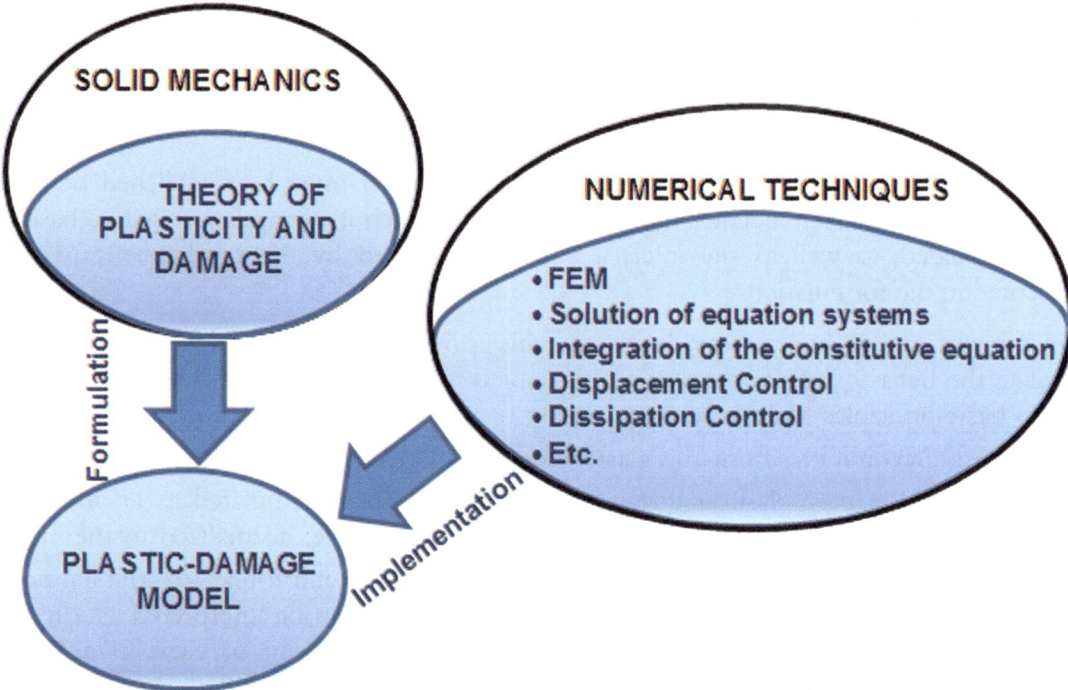

Figure 5.21 – Formulations that contribute to the plastic damage model.

5.13.1 Mechanical behavior required for the constitutive model formulation

The formulation of a constitutive model to deal with the mechanical behavior of a particular material must be carried out after assuming some simplified hypothesis in such a material. It involves creating an ideal material as closest to the real one as possible. Some of these simplified hypothesis are listed below, which are considered as premises within the *plastic damage* model.

- Frictional materials have a pronounced inelastic behavior producing permanent strain interpreted as micro cracks.
- Non-directional damage at each point is interpreted as a local isotropic behavior.
- The damaged point locus sets a macroscopic direction that will be interpreted as a crack.
- The damaged points are concentrated on a thin area called damage zone due to the strain location phenomenon. This leads to an anisotropy induced by the solid nonlinear behavior.
- The material can have a change of volume behavior, known as the dilatancy phenomenon during the inelastic process.
- The maximum strength, evolution and final deformation depend on the characteristics of the loading, tension-tension, tension-compression and compression-compression processes. In other words, the strength evolves like the process itself.
- There is continued increasing stiffness degradation, during the whole quasi static and growing monotonous loading process, including the behavior range of reversible deformations.

As observed, although all the above-mentioned items are not simple, they should be found within a mechanical formulation.

5.13.2 Some characteristics of the plastic damage model

The main mechanical characteristics of a constitutive model must be established before determining the formulation. Thus, the basic theory on which its formulation will be based must be decided, as well as the internal variables required by the mechanisms to be represented in the formulation.

The theory of plasticity provides a suitable physical-mathematical framework to formulate the behavior of frictional materials subjected to loading. From the extension of its main basic principles and the reinterpretation of its main variables, a *plastic damage model* has emerged. Accordingly, from the classic variable of plastic hardening ω^p defined in previous sections, a new plastic damage variable κ^p has been formulated as an internal variable. This variable is dealt with as a dimensionless magnitude, normalized to the unit, ranging from $0 \leq \kappa^p \leq 1$. For $\kappa^p = 0$ there is no plastic damage and for $\kappa^p = 1$ the limit of total damage of a point in a solid is defined. The latter state can be interpreted as a total loss of strength in a point of the solid, and from a physical point of view, as a mass breakup of the solid in the point of analysis (physical discontinuity). A brief presentation will be given below. For further details, check references[1,15,16].

---

[15] Lubliner, J.; Oliver J.; Oller S. and Oñate E. (1989). A Plastic-Damage Model for Concrete. International Journal of Solids and Structures, Vol.25, No.3, pp. 299,326.

The plastic yield criterion or discontinuity threshold, presented in previous sections, is now considered in this model through a mathematical expression written in the following general form,

$$\mathbb{F}(\boldsymbol{\sigma},c) = f(\boldsymbol{\sigma}) - c = 0 \qquad (5.108)$$

where $f(\boldsymbol{\sigma})$ is a first-degree homogenous scalar function in the stress tensor components that can be used to define the cohesion of c, or a scaled uniaxial stress, either as a function of a plastic hardening or as an internal variable depending on the evolution of the mechanical process.

The Mohr-Coulomb and the Drucker-Prager plasticity criteria can be represented by a similar expression to (5.108), but the approximation of the real behavior of many frictional materials is not appropriate. Many plastic yield criteria have tried to improve this approximation for cohesive-frictional materials. There are some non-homogeneous functions $f(\boldsymbol{\sigma})$ with a greater degree in the stress tensor components that cannot define a plastic c hardening by a direct physical interpretation.

Any yield criteria can be followed by the constitutive model of plastic damage; however, to improve the approximation of the concrete's particular behavior, appropriate criteria for each case must be defined[1].

Cohesion c is assumed as a scaled magnitude regarding an initial strength to uniaxial compression of the geomaterial σ_C^0 (stress discontinuity threshold), that is the stress level for which the volumetric deformation ε_V is the maximum. Therefore, the *initial cohesion*, or original material cohesion, is defined as $c^0 \propto \sigma_C^0$ for $\kappa^P = 0$, setting the initial position of the yield criterion and the *final cohesion* or cohesion of the material totally deteriorated as, $c^u = 0$ for $\kappa^P = 1$, defining the final position of the yield criterion.

Unlike the classic plasticity with isotropic hardening, the cohesion is not a simple function of the *plastic hardening variable* $c(\kappa^P)$ but an *internal variable* that depends on the evolution of the elasto-plastic process governed by an equation of evolution (differential equation).

The internal friction angle ϕ can also be defined as an internal variable depending on the elasto-plastic evolution law, but due to the physical behavior of this phenomenon in geomaterials[17], only a simple explicit function of the variable of the plastic damage $\phi(\kappa^P)$ is proposed. Based on this hypothesis, the initial friction is null $\phi^0 = 0$ even when the cohesion c^0 does not allow the mobilization of friction. The friction is maximum at the end of the elasto-plastic process, $\phi^u = \phi(\kappa^P = 1) = \phi^{max}$. In this final state, when the geomaterial or friction of component particles has lost its cohesion, the friction coincides with the sand friction.

The dilatancy angle ψ, like the internal friction angle, can be defined as an internal variable that can also be simply expressed as a function of the plastic damage variable $\psi(\kappa^P)$, because a good approximation of the real behavior of the geomaterial has been

---

[16] Oller, S.; Oñate E.; Oliver, J. and Lubliner J. (1990). Finite Element Non-Linear Analysis of Concrete Structures Using a ``Plastic-Damage Model''. *Engineering Fracture Mechanics*, Vol 35; pp 219-231.

[17] Borst, R. De and Vermeer, P. (1984). *Non associated plasticity for soils, concrete and rock*. Heron, Vol. 29, Delf. Netherlands.

obtained through this hypothesis. This angle is $\psi(\kappa^P = 0) = \psi^0 = 0$ when the plastic process starts and $\psi^{max} = \psi(\kappa^P = 1)$ at the end.

In short, for a plastic process with no stiffness degradation, the *plastic damage model* uses the following set of internal variables for its definition $\mathbf{q} = \{\boldsymbol{\varepsilon}^P, \kappa^P, \mathbf{c}\}$, their definitions and laws of evolution will be presented as part of the main equations governing the model.

1. A **deformation splits** into an elastic and a plastic part.

$$\boldsymbol{\varepsilon} = \boldsymbol{\varepsilon}^e + \boldsymbol{\varepsilon}^P = \mathbb{C}^{-1} : \boldsymbol{\sigma} + \boldsymbol{\varepsilon}^P \tag{5.109}$$

where $\mathbb{C}$ is the constitutive elastic tensor.

2. A **plastic yield criterion and a plastic potential criterion** as defined by equation (5.108),

$$\mathbb{F}(\boldsymbol{\sigma}, \mathbf{c}) = f(\boldsymbol{\sigma}) - \mathbf{c} = 0; \ \mathbb{G}(\boldsymbol{\sigma}) = g(\boldsymbol{\sigma}) - \mathrm{cte} = 0 \tag{5.110}$$

such that $f(\boldsymbol{\sigma})$ and $g(\boldsymbol{\sigma})$ are two scalar functions of tensor arguments called yield function and plastic potential respectively.

6. A **non-associated plastic flow rule** and a group of **internal variables,**

$$\mathbf{q} = \left\{ \begin{array}{c} \boldsymbol{\varepsilon}^P \\ \hline \mathbf{q}_\alpha \end{array} \right\} = \left\{ \begin{array}{c} \boldsymbol{\varepsilon}^P \\ \hline \kappa^P \\ \mathbf{c} \end{array} \right\} \quad \begin{array}{l} \text{Plastic Deformation} \\ \text{Plastic Damage Variable} \\ \text{Cohesion Variable} \end{array} \tag{5.111}$$

all of them defined by the following equations of evolution (see the classic plasticity section),

$$\frac{d\mathbf{q}}{dt} = \dot{\mathbf{q}} = \left\{ \begin{array}{c} \dot{\boldsymbol{\varepsilon}}^P \\ \hline \dot{\kappa}^P \\ \dot{\mathbf{c}} \end{array} \right\} \equiv \dot{\lambda} \cdot \mathbf{H} = \dot{\lambda} \cdot \left\{ \begin{array}{c} \dfrac{\partial \mathbb{G}}{\partial \boldsymbol{\sigma}} \\ \mathbf{h}_\kappa : \dfrac{\partial \mathbb{G}}{\partial \boldsymbol{\sigma}} \\ h_c \cdot \mathbf{h}_\kappa : \dfrac{\partial \mathbb{G}}{\partial \boldsymbol{\sigma}} \end{array} \right\} \equiv \left\{ \begin{array}{c} \dot{\boldsymbol{\varepsilon}}^P \\ \mathbf{h}_\kappa : \dot{\boldsymbol{\varepsilon}}^P \\ h_c \cdot \mathbf{h}_\kappa : \dot{\boldsymbol{\varepsilon}}^P \end{array} \right\} \tag{5.112}$$

where $\mathbf{h}_\kappa$ and h_c are a second-degree tensor and a scalar function respectively that will be defined later and which depend on the current state of the free variable $\boldsymbol{\varepsilon}^e$ and the rest of the internal variables $\mathbf{q}$. As observed in this equation, the main internal variable is the plastic deformation $\boldsymbol{\varepsilon}^P$ and the others are obtained from it. The plastic consistency factor λ is obtained as shown in the previous sections from the consistency condition of the plastic yield function.

7. **A secant and tangent constitutive equation,** defined as the classic plasticity,

$$\boldsymbol{\sigma} = \mathbb{C} : \left(\boldsymbol{\varepsilon} - \boldsymbol{\varepsilon}^{p} \right)$$

$$\dot{\boldsymbol{\sigma}} = \left\{ \mathbb{C} - \frac{\left[\mathbb{C} : \dfrac{\partial \mathbb{G}}{\partial \boldsymbol{\sigma}} \right] \otimes \left[\dfrac{\partial \mathbb{F}}{\partial \boldsymbol{\sigma}} : \mathbb{C} \right]}{- c_{k} \cdot \dfrac{\partial \mathbb{F}}{\partial \boldsymbol{\eta}} : \dfrac{\partial \mathbb{G}}{\partial \boldsymbol{\sigma}} + h_{\mathcal{K}} \, \mathbf{h}_{\kappa} : \dfrac{\partial \mathbb{G}}{\partial \boldsymbol{\sigma}} + \left(\dfrac{\partial \mathbb{F}}{\partial \boldsymbol{\sigma}} : \mathbb{C} : \dfrac{\partial \mathbb{G}}{\partial \boldsymbol{\sigma}} \right)} \right\} : \dot{\boldsymbol{\varepsilon}} \quad \Rightarrow \quad \dot{\boldsymbol{\sigma}} = \mathbb{C}^{T} : \dot{\boldsymbol{\varepsilon}} \qquad (5.113)$$

such that the notation used can be seen in the classic plasticity section. Those with a particular definition will be presented below.

The constitutive model resulting from these basic definitions shows a very good response during the fracture process. To sum up, the model has the following characteristics:

- It defines a constitutive law depending on the internal variables of cohesion and plastic damage to represent non-radial complex loading situations.
- It deals with complex states of multiaxial stress in a unified way.
- It admits that materials have different limits of maximum strength and ultimate deformation, depending on the mechanical process in progress.
- It admits different plastic yield and potential. Although this is not a characteristic of the model, it can be pre-established as one of its variables.
- It takes into consideration a non-associated plastic flow to control the dilatancy phenomenon.
- It can obtain all the information related to the point deterioration through the point mechanical information post processing.

5.14 Main variables of the plastic-damage model

The aforementioned definitions are very broad and require further detail. Hence, the expressions defining the evolution law of the internal variables of those basic definitions are presented as follows.

5.14.1 Definition of the plastic damage variable

The classic plasticity theory establishes a hardening variable as a function of the effective plastic strain $\kappa^{P} = \bar{\varepsilon}^{P}$, or also as a function of the specific plastic work $\kappa^{P} = \omega^{P} = \bar{\sigma}\,\bar{\varepsilon}^{P} = \boldsymbol{\sigma} : \boldsymbol{\varepsilon}^{P}$ (see the section about classic plasticity). These definitions are suitable for materials whose final deformation (the state of total degradation in the point) is equal in tension as in compression or for any combined process in general. However, this is not true for many materials and a plastic hardening variable with a final strain for each case must be found. Therefore, it is necessary to establish an internal variable defining a unit normalized dissipation, which is the relation between the density of the dissipated energy at a specific time of the process and the maximum dissipation of the point of the solid. Hence, it is said that *the variable of plastic damage* is a relative measure of the dissipated energy during the plastic process. For the sake of simplicity and order, it is convenient to define this variable for simple uniaxial processes and then generalize it to other multiaxial processes.

Definition of the plastic damage variable for uniaxial stress states

The curves describing stress evolution vs. the strain at every moment are obtained from uniaxial traction and compression tests. From these curves, curves $\overline{\sigma} - \overline{\varepsilon}^P$ can be obtained. The latter include the fracture energy g_f^p and crushing g_c^p, per unit of volume, respectively.

The specific dissipated energy is obtained as a function of these curves at the end of an elasto-plastic quasi-static process, which in the case of a uniaxial tension process is,

$$g_f^p = \int_{t=0}^{\infty} \sigma_T \, \dot{\varepsilon}_T^p \, dt \tag{5.114}$$

where σ_T is the uniaxial tension stress and ε_T^p is the uniaxial tension plastic strain. From this last expression *the plastic damage variable* for uniaxial quasi-static tension processes is defined as the plastic normalized dissipation between zero and the unit,

$$0 < \left[\kappa^p = \frac{1}{g_f^p} \int_{t=0}^{t} \sigma_I \, \dot{\varepsilon}_I^p \, dt \right] \leq 1 \tag{5.115}$$

As a result, there is a normalized variable with respect to the tension maximum specific energy with values ranging from $0 \leq \kappa^p \leq 1$ for the beginning and end of the plastic process, respectively. With κ^p as a variable, resulting curves of the uniaxial plastic stress-strain can be transformed into other curves depending on the plastic damage in which the following conditions must be met, $\sigma_T(\kappa^p = 0) = \sigma_T^0$ and $\sigma_T(\kappa^p = 1) = \sigma_T^{Ultima}$ in the interval $[0,1]$.

In uniaxial compression, like in traction, the plastic damage variable is obtained as

$$0 \leq \left[\kappa^p = \frac{1}{g_C^p} \int_{t=0}^{t} \sigma_C \, \dot{\varepsilon}_C^p \, dt \right] \leq 1 \tag{5.116}$$

where $g_C^p = \int_{t=0}^{\infty} \sigma_C \, \dot{\varepsilon}_C^p \, dt$ is the dissipated energy during the compression process. Like in the pure traction process, the uniaxial response of plastic stress-strain can be defined now through another curve depending on the plastic damage rather than in the plastic deformation in which the following conditions must be met, $\sigma_C(\kappa^p = 0) = \sigma_C^0$ and $\sigma_C(\kappa^p = 1) = \sigma_C^{Ultima}$ in the interval $[0,1]$ (see Figure 5.22).

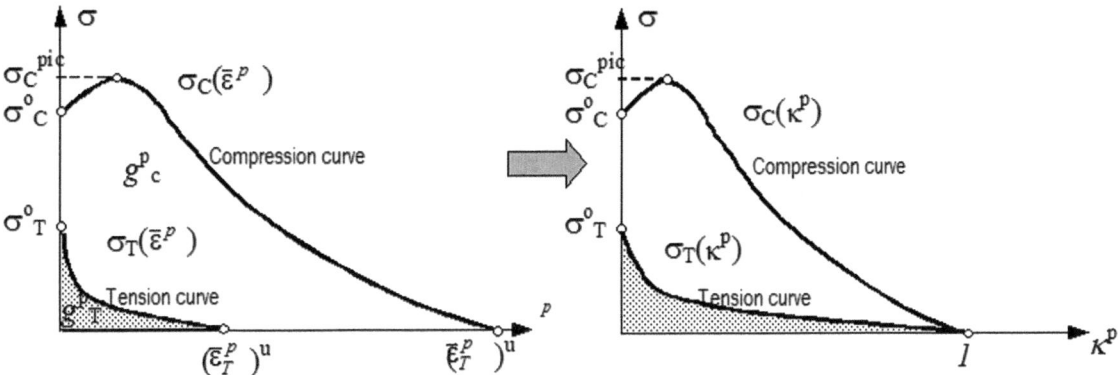

a) Uniaxial strength according to the plastic strain   b) Uniaxial strength according to the plastic damage variable

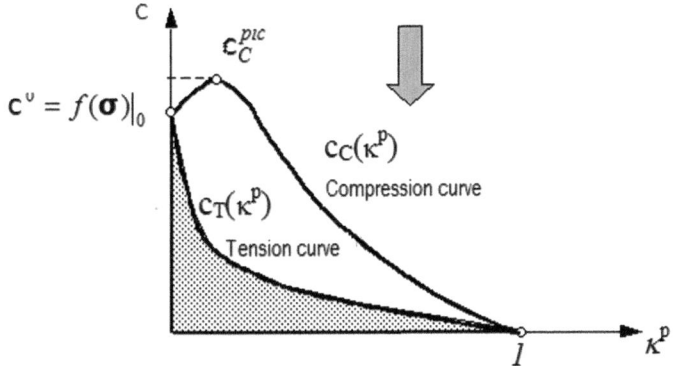

c) Cohesion according to the plastic damage variable

Figure 5.22 – Transformation of the uniaxial strength measured in the lab into the uniaxial strength used in the plastic damage model.

The plastic damage variable is objective and evolves within the same limits regardless of the mechanical process. Thus, the total damage in a point is reached when $\kappa^p = 1$, but the dissipated energy will be g_f^p in a pure tension process and g_C^p in a pure compression process.

Definition of a plastic damage variable for multiaxial stress states

Generally, for a generic loading process the plastic damage variable is defined for a multiaxial mechanical process as,

$$\dot{\kappa}^p = \mathbf{h}_\kappa : \dot{\boldsymbol{\varepsilon}}^p \tag{5.117}$$

where $\mathbf{h}_\kappa$ is a second-degree tensor that ,for uniaxial traction and compression processes, leads to a plastic damage that agrees with equations (5.115) and (5.116) respectively. For the remaining cases, the magnitude of damage depends on the loading process. To recover the plastic hardening variable of the classic plasticity theory, this tensor is equal to the stress tensor $\mathbf{h}_\kappa = \boldsymbol{\sigma}$ and it can be defined as a normalized dissipation to the unit for isotropic materials.

$$\dot{\kappa}^p = \mathbf{h}_\kappa : \dot{\boldsymbol{\varepsilon}}^p = \left[\frac{r(\boldsymbol{\sigma})}{g_f^p} + \frac{1 - r(\boldsymbol{\sigma})}{g_C^p} \right] \cdot \Xi_m \tag{5.118}$$

where $\Xi_m = \boldsymbol{\sigma} : \dot{\boldsymbol{\varepsilon}}^p$ is the plastic dissipation and $r(\boldsymbol{\sigma}) = \sum_{I=1}^{3} \langle \sigma_I \rangle / \sum_{I=1}^{3} |\sigma_I|$ is a scalar function defining the behavior states of a point as a function of the stress state, and $\langle x \rangle = 0.5 [x + |x|]$ is the Macaulay bracket. Note the following particular cases, $r(\boldsymbol{\sigma}) = 1$ for pure tension problems, $r(\boldsymbol{\sigma}) = 0$ for pure compression and $r(\boldsymbol{\sigma}) = 0.5$ for a pure shear state. Consequently, the plastic dissipation will always be normalized with respect to the maximum energy of the process at every moment.

5.14.2 Definition of the law of evolution of cohesion $c - \kappa^p$

This plastic damage model assumes that micro-cracking of frictional materials is due to an immediate *intergranular cohesion loss* produced by the relative movement among grains or particles. This intergranular cohesion loss starts at points where their effective stress gets over the cohesion limit threshold or initial cohesion (see Figure 5.18). While the loading process evolves, the amount of non-united particles grows until they form a fracture of a considerable size that leads to every solid breaking. Due to this failure mechanism, *a softening* in the strain-stress behavior occurs, non-existent at intergranular level, which can only be observed as a macroscopic effect caused by the average behavior of a set of particles.

The plastic damage constitutive model carries out the numerical analysis of an area of finite dimensions (integration point of the constitutive equation) through the finite element functional approximation technique. Therefore, every point under analysis represents infinite material points contained in its area of influence. Thus, at macroscopic level, the softening phenomenon by deformation can be considered as a material property and, in such a case, a *plastic hardening function* must be defined taking into account this *phenomenon*, which is the cohesion among particles for this constitutive model. This hardening function is represented by the cohesion, written as an internal variable to make it more general and its evolution equation for any quasi-static loading process is defined as follows,

$$\dot{c} = h_c \cdot \dot{\kappa}^p = h_c \cdot \mathbf{h}_\kappa : \dot{\boldsymbol{\varepsilon}}^p \tag{5.119}$$

where $h_c(\boldsymbol{\sigma}, \kappa^p, c)$ is a scalar function of the current state of the stress-free variable $\boldsymbol{\sigma}$ and of the internal variables κ^p and c. The expression used for the evolution law of the internal variable of cohesion is obtained by choosing the following expression for h_c,

$$h_c = c \cdot \left[\frac{r(\boldsymbol{\sigma})}{c_T} \frac{dc_T}{d\kappa^p} + \frac{1 - r(\boldsymbol{\sigma})}{c_C} \frac{dc_C}{d\kappa^p} \right] \tag{5.120}$$

where $r(\boldsymbol{\sigma}) = \sum_{I=1}^{3} \langle \sigma_I \rangle / \sum_{I=1}^{3} |\sigma_I|$ is the aforementioned function which sets the type of behavior (tension or compression or tension-compression), developed at every moment in a point of a solid. The cohesion function $c_C(\kappa^p)$ (see Figure 5.22) is obtained in explicit form and it represents the cohesion evolution during a uniaxial simple compression test. The relation between cohesion and uniaxial stress of compression is given by the following expression,

$$c_C(\kappa^p) = \frac{1}{\aleph} \sigma_C(\kappa^p) \tag{5.121}$$

such that $\aleph$ is a coefficient depending on the criterion of the discontinuity threshold and represents a scalar factor between cohesion and the uniaxial stress of compression[1]. For example, for Tresca and von-Mises its value is $\aleph = 1$, for Mohr Coulomb $\aleph = 2\sqrt{R_{Mohr}}$ where R_{Mohr} is the relation of strengths (see equation 6.111), for Drucker-Prager inscribed in the surface of Mohr-Coulomb $\aleph = 6\cos\phi/(\sin\phi - 3)$ and for Drucker-Prager circumscribed in the surface of Mohr-Coulomb $\aleph = 6\cos\phi/(3\sin\phi - 3)$. Thus, for any discontinuity threshold criterion, this coefficient must be defined.

The function $c_T(\kappa^p)$ (see Figure 5.22) can be obtained explicitly and represents the cohesion evolution during a uniaxial simple tension test. The relation between cohesion and uniaxial tension stress is given by the following expression,

$$c_T(\kappa^p) = \frac{R^0}{\aleph}\sigma_T(\kappa^p) \tag{5.122}$$

where $R^0 = \left[f_C^0 / f_T^0 \right] = \left[\sigma_C(\kappa^p = 0) / \sigma_T(\kappa^p = 0) \right]$ is the relation between uniaxial strengths. For Tresca and von-Mises its value is $R^0/\aleph = 1$, for Mohr Coulomb $R^0/\aleph = \sqrt{R_{Mohr}}/2$ where R_{Mohr} is the strengths relationship (see equation 6.111), for Drucker-Prager inscribed in the Mohr-Coulomb surface $R^0/\aleph = (3 + 3\sin\phi)/6\cos\phi$ and for Drucker-Prager circumscribed in the Mohr-Coulomb surface $R/\aleph = (3 + \sin\phi)/6\cos\phi$.

Some researchers maintain that the concrete strength curves in simple tension and compression, obtained in uniaxial experimental tests, have similar shapes[18,19], in other words, it can be stated that the scale relationship between them is a constant during the whole quasi-static process and is given by

$$R(\kappa^p) = \frac{\sigma_C(\kappa^p)}{\sigma_T(\kappa^p)} = cte. \implies R(\kappa^p) = R^0 \tag{5.123}$$

such that the explicit functions of uniaxial tension and compression cohesion coincide.

5.14.3 Definition of the variable ϕ, internal friction angle

As load is increased on a cohesive-frictional solid, inside of it the particles move among them (micro-fracturing) leading to an intergranular cohesion loss that makes the solid increasingly behave as a simple non-cohesive frictional material. Consequently, the cohesion loss involves an internal frictioning gain leading to a smoother behavior under compression due to the frictional force increase among particles and to the strength decrease under tension as a result of the loss of cohesive forces. Accordingly, the relation among the uniaxial strengths increases $R(\kappa^p) = \sigma_C(\kappa^p) / \sigma_T(\kappa^p)$ as the plastic process evolves.

[18] State of the art report on : Finite Element Analysis of Reinforced Concrete. ASCE (1982).
[19] Tasuji, E.; Slate F. and Nilson A. (1978). Stress strain response and fracture of concrete. *Journal of the structural division. ASCE – Vol 75, No 7, pp. 306-312.*

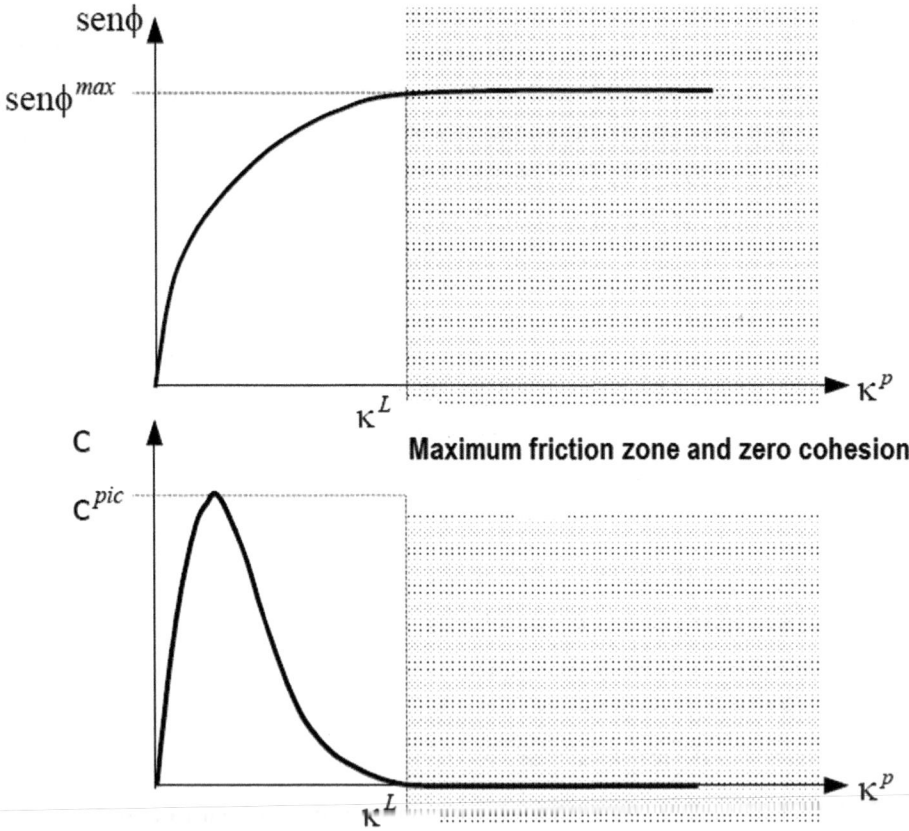

Figure 5.23 – Function of the angle evolution of internal friction as a function of the plastic damage variable and its relation to the cohesion.

Generally, an internal friction variable is formulated by a law of evolution of the type $\dot\phi = \dot\lambda\, H_\phi(\boldsymbol{\sigma};\mathbf{q}) = h_\phi(\boldsymbol{\sigma};\mathbf{q})\cdot\dot\kappa^p$, that is the above-mentioned friction increase. However, there is experimental evidence[1,20] that shows that a simple function of the plastic damage variable is enough to represent the evolution of the internal friction angle,

$$\operatorname{sen}\phi(\kappa^p) = \begin{cases} 2\dfrac{\sqrt{\kappa^p\kappa^L}}{\kappa^p + \kappa^L}\operatorname{sen}\phi^{max} & ;\ \forall\ \kappa^p \le \kappa^L \\[2ex] \operatorname{sen}\phi^{max} & ;\ \forall\ \kappa^p > \kappa^L \end{cases} \tag{5.124}$$

where κ^p is the hardening variable called internal plastic damage variable and κ^L is the damage limit from which the cohesion gets null and the internal friction keeps constant and equal to its maximum value ϕ^{max}, therefore this limit coincides with the total damage $\kappa^p = \kappa^L = 1$.

[20] Borst, R. De and Vermeer, P. (1984). *Non associated plasticity for soils, concrete and rock*. Heron. Vol. 29, Delf. Netherlands.

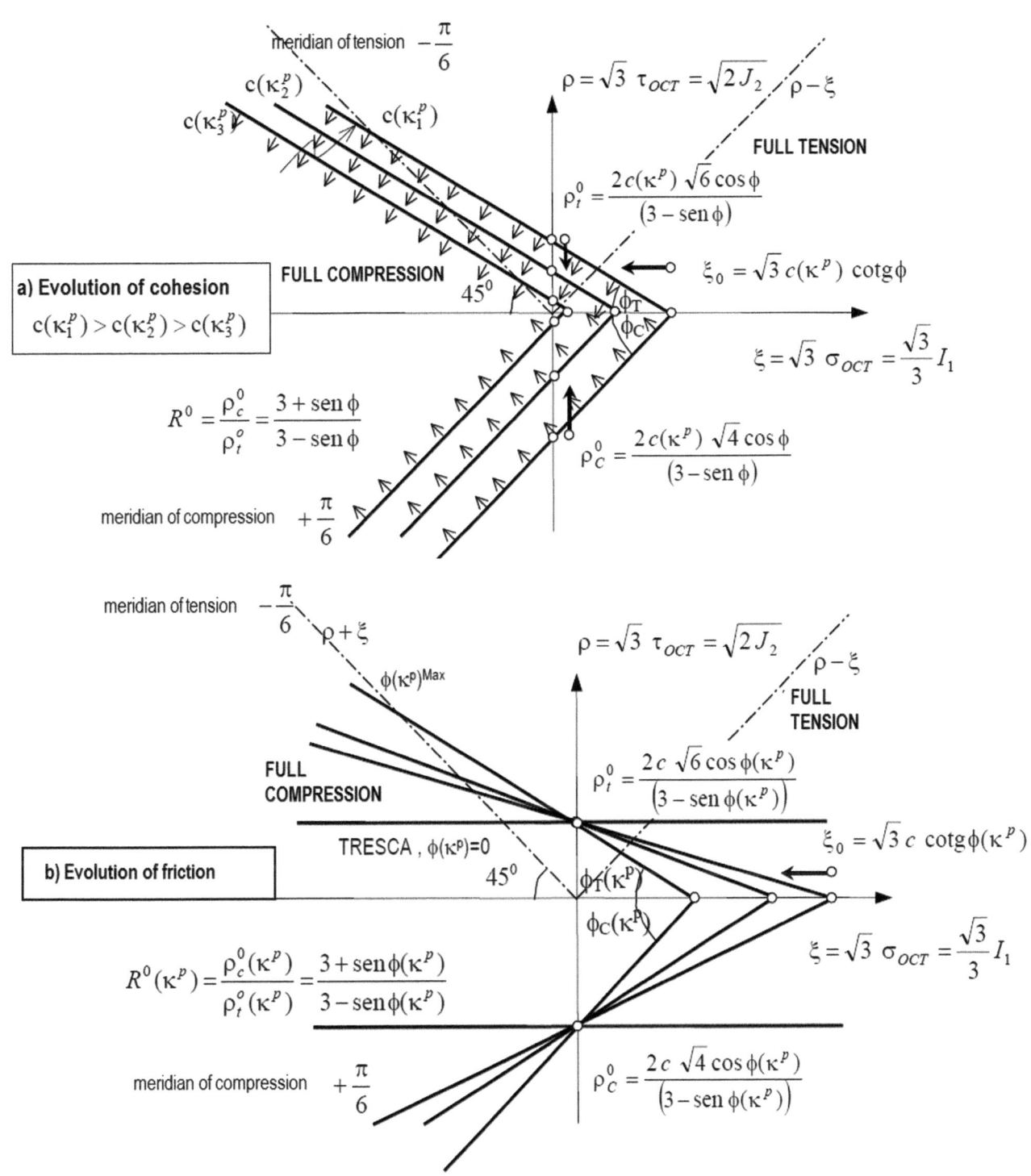

Figure 5.24 – Mohr-Coulomb's yield mobility criterion according to the meridian planes. a) Isotropic mobility by the evolution of the cohesion. b) Anisotropic mobility by internal friction change.

By using the internal friction function expressed by equation (5.124) and assuming first the hypothesis that cohesion $c = $ constant, the development of an elasto-plastic process going from an initial state $\phi = \phi^0 = 0$ can be observed, where the hydrostatic pressure is negligible, to a final state $\phi = \phi^{max}$ where the influence of pressure is very important. To illustrate the influence of the internal friction angle on the plasticity threshold, a Mohr-Coulomb type of yield surface is adopted, showing an isotropic mobility due to the cohesion evolution and another anisotropic mobility caused by the internal friction increase. This last phenomenon makes the surface move from its height towards its origin, along with its base increase (see Figures 5.24 and 5.25). This effect is presented as a hardening phenomenon for compression and softening for tensile processes.

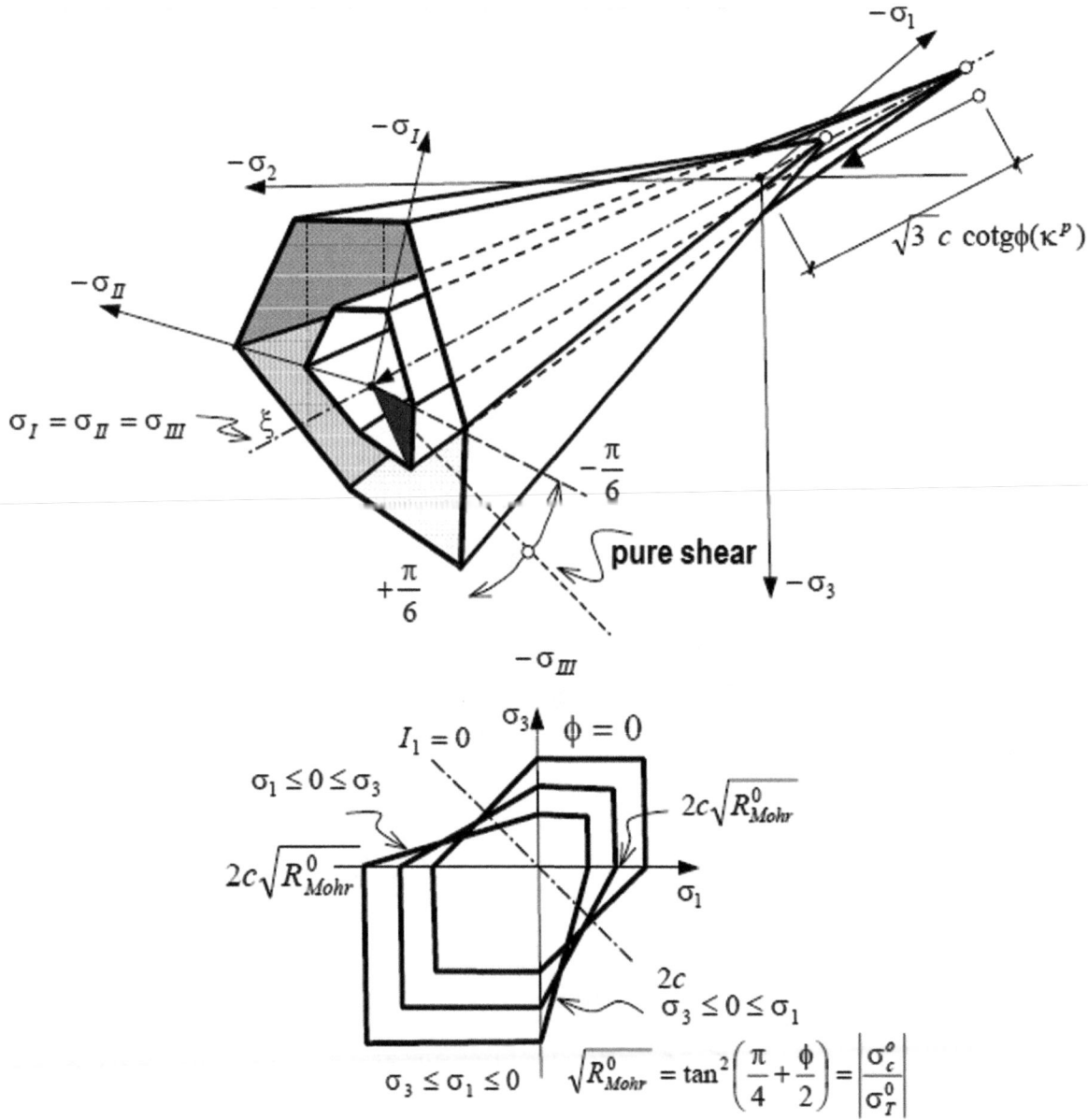

Figure 5.25 – Mohr-Coulomb yield mobility criterion when the internal friction changes. a) Function mobility in 3-D. b) Function mobility in one of the main planes.

In brittle materials with high initial cohesion such as concrete, ceramic, etc., a constant and maximum internal friction angle can be used during the whole plastic process without any problem for the resolution of multiaxial problems.

5.14.4 Variable definition ψ , dilatancy angle

A phenomenon that characterizes frictional materials is the change of the inelastic volume due to plastic deviation. This phenomenon, called dilatancy, may be due to micro-fractures in the concrete during the inelastic period.

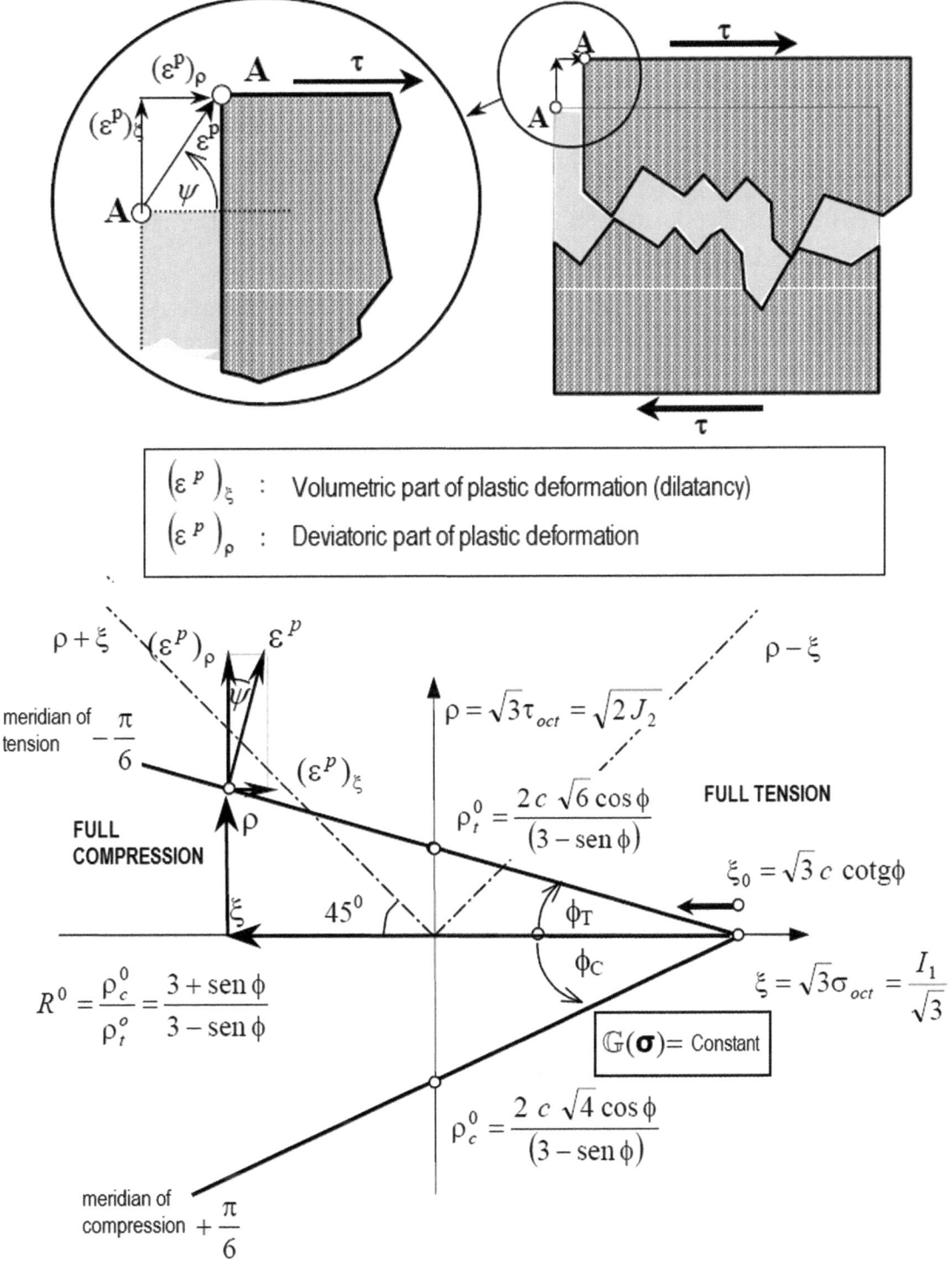

Figure 5.26 – Dilatancy graphic representation and its relation with the plastic deformation in the meridian plane.

One way to evaluate this phenomenon is through *the dilatancy angle ψ*, first introduced by B. Hansen[21], which represents the relation between the plastic increase volume and the plastic distortion.

The dilatancy can be controlled by using non-associated plasticity, in other words, when the plastic yield function is different from the plastic potential $\mathbb{F}(\boldsymbol{\sigma};\mathbf{q}) \neq \mathbb{G}(\boldsymbol{\sigma})$. Here, the plastic flow direction is orthogonal to the plastic potential surface $\mathbb{G}(\boldsymbol{\sigma})$, and the latter is responsible for the flow direction (see Figure 5.26). Thus, this flow could have a deviatoric and /or volumetric component and therefore it is possible to control these components as much as necessary to adjust the solution to the problem.

As frictional solids do not show a constant dilatancy angle during the elasto-plastic process, a function of evolution considering this magnitude must be formulated. This evolution could also be defined as an internal variable of the inelastic process, but since its variation can be described in a simple and satisfactory manner, an explicit quasi-empirical form of the plastic damage variable is adopted $\psi(\kappa^{p})$.

Among some of the possible ways to define the dilatancy evolution is P. Rowe's[22] modified equation, which can be very well adapted to the behavior of various geomaterials with cohesion.

$$\psi(\kappa^{p}) = \arcsin\left[\frac{\sin\phi(\kappa^{p}) - \sin\phi_{cv}}{1 - \sin\phi(\kappa^{p})\sin\phi_{cv}}\right] = \mathrm{atan}\frac{\left\|\left(\boldsymbol{\varepsilon}^{p}\right)_{\xi}\right\|}{\left\|\left(\boldsymbol{\varepsilon}^{p}\right)_{\rho}\right\|} \tag{5.125}$$

where $\phi(\kappa^{p})$ is the internal friction function (equation (5.124)) and ϕ_{cv} is the internal friction angle for null dilatancy, expressed as,

$$\phi_{cv} = \arcsin\left[\frac{\sin\phi^{max} - \sin\psi^{max}}{1 - \sin\phi^{max}\sin\psi^{max}}\right] \tag{5.126}$$

where ϕ^{max} is the maximum friction and ψ^{max} is the maximum dilatancy. As reference values for concrete are $\phi^{max} = 35^{0}$ and $\psi^{max} = 13^{0}$.

5.15 Generalization of the damage model with stiffness degradation

5.15.1 Introduction

Experimental results show that cohesive-frictional materials have a stiffness loss even in the elastic behavior. This effect is increased even more when decohesion among particles is produced and the elastic period starts. This evidence shows that there are two stiffness degradation phenomena acting on the material. One depends on the accumulated energy

---

[21] Hansen, B. (1958). Line ruptures regarded as narrow ruptures zone – Basic equations based on kinematics considerations. *Proc. Brussels Conf. 58 on earth pressures problems.*
[22] Rowe, P. (1972). Theoretical meaning and observed values of deformation parameter for soil. Proc. Rascoe Memorial Symp. on stress-strain behavior of soils. Cambridge.

called *elastic degradation*, and another one that depends on the friction mobility and is called *plastic degradation*. Based on these two phenomena, the above-mentioned plastic damage model will be modified in this section.

In most research studies, degradation is believed to depend only on the accumulated total energy and, in any event, this phenomenon is never due to particles decohesion (plasticity in this case).

The stiffness degradation or damage, as it is known in many references, has been formulated by Kachanov and is based on the mechanical representation of a material strength loss with recoverable deformations and stiffness decrease of the material[23,24] (the bases of this model will be presented further). In other words, the nonlinear condition of the problem is due to the irreversible loss of the stiffness properties (see Figure 5.27). This, along with the theory of plasticity, provides with a powerful tool to represent the behavior of many cohesive-frictional materials and metals where an internal porosity increase can occur.

Hence, the secant constitutive tensor $\mathbb{C}(d_i^e, d_i^p)$ will depend every moment on the internal variables of elastic damage d_i^e and of internal variables of plastic damage d_i^p, and their evolution laws are as follows,

$$\dot{d}_i^e = \Phi_i \left\langle \mathbf{k}_i^e : \dot{\boldsymbol{\varepsilon}}^e \right\rangle$$
$$\dot{d}_i^p = \dot{\lambda} H_i = \mathbf{k}_i^p : \dot{\boldsymbol{\varepsilon}}^p$$

(5.127)

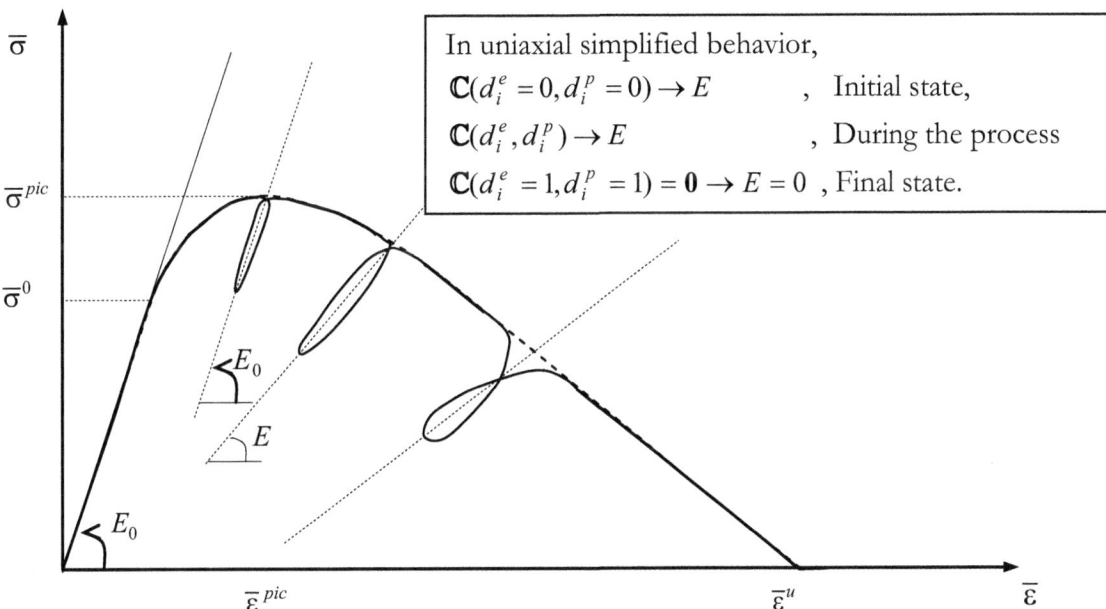

In uniaxial simplified behavior,

$\mathbb{C}(d_i^e = 0, d_i^p = 0) \to E$, Initial state,

$\mathbb{C}(d_i^e, d_i^p) \to E$, During the process

$\mathbb{C}(d_i^e = 1, d_i^p = 1) = \mathbf{0} \to E = 0$, Final state.

Figure 5.27 – Evolution of the uniaxial strength in a point due to plasticity and stiffness degradation.

where "d" is the damage index under development and the problem may be made of a finite number of different mechanisms. The tensor $\mathbf{k}_i^e$ is the direction of the elastic degradation of the ith mechanism, Φ_i is a positive scalar to be defined $H_i = \mathbf{k}_i^p : (\partial G / \partial \boldsymbol{\sigma})$

[23] Kachanov,L.M. (1958). Time Rupture Process under Creep Conditions, (in Russian). *Izv.ARad.SSSR Teckh.Nauk.,8 ,26-31.*

[24] Kachanov,L.M. (1986). Introduction to Continuum Damage Mechanics. *Martinus Nijho Publishers, Dordrecht.*

a scalar elasto-plastic function of tensorial arguments that considers the degradation direction $\mathbf{k}_i^p$ induced by the plasticity.

5.15.2 Elasto-plastic constitutive equation with stiffness degradation

The stiffness degradation phenomenon modifies the elasto-plastic constitutive equation for plasticity with small deformations. To formulate this new constitutive equation, first *a free potential energy* will be defined at a constant temperature, made of an elastic and a plastic part.

$$\Psi(\boldsymbol{\varepsilon}^e, \mathbf{q}_\alpha, \mathbf{q}_\beta) = \Psi^e(\boldsymbol{\varepsilon}^e, \mathbf{d}^p, \mathbf{q}_\beta) + \Psi^p(\mathbf{q}_\alpha) \tag{5.128}$$

where $\Psi^p(\mathbf{q}_\alpha)$ is a function of the plastic potential and $\Psi^e(\boldsymbol{\varepsilon}^e, \mathbf{d}^p, \mathbf{q}_\beta)$ is a function of the elastic potential or free energy. Additionally, the elastic deformation $\boldsymbol{\varepsilon}^e$ is a free variable of the process, $\mathbf{q}_\alpha$ is the plastic internal variables including its own plastic deformation, $\boldsymbol{\varepsilon}^p$ and $\mathbf{q}_\beta$ are the group of non-plastic internal variables including the degradation produced by the plasticity.

Based on the main principles of the mechanics, the dissipation is defined by the following simplified expression of the Clausius-Duhem's inequality (see chapter 2).

$$\Xi - \boldsymbol{\sigma} : \dot{\boldsymbol{\varepsilon}} - \dot{\Psi} \geq 0 \tag{5.129}$$

This inequality expresses the entropy balance for Cauchy's continuum and is valid for any admissible loading process. Replacing the temporal derivative of the free energy (5.128) into equation (5.129) the following expression for dissipation is obtained,

$$\Xi = \left[\boldsymbol{\sigma} - \frac{\partial \Psi}{\partial \boldsymbol{\varepsilon}^e} \right] : \dot{\boldsymbol{\varepsilon}} + \frac{\partial \Psi}{\partial \boldsymbol{\varepsilon}^e} : \dot{\boldsymbol{\varepsilon}}^p - \frac{\partial \Psi}{\partial \mathbf{q}_\alpha} \dot{\mathbf{q}}_\alpha - \frac{\partial \Psi}{\partial \mathbf{q}_\beta} \dot{\mathbf{q}}_\beta \geq 0 \tag{5.130}$$

To ensure compliance with this inequality under any deformation increase $\dot{\boldsymbol{\varepsilon}}$ for a non-degradable loading process, it must occur that,

$$\left[\boldsymbol{\sigma} - \frac{\partial \Psi}{\partial \boldsymbol{\varepsilon}^e} \right] \geq 0 \qquad \forall \; \dot{\boldsymbol{\varepsilon}} \tag{5.131}$$

from where it is assumed that the stress is,

$$\boldsymbol{\sigma} = \frac{\partial \Psi}{\partial \boldsymbol{\varepsilon}^e} \tag{5.132}$$

such that the free energy for an elasto-plastic solid with stiffness degradation can be expressed in small deformations as,

$$\Psi(\boldsymbol{\varepsilon}^e, \mathbf{q}_\alpha, \mathbf{q}_\beta) = \frac{1}{2}(\boldsymbol{\varepsilon} - \boldsymbol{\varepsilon}^p) : \mathbb{C}(d_i^e, d_i^p) : (\boldsymbol{\varepsilon} - \boldsymbol{\varepsilon}^p) + \Psi^p(\mathbf{q}_\alpha) \tag{5.133}$$

Replacing the latter into equation (5.132) then the secant constitutive equation for an elasto-plastic solid with stiffness degradation is expressed as,

$$\boldsymbol{\sigma} = \mathbb{C}(d_i^e, d_i^p) : (\boldsymbol{\varepsilon} - \boldsymbol{\varepsilon}^p) \tag{5.134}$$

From this equation, the temporal variation of the stress is obtained,

$$\dot{\boldsymbol{\sigma}} = \sum_i \left[\frac{\partial \mathbb{C}}{\partial d_i^e} d_i^e + \frac{\partial \mathbb{C}}{\partial d_i^p} d_i^p \right] : (\boldsymbol{\varepsilon} - \boldsymbol{\varepsilon}^{\text{p}}) + \mathbb{C} : (\dot{\boldsymbol{\varepsilon}} - \dot{\boldsymbol{\varepsilon}}^{\text{p}}) \tag{5.135}$$

Now a particular case of simple damage with only one damage mechanism (i=1) is considered. Then the evolution laws (5.127) can be simplified as follows,

$$\dot{d}^e = \Phi \left\langle \mathbf{k}^e : \dot{\boldsymbol{\varepsilon}} \right\rangle = \Phi \left\langle \boldsymbol{\sigma} : \dot{\boldsymbol{\varepsilon}}^{\mathbf{e}} \right\rangle$$
$$\dot{d}^p = \mathbf{k}^p : \dot{\boldsymbol{\varepsilon}}^p = \left[\frac{1 - d^p}{c} h_c \mathbf{h}_\kappa \right] : \dot{\boldsymbol{\varepsilon}}^p \tag{5.136}$$

Replacing these equations into (5.135), the following temporal evolution of the stress is obtained (see section 5.14 and reference[1]),

$$\dot{\boldsymbol{\sigma}} = \mathbb{C}_T^e(d^e, d^p) : \dot{\boldsymbol{\varepsilon}} - \mathbb{C}_T^p(d^e, d^p) : \dot{\boldsymbol{\varepsilon}}^p \begin{cases} \mathbb{C}_T^e = \mathbb{C}(d^e, d^p) - \dfrac{\Phi}{1 - d^e}(\boldsymbol{\sigma} \otimes \boldsymbol{\sigma}) \\[2mm] \mathbb{C}_T^p = \mathbb{C}(d^e, d^p) - \dfrac{h_c}{c}(\boldsymbol{\sigma} \otimes \mathbf{h}_\kappa) \end{cases} \tag{5.137}$$

where Φ is defined as constant during the whole process and represents the maximum density of elastic energy of a point in a solid,

$$d^e = \Phi \, \omega^e \tag{5.138}$$

The magnitude of Φ results from a loading process in which the plastic degradation variable is frozen and only the elastic degradation is allowed to evolve. This is,

$$\boldsymbol{\sigma} = \mathbb{C}(d^e) : \boldsymbol{\varepsilon}^e = (1 - \Phi \, \omega^e) \cdot \mathbb{C}^0 : \boldsymbol{\varepsilon}^e \implies \dot{\omega}^e = \boldsymbol{\varepsilon}^e : \left[(1 - \Phi \, \omega^e) \mathbb{C}^0 \right] : \dot{\boldsymbol{\varepsilon}}^e \tag{5.139}$$

from where then,

$$\int_{t=0}^{t} \frac{\dot{\omega}^e}{(1 - \Phi \, \omega^e)} dt = \int_{t=0}^{t} \boldsymbol{\varepsilon}^e : \left[\mathbb{C}^0 \right] : \dot{\boldsymbol{\varepsilon}}^e dt \implies d^e = \Phi \omega^e = 1 - e^{-(\Phi \, \boldsymbol{\varepsilon}^e : \mathbb{C}^0 : \boldsymbol{\varepsilon}^e)/2} \tag{5.140}$$

Once the equation is known (5.140) and the uniaxial stress-strain response is obtained in the lab, the magnitude of the constant Φ is obtained:

$$E = (1 - d^e) E^0 \implies \frac{E}{E^0} = e^{-(\Phi \, \varepsilon^e E^0 \varepsilon^e)/2} \implies \Phi = -\frac{2}{E^0 (\varepsilon^e)^2} \ln \frac{E}{E^0} \tag{5.141}$$

where E^0 represents the initial elasticity module and E the secant elasticity module for a deformation level ε^e. Formally, the elastic degradation process will be considered to be finished at this point.

The dissipated energy during this process with degradation will be obtained from equation (5.129), that is,

$$\omega^d = \int_t \Xi^e\, dt = \frac{1}{\Phi}\left[1 - e^{-(\Phi\, \boldsymbol{\varepsilon}^e : \mathbb{C}^0 : \boldsymbol{\varepsilon}^e)/2}\right] - \frac{1}{2}\boldsymbol{\varepsilon}^e : \mathbb{C}^0 : \boldsymbol{\varepsilon}^e \qquad (5.142)$$

This elastic degradation can also be adapted to other mechanical processes by establishing a degradation rate for volumetric behaviors and another for the deviatoric behavior. Examples of a formulation of this type can be seen in reference[1].

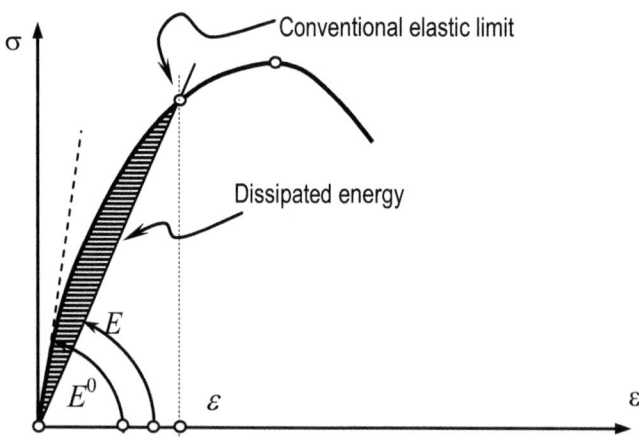

Figure 5.28 – Stress-strain curve with degradation during the elastic period.

5.15.3 Tangent constitutive equation for stiffness degradation processes

Following the same procedure established in classic plasticity, the elasto-plastic constitutive law with tangent damage without kinetic movement of the yield surface $\dot{\boldsymbol{\sigma}} = \mathbb{C}_T^{ep} : \dot{\boldsymbol{\varepsilon}}$ will be obtained. Also, the plastic consistency parameter λ is obtained by Prager's consistency condition. This is,

$$\left.\begin{array}{l} \mathbb{F}(\boldsymbol{\sigma}, c) = f(\boldsymbol{\sigma}) - c = 0 \\[2mm] \dot{\mathbb{F}} = \dfrac{\partial \mathbb{F}}{\partial \boldsymbol{\sigma}} : \dot{\boldsymbol{\sigma}} + \underbrace{\dfrac{\partial \mathbb{F}}{\partial c}}_{-1}\dot{c} = 0 \end{array}\right\} \;\Rightarrow\; \dfrac{\partial \mathbb{F}}{\partial \boldsymbol{\sigma}} : \dot{\boldsymbol{\sigma}} - \dot{c} = 0 \qquad (5.143)$$

Replacing this into equation (5.119) and equation (5.137), then

$$\frac{\partial \mathbb{F}}{\partial \boldsymbol{\sigma}} : \left[\mathbb{C}_T^e(d^e, d^p):\dot{\boldsymbol{\varepsilon}} - \mathbb{C}_T^p(d^e, d^p):\dot{\boldsymbol{\varepsilon}}^p\right] - h_c\left(\mathbf{h}_\kappa : \dot{\boldsymbol{\varepsilon}}^P\right) = 0$$

$$\left[\frac{\partial \mathbb{F}}{\partial \boldsymbol{\sigma}} : \mathbb{C}_T^e : \dot{\boldsymbol{\varepsilon}}\right] - \lambda \cdot \left[\frac{\partial \mathbb{F}}{\partial \boldsymbol{\sigma}} : \mathbb{C}_T^p : \frac{\partial \mathbb{G}}{\partial \boldsymbol{\sigma}} + h_c\, \mathbf{h}_\kappa : \frac{\partial \mathbb{G}}{\partial \boldsymbol{\sigma}}\right] = 0 \qquad (5.144)$$

The plastic consistency factor λ, which represents the magnitude (module) distance between an admissible stress state, outside the elastic domain, and the plastic loading surface, is obtained as a result of this last expression. Then,

$$\dot{\lambda} = \frac{\dfrac{\partial \mathbb{F}}{\partial \boldsymbol{\sigma}} : \mathbb{C}_T^e : \dot{\boldsymbol{\varepsilon}}}{\underbrace{\left[h_c \, \mathbf{h}_\kappa : \dfrac{\partial \mathbb{G}}{\partial \boldsymbol{\sigma}} \right] + \left(\dfrac{\partial \mathbb{F}}{\partial \boldsymbol{\sigma}} : \mathbb{C}_T^p : \dfrac{\partial \mathbb{G}}{\partial \boldsymbol{\sigma}} \right)}_{A}} \qquad \text{con} \quad \dot{\lambda} \geq 0 \tag{5.145}$$

where A is the plastic hardening parameter. Replacing equation (5.69) into the tangent constitutive equation (5.137), the following tangent elasto-plastic law with damage is obtained,

$$\dot{\boldsymbol{\sigma}} = \left\{ \mathbb{C}_T^e - \frac{\left[\mathbb{C}_T^p : \dfrac{\partial \mathbb{G}}{\partial \boldsymbol{\sigma}} \right] \otimes \left[\dfrac{\partial \mathbb{F}}{\partial \boldsymbol{\sigma}} : \mathbb{C}_T^e \right]}{h_c \, \mathbf{h}_\kappa : \dfrac{\partial \mathbb{G}}{\partial \boldsymbol{\sigma}} + \left(\dfrac{\partial \mathbb{F}}{\partial \boldsymbol{\sigma}} : \mathbb{C}_T^p : \dfrac{\partial \mathbb{G}}{\partial \boldsymbol{\sigma}} \right)} \right\} : \dot{\boldsymbol{\varepsilon}} \quad \Rightarrow \quad \dot{\boldsymbol{\sigma}} = \mathbb{C}_T^{ep} : \dot{\boldsymbol{\varepsilon}} \tag{5.146}$$

where $\mathbb{C}_T^{ep}$ is the continuous tangent constitutive tensor. As observed, the tangent stiffnesses will be symmetrical if the following proportionality is met, $\left[\mathbb{C}_T^p : \dfrac{\partial \mathbb{G}}{\partial \boldsymbol{\sigma}} \right] \propto \left[\dfrac{\partial \mathbb{F}}{\partial \boldsymbol{\sigma}} : \mathbb{C}_T^e \right]$. Therefore, it is not enough to comply with the classic non-associated flow law to ensure such symmetry.

5.15.4 Particular yield functions

As part of the classic plasticity theory generalization, it is also necessary to formulate the most suitable yield surfaces for the different behaviors of materials. A modification of the Mohr-Coulomb and the Drucker-Prager classic functions will be briefly presented as follows. Another particular formulation for concrete can be found in the reference S. Oller[1]. The limitations of each of these classic functions as well as their advantages and disadvantages will be shown.

5.15.4.1 The Mohr-Coulomb modified function

The Mohr-Coulomb function cannot be used directly in frictional cohesive materials such as concrete, which has an internal friction angle of $\phi \cong 32^0$. According to the Mohr-Coulomb classic formulation, a limit strength relation is obtained for this angle between a traction behavior and a uniaxial compression behavior of $R_{Mohr}^0 = \left\| \sigma_C^0 / \sigma_T^0 \right\|$ $= \tan\left[(\pi/4) + (\phi/2) \right] = 3{,}25$ (see (5.30)). This magnitude is very different from the concrete magnitude, $R_{Mohr}^0 = \left\| \sigma_C^0 / \sigma_T^0 \right\| \cong 10{,}0$. To solve this problem, either the friction angle can be increased causing a dilatancy excess or the original criterion[1] can be modified. By this last option, the following expression is obtained,

$$\mathbb{F}(\boldsymbol{\sigma}, \mathsf{c}, \phi) = f(\boldsymbol{\sigma}, \phi) - \mathsf{c} = 0 \tag{5.147}$$

where the stress function is expressed as,

$$f(\boldsymbol{\sigma};\phi) = \frac{1}{\cos\phi}\left\{\frac{I_1}{3}\mathbb{K}_3 + \sqrt{J_2}\left[\mathbb{K}_1\cos(\theta) - \mathbb{K}_2\frac{\sin(\theta)\sin(\phi)}{3}\right]\right\} \tag{5.148}$$

being the invariants defined in previous sections and the factors $\mathbb{K}_i$, for Mohr-Coulomb's classic function, then

$$\left.\begin{array}{l}\mathbb{K}_1 = f_1(\alpha_R;\phi)\\ \mathbb{K}_2 = f_2(\alpha_R;\phi)\\ \mathbb{K}_3 = f_3(\alpha_R;\phi)\end{array}\right\} \xrightarrow{si \quad \alpha_R=1} \left\{\begin{array}{l}\mathbb{K}_1 = 1\\ \mathbb{K}_2 =\\ \mathbb{K}_3 = \sin(\phi)\end{array}\right. \tag{5.149}$$

where $\alpha_R = R'/R_{Mohr}$ is the coefficient between the required strength R' and Mohr-Coulomb's classic-function strength relation R_{Mohr}.

Through this new Mohr-Coulomb modified function any strength relation required by the different materials can be established by only modifying $\mathbb{K}_i$, without increasing dilatancy (see the modified function of Figure 5.31). The expressions resulting from this modification and the shape obtained by the Mohr function as it modifies the relation $\alpha_R = R'/R_{Mohr}$ are shown next.

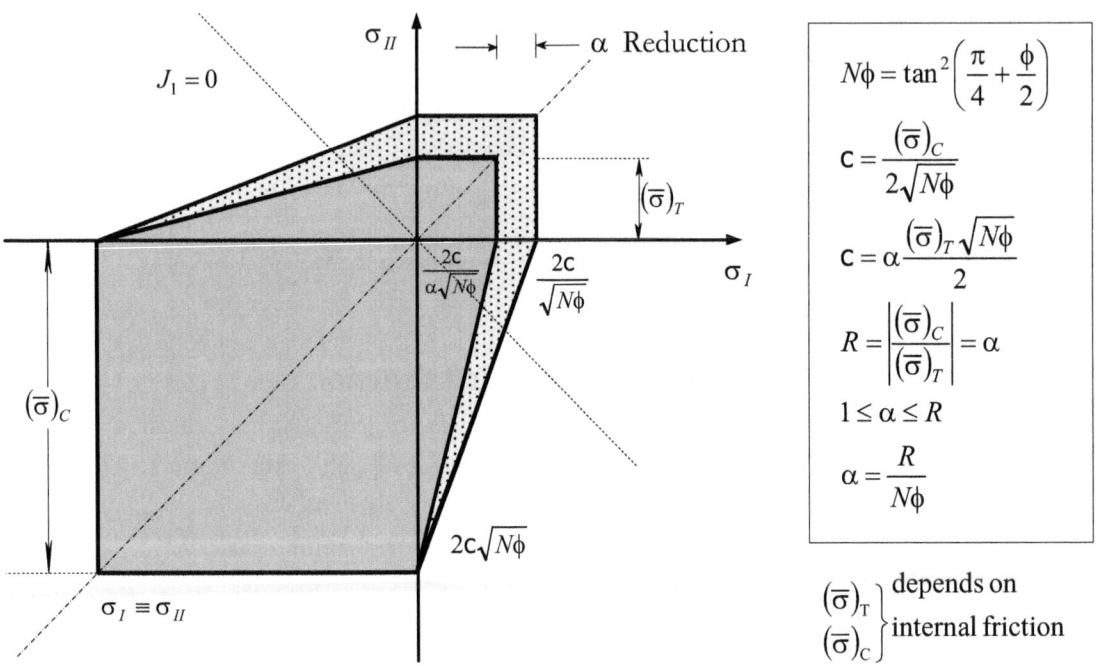

Figure 5.29 – Mohr-Coulomb function evolution as a function of the strength relation.

$$\mathbb{F}(\sigma;c;\phi) = \left\{ \frac{I_1}{3}\mathbb{K}_3 + \sqrt{J_2}\left[\mathbb{K}_1\cos(\theta) - \mathbb{K}_2\frac{\sin(\theta)\sin(\phi)}{3}\right]\right\} - c\cos(\phi) = 0 \qquad (5.150)$$

where

$$\begin{aligned}
\mathbb{K}_1 &= \left[\frac{(1+\alpha)}{2} - \frac{(1-\alpha)}{2}\sin(\phi)\right]\\[4pt]
\mathbb{K}_2 &= \left[\frac{(1+\alpha)}{2} - \frac{(1-\alpha)}{2}\frac{1}{\sin(\phi)}\right]\\[4pt]
\mathbb{K}_3 &= \left[\frac{(1+\alpha)}{2}\sin(\phi) - \frac{(1-\alpha)}{2}\right]
\end{aligned} \qquad (5.151)$$

For further details about coefficients, consulting the original reference[1] is recommended.

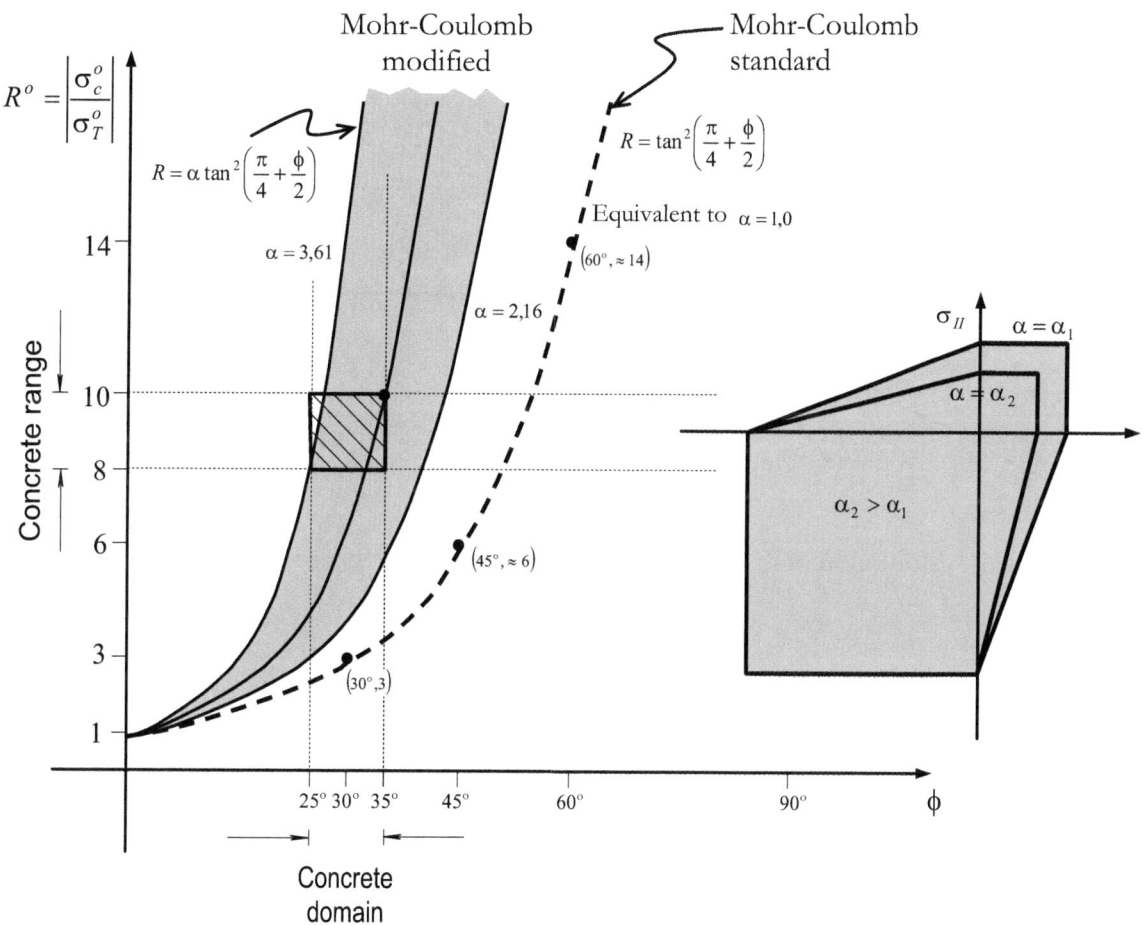

Figure 5.30 – Mohr-Coulomb´s standard and modified strength relation.

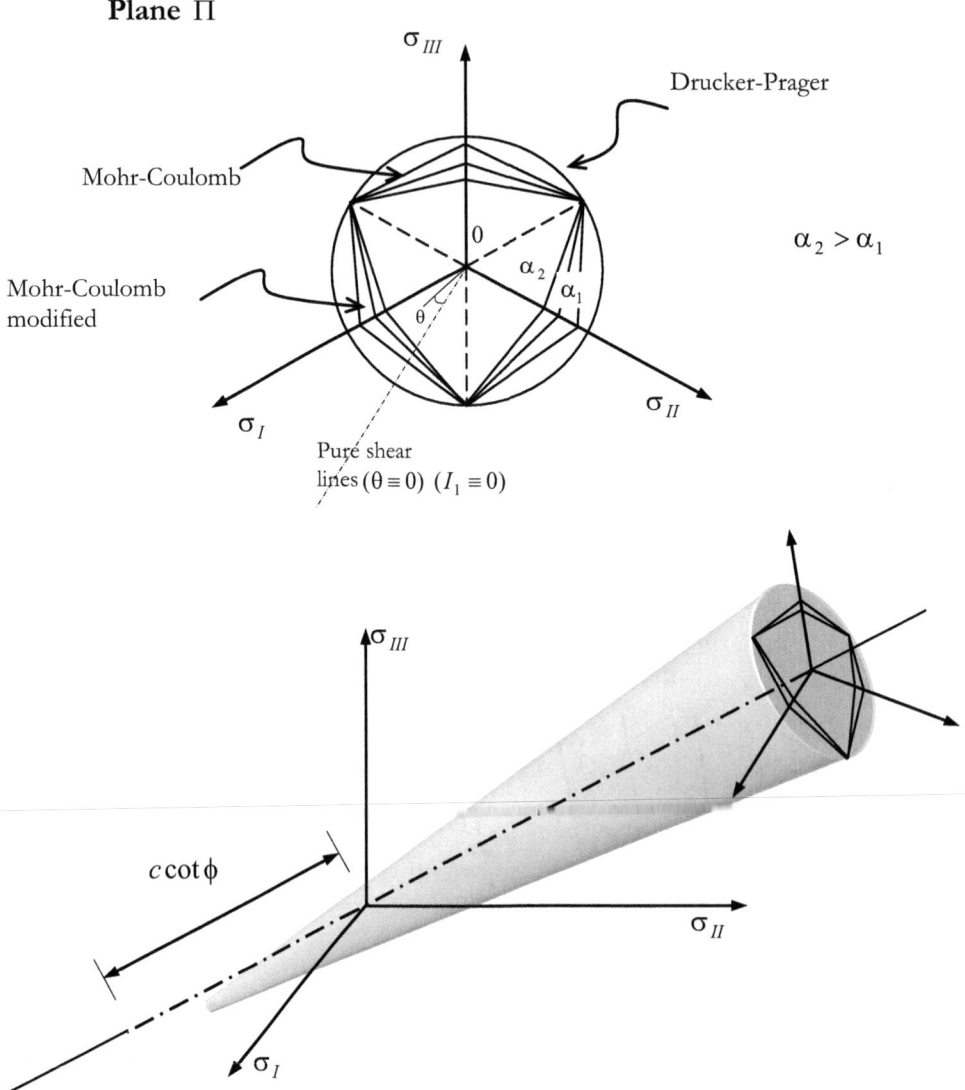

Figure 5.31 – Mohr-Coulomb's function circumscribed in Drucker Prager's function

5.15.4.2 The Drucker-Prager modified function

Drucker-Prager's standard function also has direct use limitations for materials with a strength relation between the traction and compression behavior greater than 3 (see Figure 5.32 and Figure 5.33).

Drucker-Prager's surface circumscribing Mohr-Coulomb's allows for greater strength relations with lower friction angles; however, there is an indetermination in the plane $\sigma_1 - \sigma_3$, for stresses $-\sigma_1 = -\sigma_3$ with internal friction angles of $\phi = \arcsin(3/5) \cong 36,8698\ldots$ This indetermination produces an excessive growth of the yield surface in the compression area with respect to tension[1].

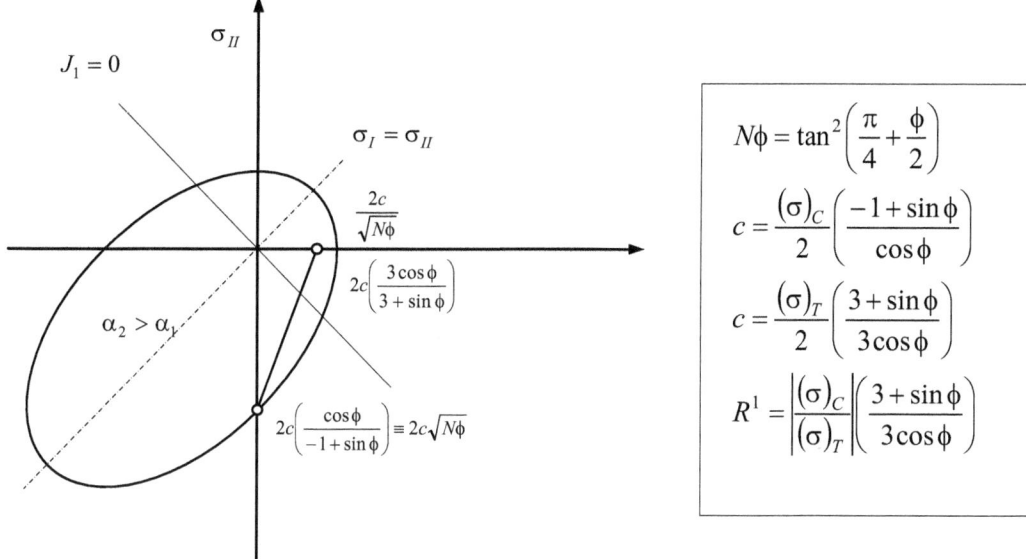

Figure 5.32 – Drucker-Prager's shape and relation with the characteristic values of Mohr-Coulomb's function.

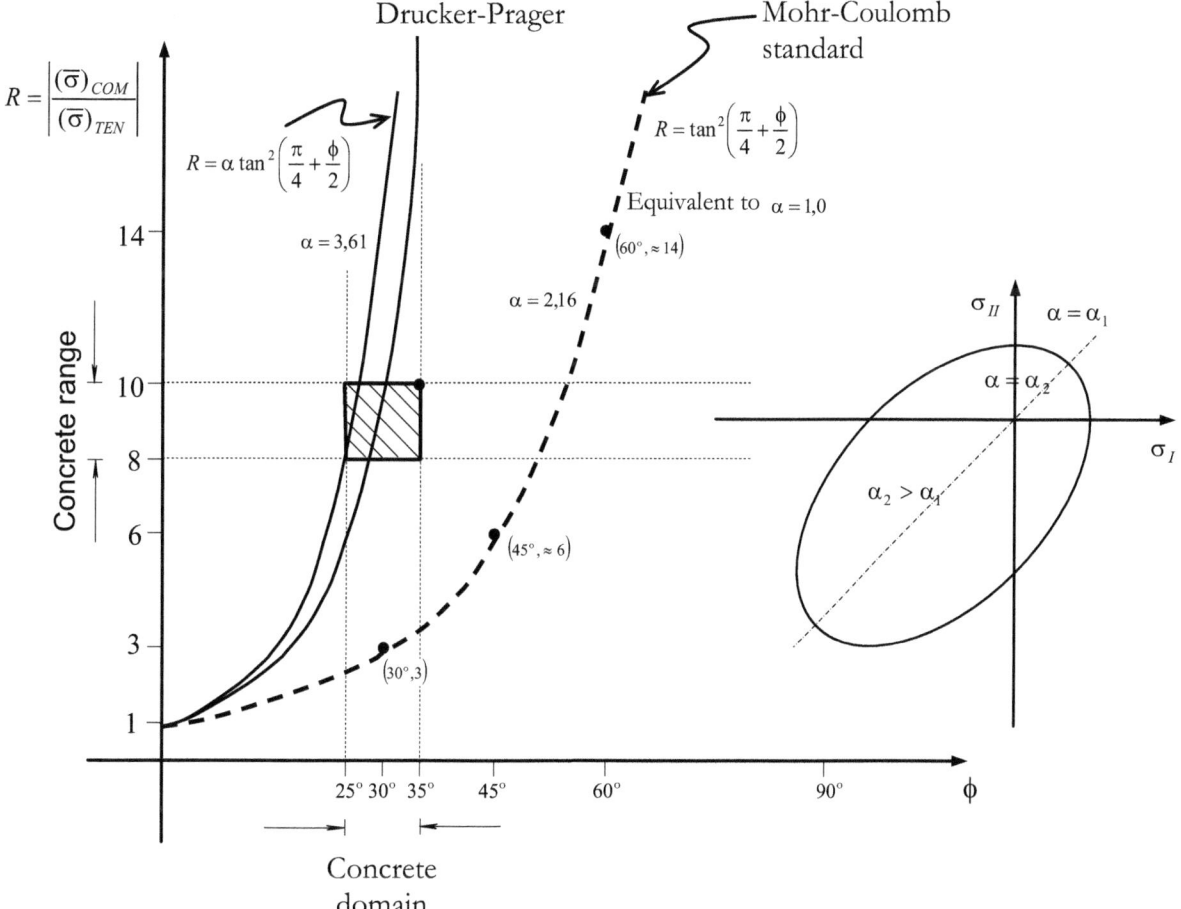

Figure 5.33 – Friction angle-strength relation.

This situation can be verified by analyzing carefully Drucker-Prager's yield function. Its mathematical form is as follows,

$$\mathbb{F}(I_1;J_2;c;\phi) = \overline{\alpha}(\phi)I_1 + \sqrt{J_2} - \overline{\mathcal{K}}(\kappa;\phi) = 0 \qquad (5.152)$$

where the hardening functions $\overline{\alpha}(\phi)$ and $\overline{\mathcal{K}}(\kappa;\phi)$, after being modified according to Mohr-Coulomb's criterion, are written as

Parameters Application	$\overline{\alpha}$	$\overline{\mathcal{K}}$
For conventional triaxial compression	$\dfrac{2\sin(\phi)}{\sqrt{3}[3-6\sin(\phi)]}$	$\dfrac{\tan(\phi)}{[9+12\tan^2(\phi)]^{1/2}}$
For plane deformation	$\dfrac{6c\cos(\phi)}{\sqrt{3}[3-6\sin(\phi)]}$	$\dfrac{3c}{[9+12\tan^2(\phi)]^{1/2}}$

5.16 Isotropic continuous damage-Introduction[*]

The damage of a continuous solid (stiffness degradation) is an alteration of the elastic properties during load application due to a decrease of the effective strength area[25]. This effective area loss is normally caused by the increase of voids and/or micro fractures.

The damage phenomenon only affects the elastic properties of the material whereas plasticity is developed as a result of the irrecoverable growth of the plastic strain. Both phenomena are complementary and it is normal to see material strength loss due to *the damage* –elasticity loss – and by *the plasticity* -inelastic deformation growth-.

The continuous damage theory was first introduced by Kachanov[26] in 1958 dealing with yield problems but it was accepted afterwards as a valid alternative for behavior simulation of different materials. Among the different possible formulations[27,28,29,30,31,32], a simple damage model with a scalar internal variable that helps to characterize the local damage will be presented in this chapter. Although this model is simple, it has great potential and can be used for the nonlinear behavior representations of both metallic materials and geomaterials.

This type of model can simulate the behavior of materials in which stiffness degradation can occur after the material damage threshold is overcome.

---

[*] This section has been written in collaboration with Dr. Eduardo Car from the Polytechnic University of Catalonia.

[25] Maugin, G. A. (1992). *The thermodynamics of plasticity and fracture*. Cambridge University Press.

[26] Kachanov, L. M. (1958). Time of rupture process under creep conditions. *Izvestia Akaademii Nauk; Otd Tech Nauk, 8* 26-31.

[27] Lemaitre, J and Chaboche, J. L. (1978). Aspects phénoménologiques de la rupture par endommagement. *J. Appl, 2, 317-365.*

[28] Chaboche, J. (1988). Continuum damage mechanics part I. General Concepts. *Journal of Applied Mechanics 55, 59-64.*

[29] Chaboche, J. (1988). Continuum damage mechanics part II. Damage Growth. *Journal of Applied Mechanics 55, 65-72.*

[30] Simo, J. and Ju, J. (1987). Strain and stress based continuum damage models – I Formulation. *Int. J. Solids Structures, 23,* 821-840.

[31] Simo, J. and Ju, J. (1987). Strain and stress based continuum damage models – II Computational aspects. *Int. J. Solids Structures, 23,* 841-869.

[32] Oliver, J.; Cervera, M.; Oller, S. and Lubliner, J. (1990). Isotropic damage models and smeared crack analysis of concrete. *Second international conference on Computer Aided Analysis and Design of Concrete Structures.*

5.16.1 Isotropic damage model

During the last years the constitutive models known as continuous damage models have been widely accepted for the simulation of the complex constitutive behavior of many materials used in engineering[4,5,6,7]. These models are characterized by their simplicity in their implementation, versatility and coherence, as they are based on the continuous damage mechanics.

Kachanov (1958)[2] introduced the concept of effective stresses to carry out fracture simulations by viscous phenomena[33]. It is also used for the representation of fatigue[34], fracture in ductile and fragile materials[35], etc.

Physically, the degradation process of the materials' properties is due to the presence and growth of small fractures and micro voids present in the structure of any material. This growth process can be simulated within the context of the mechanics of continuous media, taking into consideration the theory of internal state variables, introducing an internal variable of damage represented by a scalar, vector or tensor. This internal variable of damage characterizes the material's level of deterioration and transforms the stress real tensor into an effective stress tensor as follows,

$$\boldsymbol{\sigma}_0 = \mathbf{M}^{-1} : \boldsymbol{\sigma} \qquad (5.153)$$

where $\mathbf{M}$ is the fourth-order tensor of the anisotropic damage model. For the isotropic damage model, the material degradation is developed in all directions alike and only depends on one scalar damage variable d, so tensor $\mathbf{M}$ is reduced to $\mathbf{M} = (1-d)\mathbf{I}$ and the equation of anisotropic damage (5.153) is then

$$\boldsymbol{\sigma}_0 = \frac{\boldsymbol{\sigma}}{(1-d)} \qquad (5.154)$$

where d is the internal variable of damage, $\boldsymbol{\sigma}$ is Cauchy's stress tensor and $\boldsymbol{\sigma}_0$ is the effective stress tensor measured in the "non-damaged" space. This internal variable is a measure of the material's stiffness loss and its higher or lower limits are given by:

$$0 \leq d \leq 1 \qquad (5.155)$$

where $d=1$ is a state of the material completely damaged and defines the complete local fracture, and $d=0$ is a non-damaged material. The effective stress concept was formulated for the first time in connection with the equivalent deformation hypothesis by Lemaitre-Chaboche[3] in 1978 as follows: "... The deformation associated to a damaged state subjected to a stress $\boldsymbol{\sigma}$ is equivalent to the deformation associated to the non-damage state subjected to an effective stress $\boldsymbol{\sigma}_0$ ". A graphic representation of the effective stress hypothesis can be observed in Figure 5.34.

---

[33] Rabotnov, I. (1963). On the equation of state for creep. *Progress in Applied Mechanics. The Prager Anniversary Volume. Pp 307-315.*

[34] Salomón, O.; Oller S.; Car E.; Oñate E. (1999). *Thermomechanical fatigue analysis based on continuum mechanics. Acta del VI congreso Argentino de mecánica computacional.*

[35] Lubliner, J.; Oliver, J.; Oller, S. and Oñate (1989). A plastic damage model for concrete. Int. J. solids Structures. Vol. 25, No.3, pp. 299-326.

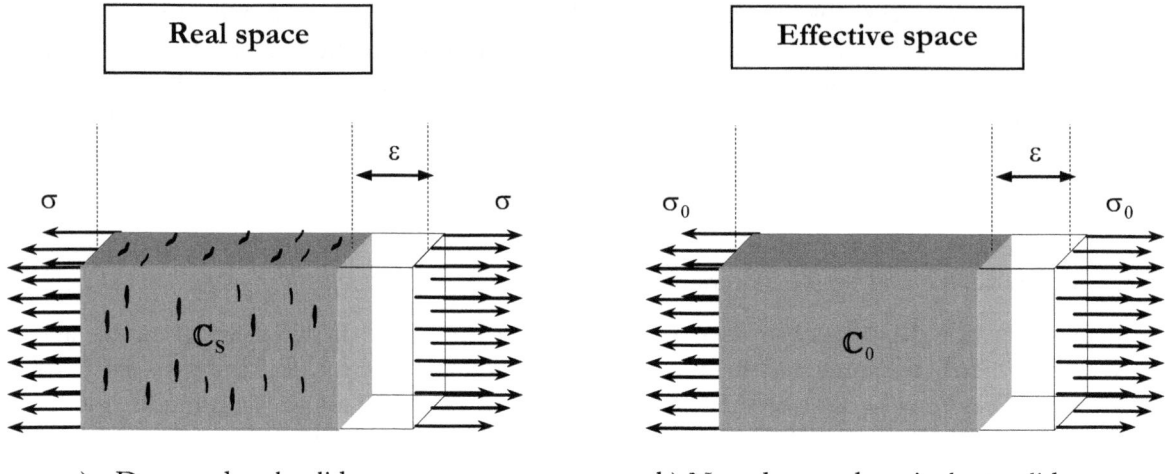

a)   Damaged real solid   b) Non-damaged equivalent solid

Figure 5.34 – Graphic representation of the effective stress hypothesis.

5.17 Helmholtz's free energy and constitutive equation

Helmholtz's free energy per unit volume for a model of isotropic damage at constant temperature is given by

$$\Psi = \Psi\left(\boldsymbol{\varepsilon}; p_i\right) \qquad \text{with} \quad p_i = \{d\}$$
$$\Psi = \Psi\left(\boldsymbol{\varepsilon}; d\right) = (1 - d)\,\Psi_0(\boldsymbol{\varepsilon}) \tag{5.156}$$

where $\Psi_0(\boldsymbol{\varepsilon})$ is Helmholtz's free initial elastic energy of the non-damaged material. In the case of small strain, the free energy can be characterized through a quadratic function of the strain of this type,

$$\Psi_0(\boldsymbol{\varepsilon}) = \frac{1}{2}\boldsymbol{\varepsilon} : \mathbb{C}_0 : \boldsymbol{\varepsilon} \tag{5.157}$$

where $\mathbb{C}_0$ is the constitutive elastic tensor of the material in a non-damaged state. For thermally stable problems the following form of the Clausius-Plank inequality is appropriate,

$$\Xi = \left(\boldsymbol{\sigma} - \frac{\partial \Psi}{\partial \boldsymbol{\varepsilon}}\right) : \dot{\boldsymbol{\varepsilon}} - \frac{\partial \Psi}{\partial d}\dot{d} \geq 0 \tag{5.158}$$

This dissipative power expression leads to the following considerations:

a. Inequality (5.158) must be met for any temporal variation of the free variable $\boldsymbol{\varepsilon}$, thus $\dot{\boldsymbol{\varepsilon}}$ must be null (Coleman's method, see Maugin[1]). This condition provides the hyper-elastic constitutive law for the scalar damage problem,

$$\boldsymbol{\sigma} = \frac{\partial \Psi}{\partial \boldsymbol{\varepsilon}} \quad , \quad \frac{\partial \Psi}{\partial d} = -\Psi_0 \leq 0 \quad \Rightarrow \quad -\Psi_0 \text{ conjugate of } d \tag{5.159}$$

b. Taking into account the constitutive law, the dissipation value of the degradation model is then,

$$\Xi = \Psi_0 \dot{d} \geq 0 \tag{5.160}$$

Assuming equation (5.159), the following form of the constitutive equation is obtained,

$$\boldsymbol{\sigma} = \frac{\partial \Psi}{\partial \boldsymbol{\varepsilon}} = (1-d)\frac{\partial \Psi_0}{\partial \boldsymbol{\varepsilon}} = (1-d)\mathbb{C}_0 : \boldsymbol{\varepsilon} \tag{5.161}$$

This is the secant constitutive equation of the damage model and it has the following characteristics:

1. The elastic degradation model is isotropic since the mechanical properties are only affected by a scalar value.
2. The integration of the constitutive equation is explicit.
3. Equation (5.161) can be interpreted as an additive decomposition of the elastic and inelastic stresses (see Figure 5.35), that is

$$\boldsymbol{\sigma} = (1-d)\mathbb{C}_0 : \boldsymbol{\varepsilon} = [\mathbb{C}_0 : \boldsymbol{\varepsilon}] - [d\,\mathbb{C}_0 : \boldsymbol{\varepsilon}] = \boldsymbol{\sigma}_0 - \boldsymbol{\sigma}_d \tag{5.162}$$

The model expressed in equation (5.161) requires knowing the damage variable at each moment of the mechanical process. Thus, it is necessary to define the evolution of this internal variable of damage. In the following sections, the necessary steps for its evaluation will be described in detail.

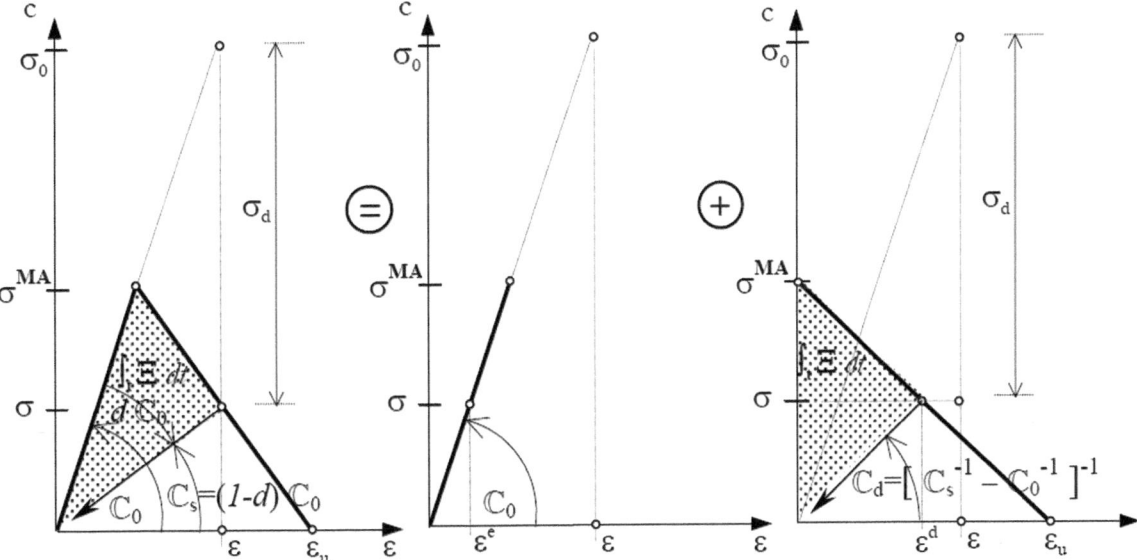

Figure 5.35 – Schematic representation of the uniaxial damage model.

5.18 Damage threshold criterion

The damage criterion makes a distinction between an elastic behavior inside the domain delimited by the damage function and another in which the degradation process of the material's properties is verified. This criterion depends on the type of material and is defined in the same way as the plasticity problems,

$$\mathbb{F}(\boldsymbol{\sigma}_0;\boldsymbol{q}) = f(\boldsymbol{\sigma}_0) - c(d) \leq 0 \quad , \quad \text{with} \quad \boldsymbol{q} \equiv \{d\} \tag{5.163}$$

where $f(\boldsymbol{\sigma}_0)$ is a function of the stress tensor $\boldsymbol{\sigma}_0 = \mathbb{C}_0 : \boldsymbol{\varepsilon}$ and $c(d)$ is the function defining the damage threshold position. This function establishes the onset of the nonlinear damage behavior and additionally defines the loading, unloading and reloading states. It is a scalar function and it must be positive and null for a non-deformed state. The initial value of the damage threshold $c(d^0) = c^{max} = \sigma^{max}$ is a property of the material and is related to its strength to compression depending on the damage threshold function chosen.

Equation (5.33) represents a limit surface in the strain space or non-damaged stresses. The damage in the material is verified when the value of $f(\boldsymbol{\sigma}_0)$ is equal or greater than $c^{max} = \sigma^{max}$ for the first time. A similar equation to (5.162) is given by the following expression,

$$\overline{\mathbb{F}}(\boldsymbol{\sigma}_0;\mathbf{q}) = G\big[f(\boldsymbol{\sigma}_0)\big] - G\big[c(d)\big] \leq 0 \quad , \quad \text{with} \quad \mathbf{q} \equiv \{d\} \tag{5.164}$$

where $G[\chi]$ is a scalar monotonically increasing function, positive invertible, with positive derivative.

5.19 Evolution law of the internal damage variable

When dealing with mechanics problems involving the internal variable formulations, the evolution law of the latter must be defined. Regarding the damage problem, the evolution law of the internal variable is given by:

$$\dot{d} = \dot{\mu}\frac{\partial\overline{\mathbb{F}}(\boldsymbol{\sigma}_0;\mathbf{q})}{\partial[f(\boldsymbol{\sigma}_0)]} \equiv \dot{\mu}\frac{\partial G[f(\boldsymbol{\sigma}_0)]}{\partial[f(\boldsymbol{\sigma}_0)]} \tag{5.165}$$

where μ is a non-negative scalar called damage consistency parameter, similar to the plastic consistency factor λ, which is used to define loading, unloading and reloading conditions through Kuhn-Tucker's conditions,

$$\dot{\mu} \geq 0 \quad ; \quad \overline{\mathbb{F}}(\boldsymbol{\sigma}_0;\mathbf{q}) \leq 0 \quad ; \quad \dot{\mu} \cdot \overline{\mathbb{F}}(\boldsymbol{\sigma}_0;\mathbf{q}) = 0 \tag{5.166}$$

The conditions expressed in the previous equation correspond to problems which have unilateral restrictions. If the value of $\overline{\mathbb{F}}(\boldsymbol{\sigma}_0;\mathbf{q}) < 0$ the damage criterion is not verified and to comply with all Kuhn-Tucker's conditions it will be necessary for $\dot{\mu} = 0$. Hence, it can be assumed from equation (5.165) that the damage temporal variation must be null $\dot{d} = 0$. Then, the material does not show damage phenomena, as it is an elastic mechanical process.

As with the plasticity theory, the consistency factor magnitude is a result of imposing Il'ushim's damage consistency conditions. From the latter and from the function properties $G[\chi]$ can be obtained,

$$\overline{\mathbb{F}}(\boldsymbol{\sigma}_0;\mathbf{q}) = 0 \Rightarrow G[f(\boldsymbol{\sigma}_0)] = G[c(d)] \Rightarrow f(\boldsymbol{\sigma}_0) = c(d) \Rightarrow \frac{\partial G[f(\boldsymbol{\sigma}_0)]}{\partial f(\boldsymbol{\sigma}_0)} = \frac{\partial G[c(d)]}{\partial c(d)} \tag{5.167}$$

From the condition of permanence over the damage threshold surface it can be assumed that,

$$\dot{\mathbb{F}}(\boldsymbol{\sigma}_0;\mathbf{q})=0 \quad \Rightarrow \quad \frac{\partial G[f(\boldsymbol{\sigma}_0)]}{\partial f(\boldsymbol{\sigma}_0)}\dot{f}(\boldsymbol{\sigma}_0)-\frac{\partial G[c(d)]}{\partial c(d)}\dot{c}(d)=0 \quad \Rightarrow \quad \dot{f}(\boldsymbol{\sigma}_0)=\dot{c}(d) \tag{5.168}$$

As observed with the temporal variation of $\partial G[f(\boldsymbol{\sigma}_0)]/\partial t = \dot{G}[f(\boldsymbol{\sigma}_0)]$ (equation (5.168)) and making an analogy with the evolution law of the internal variable $\dot{d}$ (equation (5.165)), the damage consistency parameter is obtained as follows,

$$\left.\begin{array}{l} \dot{G}[f(\boldsymbol{\sigma}_0)]=\dfrac{\partial G[f(\boldsymbol{\sigma}_0)]}{\partial f(\boldsymbol{\sigma}_0)}\dot{f}(\boldsymbol{\sigma}_0) \\[4mm] \dot{d}=\dot{\mu}\,\dfrac{\partial G[f(\boldsymbol{\sigma}_0)]}{\partial[f(\boldsymbol{\sigma}_0)]} \end{array}\right\} \quad \Rightarrow \quad \dot{d}\equiv\dot{G}[f(\boldsymbol{\sigma}_0)] \quad \Rightarrow \quad \dot{\mu}\equiv\dot{f}(\boldsymbol{\sigma}_0) \tag{5.169}$$

Developing further the formulation of the consistency parameter, it can be expressed that,

$$\dot{\mu}=\dot{f}(\boldsymbol{\sigma}_0)=\dot{c}(d)=\frac{\partial f(\boldsymbol{\sigma}_0)}{\partial\boldsymbol{\sigma}_0}:\dot{\boldsymbol{\sigma}}_0=\frac{\partial f(\boldsymbol{\sigma}_0)}{\partial\boldsymbol{\sigma}_0}:\mathbb{C}_0:\dot{\boldsymbol{\varepsilon}} \tag{5.170}$$

By integrating in time the damage temporal variation (equation (5.169)), the following explicit form for the damage representation of a point in a solid is obtained,

$$d=\int_t\dot{d}\ dt=\int_t\dot{G}[f(\boldsymbol{\sigma}_0)]\ dt=G[f(\boldsymbol{\sigma}_0)] \tag{5.171}$$

Substituting this equation into the dissipation (equation (5.160)), the expression describing the dissipation evolution in time is obtained,

$$\Xi=\Psi_0\,\dot{G}[f(\boldsymbol{\sigma}_0)]=\Psi_0\,\frac{\partial G[f(\boldsymbol{\sigma}_0)]}{\partial f(\boldsymbol{\sigma}_0)}\frac{\partial f(\boldsymbol{\sigma}_0)}{\partial\boldsymbol{\sigma}_0}:\mathbb{C}_0:\dot{\boldsymbol{\varepsilon}} \tag{5.172}$$

From the previously presented definitions, the damage threshold c at a time $s=t$ can be obtained,

$$c=\max\left\{c^{max},\max\left\{f(\boldsymbol{\sigma}_0)\big|_s\right\}\right\} \quad \forall \ \ 0\le s\le t \tag{5.173}$$

5.20 Tangent constitutive tensor of damage

The constitutive tensor of the tangent damage is obtained by the temporal variation of the constitutive secant equation (5.161).

$$\dot{\boldsymbol{\sigma}}=(1-d)\mathbb{C}_0:\dot{\boldsymbol{\varepsilon}}-\dot{d}\,\mathbb{C}_0:\boldsymbol{\varepsilon} \tag{5.174}$$

Replacing the evolution law of the damage internal variable $\dot{d}$ given by equation (5.165) into the previous equation, the following is obtained:

$$\dot{\boldsymbol{\sigma}}=(1-d)\mathbb{C}_0:\dot{\boldsymbol{\varepsilon}}-\frac{\partial G[f(\boldsymbol{\sigma}_0)]}{\partial[f(\boldsymbol{\sigma}_0)]}\dot{f}(\boldsymbol{\sigma}_0)\cdot[\mathbb{C}_0:\boldsymbol{\varepsilon}] \tag{5.175}$$

Taking into consideration that the temporal variation of the threshold function is expressed as,

$$\dot{f}(\pmb{\sigma}_0) = \frac{\partial f(\pmb{\sigma}_0)}{\partial \pmb{\sigma}_0} : \dot{\pmb{\sigma}}_0 = \frac{\partial f(\mathbb{C}_0 : \pmb{\varepsilon})}{\partial \pmb{\varepsilon}} : \dot{\pmb{\varepsilon}} \qquad (5.176)$$

Substituting this equation into equation (5.175) then:

$$\dot{\pmb{\sigma}} = (1-d)\,\mathbb{C}_0 : \dot{\pmb{\varepsilon}} - \frac{\partial G[f(\pmb{\sigma}_0)]}{\partial [f(\pmb{\sigma}_0)]} \left[\frac{\partial f(\mathbb{C}_0 : \pmb{\varepsilon})}{\partial \pmb{\varepsilon}} : \dot{\pmb{\varepsilon}} \right] \cdot [\mathbb{C}_0 : \pmb{\varepsilon}] \qquad (5.177)$$

From the previous equation the damage tangent tensor is obtained as,

$$\mathbb{C}^T = (1-d)\,\mathbb{C}_0 - \frac{\partial G[f(\pmb{\sigma}_0)]}{\partial [f(\pmb{\sigma}_0)]} [\mathbb{C}_0 : \pmb{\varepsilon}] \otimes \left[\frac{\partial f(\mathbb{C}_0 : \pmb{\varepsilon})}{\partial \pmb{\varepsilon}} \right] \qquad (5.178)$$

5.21 Particularization of the damage criterion

The type of softening to be defined in the general damage criterion depends on the problem to be solved. One general and two particular cases will be presented below; the two latter correspond to an exponential softening and a linear softening.

5.21.1 General softening

The scalar function $G[\chi]$ defining the evolution of the damage threshold must be monotonous and with a value ranging from 0 to 1. One way of expressing the damage threshold evolution is through an auxiliary variable κ, that will be called *dissipation variable* normalized to the unit and its expression is similar to the plasticity equation (5.118)

$$\dot{\kappa} = K(\pmb{\sigma}_0) \cdot \Xi_m = \left[\frac{r(\pmb{\sigma}_0)}{g_f} + \frac{1-r(\pmb{\sigma}_0)}{g_C} \right] \cdot \Xi_m \qquad (5.179)$$

where $\Xi_m = \Psi_0\,\dot{d}$ is the damage dissipation and $r(\pmb{\sigma}) = \sum_{I=1}^{3} \langle \sigma_I \rangle / \sum_{I=1}^{3} |\sigma_I|$ is a scalar function defining the behavior states of a point as a function of the stress state, where $\langle x \rangle = 0{,}5\,[x+|x|]$ is Macaulay's brackets. Magnitudes g_f and g_c are the maximum dissipation of a point subjected to tension and compression, respectively. Accordingly, the damage dissipation will always be normalized with respect to the maximum energy of the mechanical process at every moment.

Using variable κ as an auxiliary variable, now $G[\chi]$ can be defined as the following general form,

$$G[c(\kappa)] = 1 - \frac{c(\kappa)}{f(\pmb{\sigma}_0)} \qquad (5.180)$$

In this formulation the value of $f^0(\pmb{\sigma}_0) = c^{max}$ must be met and in this case the damage criterion is satisfied for the first degradation threshold. Moreover, and according to equations (5.169) and (5.171) the evolution of the damage threshold will be

$$d \equiv G[c(\kappa)] = 1 - \frac{c(\kappa)}{f(\pmb{\sigma}_0)} \quad .$$

In Table 5.2 the algorithm to obtain the stress through this general damage model, suitable for any $f(\boldsymbol{\sigma}_0)$ and $c(\kappa)$, is observed. This general model needs to be integrated numerically.

---

1. Calculation of the predictive stress and the internal variables for current time "$t + \Delta t$", equilibrium iteration "i", convergence counter of the constitutive model "$k = 1$"

$$[\boldsymbol{\sigma}_0]^{t+\Delta t} = \mathbb{C}_0 : [\boldsymbol{\varepsilon}]^{t+\Delta t}$$

$$\substack{i \\ k-1}[\mathbf{q}]^{t+\Delta t} = {}^{i}[\mathbf{q}]^{t+\Delta t} = {}^{i}\{\kappa, d\}^{t+\Delta t}$$

2. Damage threshold condition verification:

   a. If: $\substack{i \\ k-1}\big[\mathbb{F}(\boldsymbol{\sigma}_0;\mathbf{q})\big]^{t+\Delta t} = \substack{i \\ k-1}\big[G[f(\boldsymbol{\sigma}_0)] - G[c(\kappa)]\big]^{t+\Delta t} \leq 0$,

   Then $\left\{ \begin{array}{l} {}^{i}[\boldsymbol{\sigma}]^{t+\Delta t} = [\boldsymbol{\sigma}_0]^{t+\Delta t} \\ {}^{i}[\mathbf{q}]^{t+\Delta t} = \substack{i \\ k-1}[\mathbf{q}]^{t+\Delta t} \end{array} \right\}$ and goes to the **EXIT**

   b. If: $\substack{i \\ k-1}\big[\mathbb{F}(\boldsymbol{\sigma}_0;\mathbf{q})\big]^{t+\Delta t} = \substack{i \\ k-1}\big[G[f(\boldsymbol{\sigma}_0)] - G[c(\kappa)]\big]^{t+\Delta t} > 0$,

   Then the integration of the constitutive equation starts

3. Integration of the equation itself,

$$\substack{i \\ k}[d]^{t+\Delta t} = 1 - \frac{\substack{i \\ k-1}[c(\kappa)]^{t+\Delta t}}{\substack{i \\ k-1}[f(\boldsymbol{\sigma}_0)]^{t+\Delta t}}$$

$$\substack{i \\ k}[\delta d] = {}_{k}^{i}\left[\delta\mu \frac{\partial \overline{\mathbb{F}}(\boldsymbol{\sigma}_0;\mathbf{q})}{\partial[f(\boldsymbol{\sigma}_0)]}\right]^{t+\Delta t} \equiv {}_{k}^{i}[\delta\mu]^{t+\Delta t}\underbrace{\frac{\partial G[f(\boldsymbol{\sigma}_0)]}{\partial[f(\boldsymbol{\sigma}_0)]}}_{1} \Rightarrow \substack{i \\ k}[\delta d] = \substack{i \\ k}[d]^{t+\Delta t} - \substack{i \\ k-1}[d]^{t+\Delta t}$$

$$\substack{i \\ k}[\delta\Xi] = \underbrace{\left[\frac{1}{2}\boldsymbol{\varepsilon} : \mathbb{C}_0 : \boldsymbol{\varepsilon}\right]^{t+\Delta t}}_{\Psi_0^{t+\Delta t}} \cdot \substack{i \\ k}[\delta d] \Rightarrow \text{using } (5.115): \substack{i \\ k}[\delta\kappa]^{t+\Delta t} = \mathcal{K}(\boldsymbol{\sigma}_0)^{t+\Delta t} \cdot \substack{i \\ k}[\delta\Xi]$$

4. Update the stress, the tangent constitutive tensor, the normalized auxiliary internal variable of dissipation κ and the normalized dissipation Ξ,

$$\substack{i \\ k}[\Xi]^{t+\Delta t} = \substack{i \\ k-1}[\Xi]^{t+\Delta t} + \substack{i \\ k}[\Xi]^{t+\Delta t}$$

$$\substack{i \\ k}[\kappa]^{t+\Delta t} = \substack{i \\ k-1}[\kappa]^{t+\Delta t} + \substack{i \\ k}[\kappa]^{t+\Delta t}$$

$${}^{i}[\boldsymbol{\sigma}]^{t+\Delta t} = (1 - \substack{i \\ k}[d]^{t+\Delta t})[\boldsymbol{\sigma}_0]^{t+\Delta t}$$

$${}^{i}[\mathbb{C}^T]^{t+\Delta t} = {}^{i}\left[(1-d)\mathbb{C}_0 - \frac{\partial G[f(\boldsymbol{\sigma}_0)]}{\partial[f(\boldsymbol{\sigma}_0)]}[\mathbb{C}_0 : \boldsymbol{\varepsilon}] \otimes \left[\frac{\partial f(\mathbb{C}_0 : \boldsymbol{\varepsilon})}{\partial \boldsymbol{\varepsilon}}\right]\right]^{t+\Delta t}$$

5. Update $k = k + 1$ and return to point 2

Table 5.2 – Integration of the general damage constitutive equation

5.21.2 Exponential softening

The scalar function $G[\chi]$ defining the evolution of the damage threshold must be monotonous and with a value ranging from 0 to 1. In various publications about the scalar damage problem, the stress behavior with softening is represented in a variety of forms. Particularly, in Oliver's *et al.* (1990)[8] work the following function is proposed,

$$G\big[c(d)\big]=1-\frac{c^{\max}}{c(d)}\,\mathrm{e}^{A\left(1-\frac{c(d)}{c^{\max}}\right)} \quad \text{with} \quad 0\le c^{\max}\le c(d) \tag{5.181}$$

Or it can also be expressed as,

$$G\big[f(\boldsymbol{\sigma}_0)\big]=1-\frac{f^0(\boldsymbol{\sigma}_0)}{f(\boldsymbol{\sigma}_0)}\,\mathrm{e}^{A\left(1-\frac{f(\boldsymbol{\sigma}_0)}{f^0(\boldsymbol{\sigma}_0)}\right)} \quad \text{with} \quad f^0(\boldsymbol{\sigma}_0)=c^{\max} \tag{5.182}$$

where A is a parameter that depends on the fracture energy of the material[8]. The value of $f^0(\boldsymbol{\sigma}_0)=c^{max}$ is obtained from the fulfillment of the damage criterion for the first threshold of degradation; this complies with $G\big[f^0(\boldsymbol{\sigma}_0)\big]-G\big[c^{\max}\big]=0$ and $G\big[f^0(\boldsymbol{\sigma}_0)\big]=G\big[c^{\max}\big]\equiv 0$.

In Table 5.3 the algorithm of this already integrated model is presented. It is easier to use but less general because only exponential softening is used.

---

1. Calculation of the predictive stress and internal damage variable for the current time "$t+\Delta t$", equilibrium iteration "i",

$$\big[\boldsymbol{\sigma}_0\big]^{t+\Delta t}=\mathbb{C}_0:\big[\boldsymbol{\varepsilon}\big]^{t+\Delta t}$$

$${}^i\big[d\big]^{t+\Delta t} \;;\; \tau={}^i\big[G[f(\boldsymbol{\sigma}_0)]\big]^{t+\Delta t}$$

2. Verification of the damage threshold condition:

    a. If: $\tau-\tau^{\max}\le 0$

$$\text{Then}\;\begin{cases}{}^i\big[\boldsymbol{\sigma}\big]^{t+\Delta t}=\big[\boldsymbol{\sigma}_0\big]^{t+\Delta t}\\{}^i\big[d\big]^{t+\Delta t}\;;\;\tau^{\max}=\tau\end{cases}\;\text{and goes to \textbf{the EXIT}}$$

    b. If: $\tau-\tau^{\max}>0$ Then starts the integration of the constitutive equation.

3. Integration of the equation itself,

$$\tau^{\max}=\tau$$

$${}^i\big[d\big]^{t+\Delta t}=1-{}^i\left[\frac{f^0(\boldsymbol{\sigma}_0)}{\tau}\,\mathrm{e}^{A\left(1-\frac{\tau}{f^0(\boldsymbol{\sigma}_0)}\right)}\right]^{t+\Delta t}$$

4. Update the stress and the tangent constitutive tensor.

$${}^i\big[\boldsymbol{\sigma}\big]^{t+\Delta t}=(1-{}^i\big[d\big]^{t+\Delta t})\big[\boldsymbol{\sigma}_0\big]^{t+\Delta t}$$

$${}^i\big[\mathbb{C}^T\big]^{t+\Delta t}={}^i\left[(1-d)\mathbb{C}_0-\frac{\partial G[f(\boldsymbol{\sigma}_0)]}{\partial[f(\boldsymbol{\sigma}_0)]}[\mathbb{C}_0:\boldsymbol{\varepsilon}]\otimes\left[\frac{\partial f(\mathbb{C}_0:\boldsymbol{\varepsilon})}{\partial\boldsymbol{\varepsilon}}\right]\right]^{t+\Delta t}$$

Table 5.3 – Integration of the damage constitutive equation with exponential softening.

5.21.3 Linear softening

For linear softening a new definition of the scalar function $G[\chi]$ is given for the damage threshold. As in the previous section, this function must be monotone increasing and ranging from 0 to 1. This is,

$$G[c(d)] = \frac{1 - \dfrac{c^{max}}{c(d)}}{1 + A} \qquad \text{with} \quad 0 \le c^{max} \le c(d) \qquad (5.183)$$

or also as,

$$G[f(\boldsymbol{\sigma}_0)] = \frac{1 - \dfrac{f^0(\boldsymbol{\sigma}_0)}{f(\boldsymbol{\sigma}_0)}}{1 + A} \qquad \text{with} \quad f^0(\boldsymbol{\sigma}_0) = c^{max} \qquad (5.184)$$

where A is a parameter that depends on the fracture energy of the material. The initial value of $f^0(\boldsymbol{\sigma}_0)$ is obtained from the damage criterion expressed in equations (5.33) or (5.164) for the first threshold of degradation, so it is necessary for $G[f^0(\boldsymbol{\sigma}_0)] - G[c^{max}] = 0$ and also for $G[f^0(\boldsymbol{\sigma}_0)] = G[c^{max}] \equiv 0$.

5.22 Particularization of the stress threshold function

5.22.1 The Simo-Ju model

This model[6,7], formulated in 1987, is one of the most widespread models and the sources are recommended to deepen into its concepts. Here only the model defining the damage threshold function expressed in the stress space will be mentioned.

$$\tau = f(\boldsymbol{\sigma}_0) = \sqrt{2\,\Psi_0(\boldsymbol{\varepsilon})} = \sqrt{\boldsymbol{\varepsilon} : \mathbb{C}_0 : \boldsymbol{\varepsilon}} \qquad (5.185)$$

In this case the tangent constitutive tensor is obtained by equation (5.178), then:

$$\mathbb{C}^T = (1-d)\mathbb{C}_0 - \left[\frac{\partial G[\tau]}{\partial \tau} \frac{1}{\tau} [\mathbb{C}_0 : \boldsymbol{\varepsilon}] \right] \otimes [\mathbb{C}_0 : \boldsymbol{\varepsilon}] \qquad (5.186)$$

5.22.1.1 Setting of the A parameter for the Simo-Ju model

Parameter A is determined from the dissipation expression (5.160), particularized for an uniaxial process subjected to a monotone growing load. Taking into consideration the damage threshold function $\tau = f(\boldsymbol{\sigma}_0)$ proposed by Simo and Ju (1987), the first threshold is obtained:

$$\tau = f(\sigma_0) = \sqrt{2\,\Psi_0(\varepsilon)} = \sqrt{\varepsilon\, C_0\, \varepsilon} = \frac{\sigma_t^{max}}{\sqrt{C_0}} \quad \Rightarrow \quad \sigma_t^{max} = \tau\sqrt{C_0} \qquad (5.187)$$

where σ_t^{max} is the stress in the threshold of strength to tension. Replacing this into Helmholtz's free energy expression (equation (5.157)) then:

$$\Psi_0 = \frac{1}{2}\varepsilon\, C_0\,\varepsilon = \frac{1}{2}\sigma_t^{max}\,\varepsilon = \frac{1}{2}\frac{\left(\sigma_t^{max}\right)^2}{C_0} = \frac{\left(\tau\sqrt{C_0}\right)^2}{C_0} = \frac{1}{2}\tau^2 \qquad (5.188)$$

The total dissipation is obtained by integrating the dissipation expression and taking into account equations (5.171) and (5.172),

$$\int_{t=0}^{\infty}\Xi\,dt = \int_{t=0}^{\infty}\Psi_0\,\dot{d}\,dt = \int_{\tau^0}^{\infty}\frac{1}{2}\tau^2\,\frac{\partial G[\tau]}{\partial\tau}\,d\tau \qquad (5.189)$$

Taking into consideration the definition of $G[f(\boldsymbol{\sigma}_0)] = G[\tau]$ in the exponential form given in equation (5.181) and applying the concept of integrating by parts,

$$\begin{aligned}
\int_{\tau^0}^{\infty}\frac{1}{2}\tau^2\,dG[\tau] &= \left[\frac{1}{2}\tau^2\,G[\tau]\right]_{\tau^0}^{\infty} - \int_{\tau^0}^{\infty}\tau\,G[\tau]\,d\tau \\
&= \left[\frac{1}{2}\tau^2\,G[\tau]\right]_{\tau^0}^{\infty} - \int_{\tau^0}^{\infty}\left(\tau - \tau^0\ e^{A\left(1-\frac{\tau}{\tau^0}\right)}\right)d\tau \\
&= \frac{1}{2}\left(\tau^2 - \tau\tau^0\ e^{A\left(1-\frac{\tau}{\tau^0}\right)}\right)\Big|_{\tau^0}^{\infty} - \left[\frac{\tau^2}{2}\Big|_{\tau^0}^{\infty} + \frac{1}{A}\left(\tau^0\right)^2 e^{A\left(1-\frac{\tau}{\tau^0}\right)}\Big|_{\tau^0}^{\infty}\right] \\
&= \tau^2 - \left(\frac{\tau^2}{2} - \frac{\left(\tau^0\right)^2}{2} - \frac{\left(\tau^0\right)^2}{A}\right) - \left(\tau^0\right)^2\left(\frac{1}{2} + \frac{1}{A}\right)
\end{aligned} \qquad (5.190)$$

where $\tau^0 = f^0(\boldsymbol{\sigma}_0)$ is the initial value of the damage criterion (equation (5.33) or (5.164)) for the first threshold of degradation.

The maximum dissipated energy for each point will be g_f and from here,

$$\left(\tau^0\right)^2\left(\frac{1}{2} + \frac{1}{A}\right) = g_f \ \Rightarrow\ A = \frac{1}{\dfrac{g_f}{\left(\tau^0\right)^2} - \dfrac{1}{2}} \qquad (5.191)$$

When $G(\tau)$ is a linear function (see equation (5.183)), its top value is obtained considering the maximum value of the function (5.171), then

$$G[\tau\to\infty] = 1 = \frac{\left(1 - \dfrac{\tau^0}{\tau}\right)}{(1+A)} \ \Rightarrow\ \tau = -\frac{\tau^0}{A} \qquad (5.192)$$

The total dissipation in this case is then,

$$\int_{t=0}^{\infty} \Xi \, dt = \int_{\tau^0}^{-\frac{\tau^0}{A}} \frac{1}{2} \tau^2 \frac{\partial G[\tau]}{\partial \tau} d\tau$$

$$= \left[\frac{1}{2} \tau^2 G[\tau] \right]_{\tau^0}^{-\frac{\tau^0}{A}} - \int_{\tau^0}^{-\frac{\tau^0}{A}} \frac{\tau - \tau^0}{1 + A} d\tau$$

$$= \left[\frac{1}{2} \tau^2 G[\tau] \right]_{\tau^0}^{-\frac{\tau^0}{A}} - \frac{1}{1 + A} \left(\frac{\tau^2}{2} - \tau^0 \tau \right) \Bigg|_{\tau^0}^{-\frac{\tau^0}{A}}$$

$$= \frac{1}{2} \frac{\left(\tau^0 \right)^2}{A^2} - \frac{1}{2} (1 + A) \frac{\left(\tau^0 \right)^2}{A^2} = -\frac{1}{2} \frac{\left(\tau^0 \right)^2}{A}$$

(5.193)

Proceeding as before, it is assumed that the maximum energy to be dissipated by a point of the solid will be g_f, thus matching the maximum dissipation equation (5.193)). Then

$$-\frac{1}{2} \frac{\left(\tau^0 \right)^2}{A} = g_f \quad \Rightarrow \quad A = -\frac{1}{2} \frac{\left(\tau^0 \right)^2}{g_f}$$

(5.194)

Assuming a post-peak behavior (for $\tau \geq \tau^0$) exponential function (see equation (5.181)) or linear (see equation (5.183)), the tangent constitutive tensor for the model proposed by Simo and Ju (1987), equation (5.186), will then be

$$\begin{cases} \mathbb{C}^T = (1 - d)\mathbb{C}_0 - e^{A\left(1 - \frac{\tau}{\tau^0}\right)} \frac{\tau^0 + A\tau}{\tau^2} \left[\frac{1}{\tau} [\mathbb{C}_0 : \boldsymbol{\varepsilon}] \right] \otimes [\mathbb{C}_0 : \boldsymbol{\varepsilon}] , & \text{Exponential softening} \\[3mm] \mathbb{C}^T = (1 - d)\mathbb{C}_0 - \frac{\tau^0}{\tau^2 (1 + A)} \left[\frac{1}{\tau} [\mathbb{C}_0 : \boldsymbol{\varepsilon}] \right] \otimes [\mathbb{C}_0 : \boldsymbol{\varepsilon}] & , \text{Linear softening} \end{cases}$$

(5.195)

Both tangent constitutive tensor shapes are symmetrical. The symmetry of this tensor depends on the norm $\tau = f(\boldsymbol{\sigma}_0)$.

5.22.2 The Lemaitre and Mazars model

Lemaitre and Mazars's model is different from Simo-Ju's in the definition of the damage threshold function. This model uses the norm based on the deformation tensor given by:

$$\tau = f(\boldsymbol{\sigma}_0) = \sqrt{\boldsymbol{\varepsilon} : \boldsymbol{\varepsilon}}$$

(5.196)

This norm leads to a tangent non-symmetrical constitutive tensor, that for a post peak behavior case ($\tau \geq \tau^0$) there is an exponential function (equation (5.181)) or linear function (equation (5.183)), then

$$\begin{cases} \mathbb{C}^T = (1 - d)\mathbb{C}_0 - e^{A\left(1 - \frac{\tau}{\tau^0}\right)} \frac{\tau^0 + A\tau}{\tau^2} \left[\frac{1}{\tau} [\mathbb{C}_0 : \boldsymbol{\varepsilon}] \right] \otimes [\boldsymbol{\varepsilon}] , & \text{Exponential softening} \\[3mm] \mathbb{C}^T = (1 - d)\mathbb{C}_0 - \frac{\tau^0}{\tau^2 (1 + A)} \left[\frac{1}{\tau} [\mathbb{C}_0 : \boldsymbol{\varepsilon}] \right] \otimes [\boldsymbol{\varepsilon}] & , \text{Linear softening} \end{cases}$$

(5.197)

5.22.3 General model for different damage surfaces

This model's formulation was started during B. Luccioni's PhD thesis[36] (1993) and concluded during E. Car's PhD thesis[37] (2000) with the following characteristics: it admits, as a damage threshold function, any scalar function of tensorial arguments, homogenous and of first-order degree in stresses, such as the yield functions frequently used in plasticity. Examples of these are Rankine, von Mises, Mohr-Coulomb, Tresca and Drucker-Prager's functions, among others, presented in the classic plasticity section.

5.22.4 Setting of the A parameter

Similarly to Simó-Ju's model, presented in the previous section, the A parameter is calculated from the dissipation expression shown in equation (5.160), particularized for an uniaxial process subjected to a growing monotone load. Assuming any damage threshold function, $\tau = f(\boldsymbol{\sigma}_0)$, the first damage threshold is obtained:

$$\tau = f(\sigma_0) = \sigma_t^{max} \tag{5.198}$$

where σ_t^{max} is the stress in the traction strength threshold. Replacing this into Helmhotz's free energy equation (5.157)) then:

$$\Psi_0 = \frac{1}{2}\varepsilon\, C_0\, \varepsilon = \frac{1}{2}\sigma_t^{max}\, \varepsilon = \frac{1}{2}\frac{\left(\sigma_t^{max}\right)^2}{C_0} = \frac{1}{2}\frac{(\tau)^2}{C_0} \tag{5.199}$$

The total dissipation is obtained by integrating the dissipation expression in time, then considering equation (5.171) and (5.172),

$$\int_{t=0}^{\infty} \Xi\, dt = \int_{t=0}^{\infty} \Psi_0\, \dot{d}\, dt = \int_{\tau^0}^{\infty} \frac{1}{2}\frac{(\tau)^2}{C_0}\frac{\partial G[\tau]}{\partial \tau}d\tau \tag{5.200}$$

Additionally, by the exponential definition of $G[f(\boldsymbol{\sigma}_0)] = G[\tau]$ given by equation (5.181), and by integration by parts, we have:

$$
\begin{aligned}
\int_{\tau^0}^{\infty} \frac{1}{2}\frac{(\tau)^2}{C_0}dG[\tau] &= \left[\frac{1}{2}\frac{(\tau)^2}{C_0}G[\tau]\right]_{\tau^0}^{\infty} - \int_{\tau^0}^{\infty}\frac{(\tau)^2}{C_0}G[\tau]\,d\tau \\
&= \left[\frac{1}{2}\frac{(\tau)^2}{C_0}G[\tau]\right]_{\tau^0}^{\infty} - \int_{\tau^0}^{\infty}\frac{1}{C_0}\left(\tau - \tau^0\ e^{A\left(1-\frac{\tau}{\tau^0}\right)}\right)d\tau \\
&= \frac{1}{2C_0}\left(\tau^2 - \tau\tau^0\ e^{A\left(1-\frac{\tau}{\tau^0}\right)}\right)\Big|_{\tau^0}^{\infty} - \frac{1}{C_0}\left[\frac{\tau^2}{2}\Big|_{\tau^0}^{\infty} + \frac{1}{A}\left(\tau^0\right)^2 e^{A\left(1-\frac{\tau}{\tau^0}\right)}\Big|_{\tau^0}^{\infty}\right] \\
&= \frac{1}{C_0}\tau^2 - \frac{1}{C_0}\left(\frac{\tau^2}{2} - \frac{\left(\tau^0\right)^2}{2} - \frac{\left(\tau^0\right)^2}{A}\right) = \frac{\left(\tau^0\right)^2}{C_0}\left(\frac{1}{2} + \frac{1}{A}\right)
\end{aligned} \tag{5.201}
$$

[36] Luccioni, B. (1993). *Formulación de un modelo constitutivo para materiales ortótropos* - Tesis Doctoral, Universidad Nacional de Tucumán. Argentina.(In Spanish).

[37] Car, E. (2000). *Modelo Constitutivo para el Estudio del Comportamiento Mecánico de los Materiales Compuestos*. Tesis Doctoral, Universidad Politécnica de Cataluña. Barcelona. España. (In Spanish).

where $\tau^0 = f^0(\boldsymbol{\sigma}_0)$ is the initial value of the damage criterion (equations (5.33) or (5.164)) for the first threshold of degradation.

Like other models, the maximum dissipated energy for each point will be g_f; from here the A parameter is obtained to guarantee a controlled dissipation,

$$\frac{\left(\tau^0\right)^2}{C_0}\left(\frac{1}{2}+\frac{1}{A}\right)=g_f \quad\Rightarrow\quad A=\frac{1}{\dfrac{C_0\,g_f}{\left(\tau^0\right)^2}-\dfrac{1}{2}} \tag{5.202}$$

When $G(\tau)$ is a linear function (see equation (5.183)), the maximum value of it is obtained by considering the maximum value of function (5.171), similarly to equation (5.192),

$$G[\tau\to\infty]=1=\frac{\left(1-\dfrac{\tau^0}{\tau}\right)}{(1+A)} \quad\Rightarrow\quad \tau=-\frac{\tau^0}{A} \tag{5.203}$$

The total dissipation in this case is,

$$\begin{aligned}
\int_{t=0}^{\infty}\Xi\,dt &= \int_{\tau^0}^{-\frac{\tau^0}{A}}\frac{1}{2}\frac{\left(\tau\right)^2}{C_0}\frac{\partial G[\tau]}{\partial\tau}d\tau \\
&= \left[\frac{1}{2}\frac{\left(\tau\right)^2}{C_0}G[\tau]\right]_{\tau^0}^{-\frac{\tau^0}{A}} - \int_{\tau^0}^{-\frac{\tau^0}{A}}\frac{1}{C_0}\frac{\tau-\tau^0}{1+A}\,d\tau \\
&= \left[\frac{1}{2}\frac{\left(\tau\right)^2}{C_0}G[\tau]\right]_{\tau^0}^{-\frac{\tau^0}{A}} - \frac{1}{C_0(1+A)}\left(\frac{\tau^2}{2}-\tau^0\,\tau\right)\Bigg|_{\tau^0}^{-\frac{\tau^0}{A}} \\
&= \frac{1}{2}\frac{\left(\tau^0\right)^2}{A^2 C_0} - \frac{1}{2}(1+A)\frac{\left(\tau^0\right)^2}{A^2 C_0} = -\frac{1}{2}\frac{\left(\tau^0\right)^2}{A C_0}
\end{aligned} \tag{5.204}$$

As proceeded before, it is assumed that the maximum energy to be dissipated by a single point in a solid will be g_f. Then, matching with the maximum dissipation (equation (5.193)), we have

$$-\frac{1}{2}\frac{\left(\tau^0\right)^2}{A C_0}=g_f \quad\Rightarrow\quad A=-\frac{1}{2}\frac{\left(\tau^0\right)^2}{g_f\,C_0} \tag{5.205}$$

In this case the tangent constitutive tensor is obtained by considering that,

$$\frac{\partial\tau}{\partial\boldsymbol{\varepsilon}}=\frac{\partial\tau}{\partial\boldsymbol{\sigma}^0}:\frac{\partial\boldsymbol{\sigma}^0}{\partial\boldsymbol{\varepsilon}}=\mathbb{C}_0:\frac{\partial\boldsymbol{\sigma}^0}{\partial\boldsymbol{\varepsilon}} \tag{5.206}$$

From this expression, the tangent constitutive tensor is obtained either considering a post-peak ($\tau\geq\tau^0$) exponential equation (5.181)) or linear equation (5.183)). This is,

$$\begin{cases} \mathbb{C}^T = (1-d)\mathbb{C}_0 - \mathrm{e}^{A\left(1-\frac{\tau}{\tau^0}\right)} \dfrac{\tau^0 + A\tau}{\tau^2} \left[\mathbb{C}_0 : \boldsymbol{\varepsilon}\right] \otimes \left[\mathbb{C}_0 : \dfrac{\partial \boldsymbol{\sigma}^0}{\partial \boldsymbol{\varepsilon}}\right] & , \ \text{Exponencial softening} \\[3mm] \mathbb{C}^T = (1-d)\mathbb{C}_0 - \dfrac{\tau^0}{\tau^2(1+A)} \left[\mathbb{C}_0 : \boldsymbol{\varepsilon}\right] \otimes \left[\mathbb{C}_0 : \dfrac{\partial \boldsymbol{\sigma}^0}{\partial \boldsymbol{\varepsilon}}\right] & , \ \text{Linear softening} \end{cases} \qquad (5.207)$$

Neither form of the tangent constitutive tensor is symmetrical.

6 Time-dependent Models

6.1 Introduction

As shown in Chapters 3 and 4, the nonlinearity in dynamics is caused by changes in the direction of external forces due to large movements and to the nonlinearity of internal forces, caused by non-time-dependent phenomena studied in Chapter 5 and time-dependent phenomena that will be presented in this chapter.

One of the behaviors responsible for the nonlinearity in the materials' response over the time field is due to viscoelasticity. Viscoelasticity studies the rheological behavior of materials, in other words, behaviors affected by the course of time. There is extensive literature available on this subject, especially in books that deal with the influence of time on these materials[1,2].

The first part of this chapter will focus on the states described by a single stress and deformation component. This simplified form will allow us to introduce the concept and later extend the formulation to multiaxial behavior. A simile is used in these simplified models, such that the force is the stress and the displacement is the strain, thus the phenomenological concept describing the behavior equation can be explained in a simple form.

6.2 Constitutive equations based on spring-damping analogies

There are two types of time-dependent elasticity models:

1.   One of them is known as *delayed elasticity or creep model* in which *the stress is the free variable of the problem*. The Kelvin viscoelastic model is a good example of this type of model (see Figure 6.1).
2.   The other one, in which the *free variable is the strain*, is called *relaxation model*. The Maxwell viscoelastic model is a good example of this type of viscoelastic model (see Figure 6.2).

These models have non-invertible constitutive laws; however, each one represents the implicit inverse form of the other, in other words, a delayed elasticity model is the inverse form of a relaxation model representation.

---

[1] Creus G. (1986). Viscoelasticity – Basic theory and applications to concrete structures. Ed. By C. Brebbia and S. Orszag. Springer-Verlag. Berlin.
[2] Christensen R. M. (1982). Theory of viscoelasticity. An introduction. Academic Press, Inc. N. York.

6.2.1   The simplified Kelvin model

In this delayed elasticity model, or creep model, the stress is assumed as the free variable. Therefore, a model in parallel with strain compatibility is used to write the governing equation of the problem. The stress is then obtained in the following additive form

$$\sigma(t) = \sigma^e(t) + \sigma^{vis}(t) = \mathbb{C}\,\varepsilon(t) + \xi\,\dot{\varepsilon}(t) \tag{6.1}$$

where σ^e and σ^{vis} are the elastic stress in the spring and the viscous stress in the damping respectively; ε and $\dot{\varepsilon}$ are the strain in the spring and the strain velocity in the damping; $\mathbb{C}$ is the spring elastic constant and ξ is the damping viscous constant. The following relationship is satisfied in the deformation at each moment (see Figure 6.1),

$$\varepsilon(t) = \varepsilon^e(t) = \varepsilon^{vis}(t) \tag{6.2}$$

where ε^e is the elastic strain and ε^{vis} is the viscous strain. A stress applied during a period of time from τ_0 leads to the next field of transitory strain obtained from equation (6.1),

$$\varepsilon(t) = \frac{\sigma(t) - \xi\,\dot{\varepsilon}(t)}{\mathbb{C}} = \frac{\sigma(t)}{\mathbb{C}} - \frac{\xi}{\mathbb{C}}\dot{\varepsilon}(t) = \varepsilon_0 - r\,\dot{\varepsilon}(t) \tag{6.3}$$

where ε_0 is the model strain for infinite time or steady state and r is the so-called delayed time which, as its name suggests, is the time-delay in the model response due to the influence of viscosity. This delay is measured by the instantaneous response that an ideal material would undergo without viscosity (see Figure 6.1).

By solving this first-order differential equation (6.3) a convolution integral is obtained describing the following strain,

$$\begin{cases} \varepsilon(t) = 0 & \forall \ \tau < \tau_0 \\[2mm] \varepsilon(t) = \displaystyle\int_{-\infty}^{t} \frac{1}{\xi} e^{-(t-s)/r} \cdot \sigma(s)\, ds & \forall \ \tau \geq \tau_0 \end{cases} \tag{6.4}$$

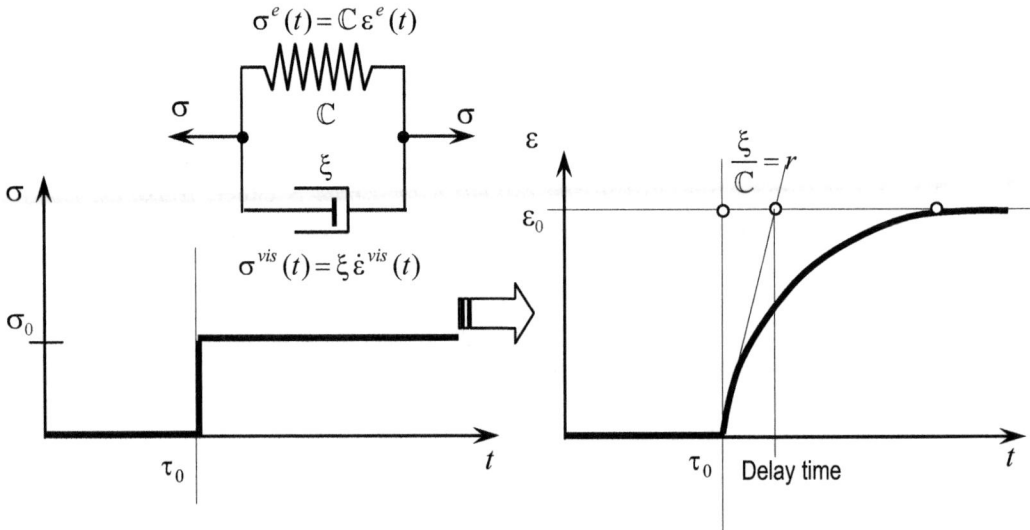

Figure 6.1 – Simplified form of the viscous behavior of the Kelvin model.

6.2.2 The simplified Maxwell model

The Maxwell formulation is also known as a relaxation model where the strain is the free variable. Here the spring and the damping are arranged in series, so the total strain is the result of both an elastic part ε^e and a viscous one ε^{vis},

$$\varepsilon(t) = \varepsilon^e(t) + \varepsilon^{vis}(t) = \frac{\sigma}{\mathbb{C}} + \int_{t_0}^{t} \frac{\sigma(s)}{\xi} ds \tag{6.5}$$

Meanwhile, to guarantee the equilibrium condition in the point, the stress is only one and therefore it follows the following relationship,

$$\sigma(t) = \sigma^e(t) = \sigma^{vis}(t) \tag{6.6}$$

The solution of equation (6.5) leads to the following stress expression

$$\begin{cases} \sigma(t) = 0 & \forall \ t < \tau_0 \\ \sigma(t) = \int_{-\infty}^{t} \mathbb{C} \ e^{-(t-s)/\tau} \ \varepsilon(s) \, ds & \forall \ t \geq \tau_0 \end{cases} \tag{6.7}$$

Like equation (6.4), this is a convolution equation over time and its solution involves a high computational cost.

As observed, the model's response gets relaxed over time under steady strain (see Figure 6.2). Thus, there is a time called relaxation time r, and its expression is the same as the delayed time introduced by the Kelvin model. Although it is not evident in its formulation, as mentioned in the introduction, the Maxwell model sets a mechanical form inverse to the Kelvin model that could be solved numerically.

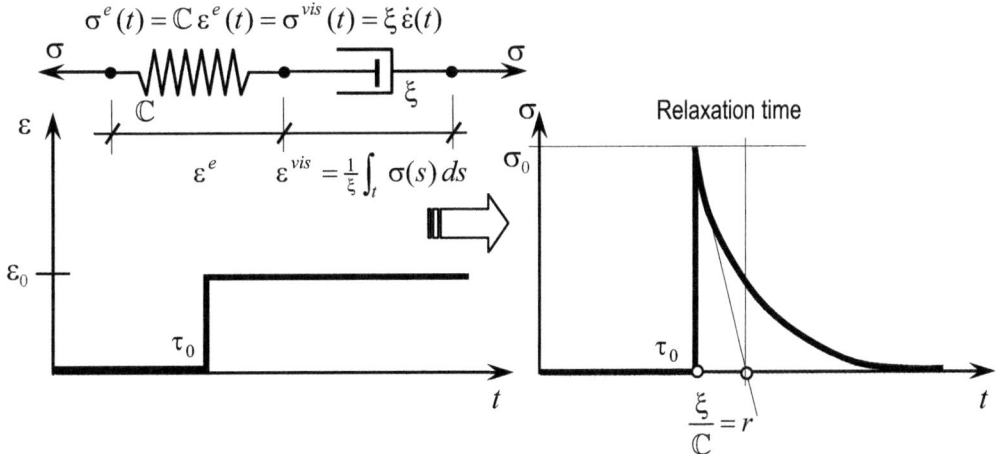

Figure 6.2 – Simplified form of the viscous behavior of the Maxwell model.

6.2.3 The generalized Kelvin model

This model sums up the Kelvin and the Maxwell models previously presented. Its main characteristic is its capacity to tend to the Maxwell model when $\mathbb{C}_1 \to 0$ and to the Kelvin model when $\mathbb{C}_0 \to \infty$ (see Figure 6.3). The characteristic of the "generalized Kelvin model" makes its formulation useful and versatile to represent different types of viscous behaviors of solids.

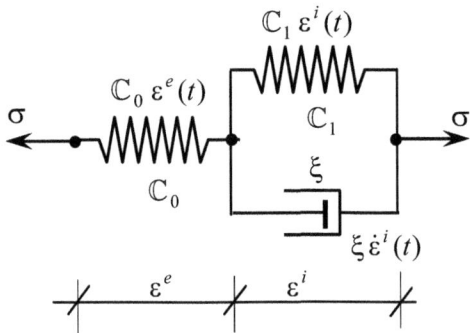

Figure 6.3 – The generalized Kelvin model.

The stress state at any moment is expressed as

$$\begin{cases} \sigma(t) = \mathbb{C}_0\,\varepsilon^e(t) & , \text{ initial response, with } t \to 0 \\ \sigma(t) = \mathbb{C}_1\,\varepsilon^i(t) + \xi\,\dot{\varepsilon}^i(t) \end{cases} \tag{6.8}$$

And the equilibrium condition requires both stresses to be the same at any instant of time,

$$\sigma(t) = \mathbb{C}_0\,\varepsilon^e(t) = \mathbb{C}_1\,\varepsilon^i(t) + \xi\,\dot{\varepsilon}^i(t) \;\Rightarrow\; \dot{\varepsilon}^i(t) = \frac{\sigma(t) - \mathbb{C}_1\,\varepsilon^i(t)}{\xi} = \frac{\mathbb{C}_0\,\varepsilon^e(t) - \mathbb{C}_1\,\varepsilon^i(t)}{\xi} \tag{6.9}$$

where the inelastic strain ε^i can be assumed as an internal variable of the constitutive model. This strain is obtained from the solution of the differential equation (6.9), under a stress applied $\sigma(t)$ from a time $t \geq \tau_0$. This is,

$$\begin{cases} \varepsilon^i(t) = 0 & \forall\ \tau < \tau_0 \\ \varepsilon^i(t) = \int_{-\infty}^{t} \frac{1}{\xi} e^{-(t-s)/r_1} \cdot \sigma(s)\,ds & \forall\ \tau \geq \tau_0 \end{cases} \tag{6.10}$$

Compared to equation (6.4), it can be observed that the classic Kelvin model represents now, conventionally, the inelastic part of the strain of this more general model. As in that expression, $r_1 = \xi/\mathbb{C}_1$ is the delayed time.

The total strain is then,

$$\begin{cases} \varepsilon(t) = 0 & \forall\ \tau < \tau_0 \\ \varepsilon(t) = \varepsilon^e(t) + \varepsilon^i(t) = \dfrac{\sigma(t)}{\mathbb{C}_0} + \displaystyle\int_{-\infty}^{t} \frac{1}{\xi} e^{-(t-s)/r_1} \cdot \sigma(s)\,ds & \forall\ \tau \geq \tau_0 \end{cases} \tag{6.11}$$

Defining now the uniaxial creep function $J(t)$ as,

$$J(t) = \frac{1}{C_0} + \frac{1}{C_1}\left[1 - e^{-t/r_1}\right] \tag{6.12}$$

And carrying out the integration by parts (6.11), assuming also that $\sigma(-\infty) = 0$, 1 the total strain can be expressed as,

$$\begin{cases} \varepsilon(t) = 0 & \forall \ \tau < \tau_0 \\ \varepsilon(t) = \displaystyle\int_{-\infty}^{t} J(t-s) \cdot \frac{d\sigma(s)}{ds} ds & \forall \ \tau \geq \tau_0 \end{cases} \tag{6.13}$$

In case the applied stress $\sigma(t)$ complies with the following simplifications (see Figure 6.4),

$$\begin{cases} \sigma(t) = 0 & \forall \ t < \tau_0 = 0 \\ \sigma(t) = \sigma_0 & \forall \ t \geq \tau_0 = 0 \end{cases} \tag{6.14}$$

The expression of the inelastic strain (6.10) and the total is reduced to the following simple form,

$$\begin{cases} \varepsilon(t) = 0 & \forall \ \tau < \tau_0 \\ \varepsilon(t) = \varepsilon^e(t) + \varepsilon^i(t) = \dfrac{\sigma_0}{C_0} + \underbrace{\dfrac{1}{C_1}(1 - e^{-t/r_1}) \cdot \sigma_0}_{\varepsilon^i(t)} & \forall \ \tau \geq \tau_0 \end{cases} \tag{6.15}$$

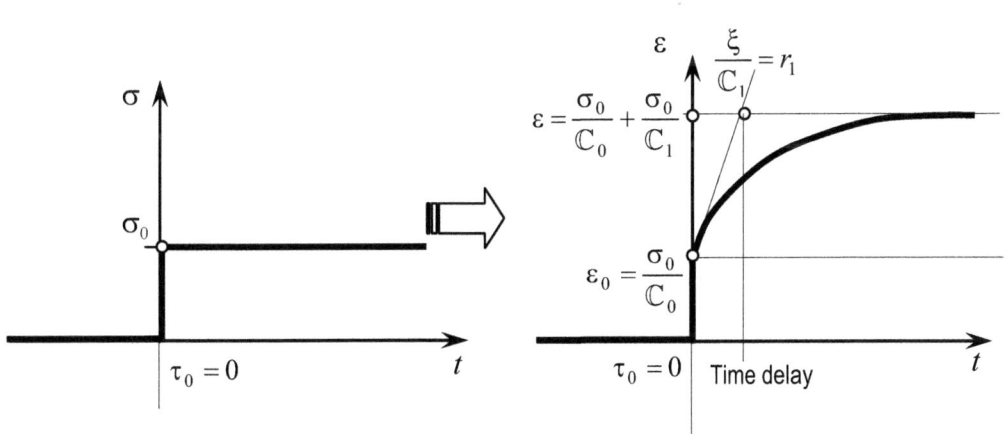

Figure 6.4 – Response of the generalized Kelvin model under a constant stress.

6.2.4 The generalized multiple Kelvin model

This model is obtained by combining the Kelvin generalized models (see Figure 6.5) and it helps to keep better control of the materials' viscous behavior. In this model, the total inelastic strain ε^i results from the addition of the inelastic strains ε^i_α in each sub-model "α" of the Kelvin "n" models,

$$\begin{cases} \varepsilon^i(t) = 0 & \forall \quad \tau < \tau_0 \\ \varepsilon^i(t) = \sum_{\alpha=1}^{n-Kelvin} \varepsilon_\alpha^i(t) & \forall \quad \tau \geq \tau_0 \end{cases} \tag{6.16}$$

By analogy with the previously obtained expressions, particularly with expression (6.9), there is the time variation of the inelastic strain for any sub-model α. This is,

$$\sigma(t) = \mathbb{C}_0 \, \varepsilon^e(t) = \mathbb{C}_\alpha \, \varepsilon_\alpha^i(t) + \xi_\alpha \, \dot{\varepsilon}_\alpha^i(t) \quad \Rightarrow \quad \dot{\varepsilon}_\alpha^i(t) = \frac{\sigma(t) - \mathbb{C}_\alpha \, \varepsilon_\alpha^i(t)}{\xi_\alpha} \tag{6.17}$$

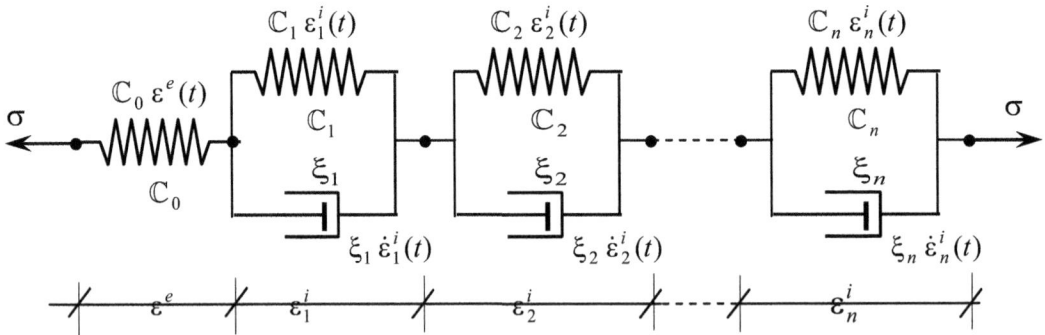

Figure 6.5 – Simplified representation of the generalized multiple Kelvin model.

By solving this differential equation, the expression for an inelastic strain of any sub-model α is obtained,

$$\begin{cases} \varepsilon_\alpha^i(t) = 0 & \forall \quad \tau < \tau_0 \\ \varepsilon_\alpha^i(t) = \int_{-\infty}^{t} \frac{1}{\xi_\alpha} e^{-(t-s)/r_\alpha} \cdot \sigma(s)\,ds & \forall \quad \tau \geq \tau_0 \end{cases} \tag{6.18}$$

Thus, the total strain is expressed as follows,

$$\begin{cases} \varepsilon(t) = 0 & \forall \quad \tau < \tau_0 \\ \varepsilon(t) = \varepsilon^e(t) + \sum_{\alpha=1}^{n-Kelvin} \varepsilon_\alpha^i(t) = \frac{\sigma(t)}{\mathbb{C}_0} + \int_{-\infty}^{t} \left[\sum_{\alpha=1}^{n-Kelvin} \frac{1}{\xi_\alpha} e^{-(t-s)/r_\alpha} \right] \cdot \sigma(s)\,ds & \forall \quad \tau \geq \tau_0 \end{cases} \tag{6.19}$$

Defining now the *creep uniaxial function* $J(t)$ for this general problem as,

$$J(t) = \frac{1}{\mathbb{C}_0} + \sum_{\alpha=1}^{n-Kelvin} \frac{1}{\mathbb{C}_\alpha} \left[1 - e^{-t/r_\alpha} \right] \tag{6.20}$$

and integrating equation (6.19) by parts, assuming that $\sigma(-\infty) = 0$, the total strain can be expressed as,

$$\begin{cases} \varepsilon(t) = 0 & \forall \quad \tau < \tau_0 \\ \varepsilon(t) = \int_{-\infty}^{t} J(t-s) \cdot \frac{d\sigma(s)}{ds}\,ds & \forall \quad \tau \geq \tau_0 \end{cases} \tag{6.21}$$

As observed, this equation coincides with the general form of equation (6.13), but the creep function $J(t)$ here is different from the aforementioned equation. When the applied stress $\sigma(t)$ meets the following simplification,

$$\begin{cases} \sigma(t) = 0 & \forall \quad t < \tau_0 = 0 \\ \sigma(t) = \sigma_0 & \forall \quad t \geq \tau_0 = 0 \end{cases} \tag{6.22}$$

The following simplified form is obtained for the total strain,

$$\begin{cases} \varepsilon(t) = 0 & \forall \quad \tau < \tau_0 \\ \varepsilon(t) = J(t)\sigma_0 & \forall \quad \tau \geq \tau_0 \end{cases} \tag{6.23}$$

6.2.5 The generalized Maxwell model

Like the generalized Kelvin model previously described, this model is an alternative general form to summarize the Kelvin and the Maxwell simplified models in a single formulation. Thus, this model tends to the basic Kelvin model when $\mathbb{C}_1 \to \infty$ and transforms itself into the basic Maxwell model when $\mathbb{C}_\infty \to 0$ (see Figure 6.3). Like the generalized Kelvin model (6.23), this formulation is useful and suitable for the representation of different types of viscous behaviors in solids. Its inverse formulation fits within the standard Kelvin models, although not quite as the model presented in section 6.2.3.

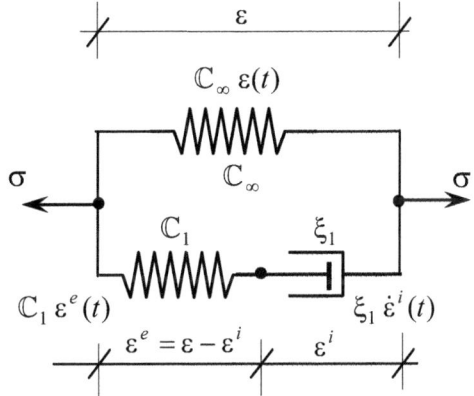

Figure 6.6 – The generalized Maxwell model.

The stress state at any time is expressed as

$$\begin{cases} \sigma^\infty(t) = \mathbb{C}_\infty \, \varepsilon(t) \\ \sigma^i(t) = \mathbb{C}_1 \left(\varepsilon(t) - \varepsilon^i(t) \right) = \xi_1 \, \dot{\varepsilon}^i(t) \end{cases} \tag{6.24}$$

The equilibrium condition must meet the following relation,

$$\sigma(t) = \sigma^i(t) + \sigma^\infty(t) = \mathbb{C}_1 \left(\varepsilon(t) - \varepsilon^i(t) \right) + \mathbb{C}_\infty \varepsilon(t) = \xi_1 \, \dot{\varepsilon}^i(t) + \mathbb{C}_\infty \varepsilon(t) \tag{6.25}$$

Stating that $\mathbb{C}_0 = \mathbb{C}_\infty + \mathbb{C}_1$ and operating algebraically in the last equation, the stress equation is obtained as,

$$\sigma(t) = \mathbb{C}_0 \varepsilon(t) - \mathbb{C}_1 \, \varepsilon^i(t) \tag{6.26}$$

Using the second equation in (6.24), the following differential equation is obtained in the inelastic strain,

$$\mathbb{C}_1\, \varepsilon(t) = \mathbb{C}_1\, \varepsilon^i(t) + \xi_1\, \dot{\varepsilon}^i(t) \quad \Rightarrow \quad \frac{\varepsilon(t)}{r_1} = \frac{\varepsilon^i(t)}{r_1} + \dot{\varepsilon}^i(t) \tag{6.27}$$

Applying a strain $\varepsilon(t)$, from $t \geq \tau_0$, this differential equation can be solved for $\varepsilon^i(t)$ as,

$$\begin{cases} \varepsilon^i(t) = 0 & \forall\ \tau < \tau_0 \\[2mm] \varepsilon^i(t) = \displaystyle\int_{-\infty}^{t} \frac{1}{r_1} e^{-(t-s)/r_1} \cdot \varepsilon(s)\, ds & \forall\ \tau \geq \tau_0 \end{cases} \tag{6.28}$$

Substituting this equation into (6.26), the following stress expression is obtained,

$$\begin{cases} \sigma(t) = 0 & \forall\ \tau < \tau_0 \\[2mm] \sigma(t) = \mathbb{C}_0\varepsilon(t) - \dfrac{\mathbb{C}_1}{r_1}\displaystyle\int_{-\infty}^{t} e^{-(t-s)/r_1} \cdot \varepsilon(s)\, ds & \forall\ \tau \geq \tau_0 \end{cases} \tag{6.29}$$

Defining now the *uniaxial relaxation function* $G(t)$, as the inverse of the *uniaxial creep function*,

$$G(t) = \left[J(t)\right]^{-1} = \mathbb{C}_\infty + \mathbb{C}_1\, e^{-t/r_1} \tag{6.30}$$

Assuming the relaxation function inversion for this particular model, the following uniaxial creep function is obtained,

$$J(t) = \frac{1}{\mathbb{C}_\infty}\left[1 - \frac{\mathbb{C}_1}{\mathbb{C}_0} e^{-\left(\frac{\mathbb{C}_\infty}{r_1\mathbb{C}_0}\right)t} \right] \tag{6.31}$$

And integrating equation (6.29) by parts, the stress can be written as the following compact form,

$$\begin{cases} \sigma(t) = 0 & \forall\ \tau < \tau_0 \\[2mm] \sigma(t) = \displaystyle\int_{-\infty}^{t} G(t-s) \cdot \dfrac{d\varepsilon(s)}{ds}\, ds & \forall\ \tau \geq \tau_0 \end{cases} \tag{6.32}$$

The stress expressed as shown above is the strain inverse function obtained in equation[3] (6.13), provided that $G(t)$ is twice differentiable and that $G(t = 0) \neq 0$. Then,

$$\sigma(t) = \int_{-\infty}^{t} G(t-s) \cdot \frac{d\varepsilon(s)}{ds}\, ds \quad \Rightarrow \quad \varepsilon(t) = \left[\sigma(t)\right]^{-1} = \int_{-\infty}^{t} J(t-s) \cdot \frac{d\sigma(s)}{ds}\, ds \tag{6.33}$$

In the particular case that the imposed strain $\varepsilon(t)$ meets the following simplification (see Figure 6.4),

---

[3] Fung Y. C. (1965). Foundations of solid mechanics. Prentice-Hall international series in dynamics.

$$\begin{cases} \varepsilon(t) = 0 & \forall \quad t < \tau_0 = 0 \\ \varepsilon(t) = \varepsilon_0 & \forall \quad t \geq \tau_0 = 0 \end{cases} \tag{6.34}$$

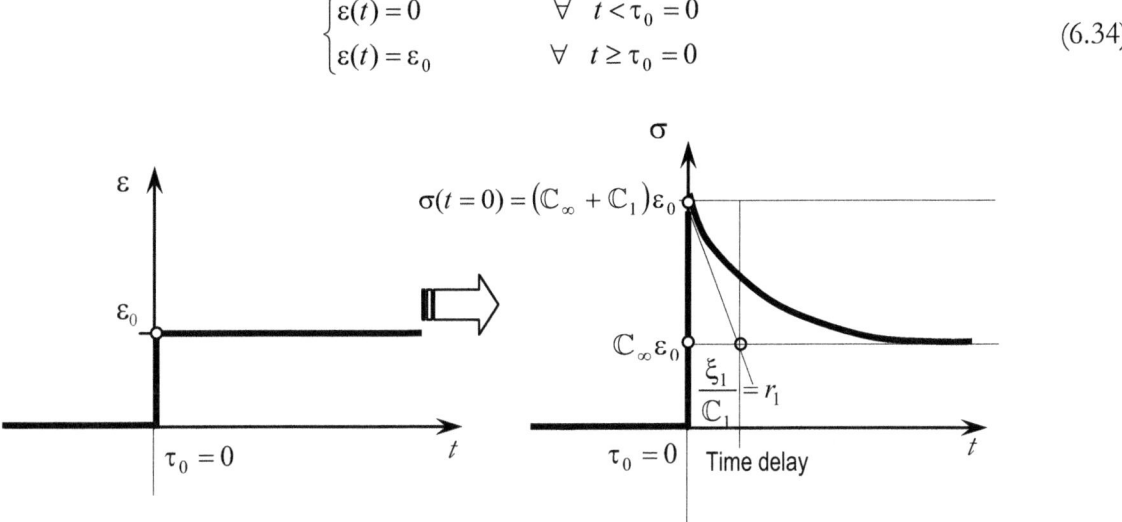

Figure 6.7 – Response of the generalized Maxwell model under constant deformation.

The stress expression (6.11) is reduced to the following simple form,

$$\begin{cases} \sigma(t) = 0 & \forall \quad \tau < \tau_0 \\ \sigma(t) = \left(\mathbb{C}_\infty + \mathbb{C}_1 e^{-t/r_1} \right) \varepsilon_0 & \forall \quad \tau \geq \tau_0 \end{cases} \tag{6.35}$$

6.2.6 The generalized multiple Maxwell model

This model is obtained by combining several generalized Maxwell models (see section 6.2.5). Like the aforementioned generalized Kelvin model, this general form to formulate the Maxwell model helps to keep better control of the material's viscous behavior. The total stress is obtained by each contribution of the Maxwell chains in addition to the stress on the spring, which gives the response in infinite time or by following the steps of the previous section for a single chain (equation (6.26)). This is,

$$\begin{cases} \sigma(t) = 0 & \forall \quad \tau < \tau_0 \\ \sigma(t) = \mathbb{C}_\infty \varepsilon(t) + \displaystyle\sum_{\alpha=1}^{n-Maxw} \mathbb{C}_\alpha \left[\varepsilon(t) - \varepsilon_\alpha^i(t) \right] = \\ \qquad = \mathbb{C}_0 \varepsilon(t) - \displaystyle\sum_{\alpha=1}^{n-Maxw} \mathbb{C}_\alpha \varepsilon_\alpha^i(t) & \forall \quad \tau \geq \tau_0 \end{cases} \tag{6.36}$$

where $\mathbb{C}_0 = \mathbb{C}_\infty + \sum_{\alpha=1}^{n-Maxw} \mathbb{C}_\alpha$. On the other hand, the strain ε is only one for all the Maxwell chains and the inelastic strain in the chain α^{th} is ε_α^i, and their expression results from solving the differential equation (see equation (6.27)) deriving from the equilibrium in each Maxwell chain,

$$\mathbb{C}_\alpha \, \varepsilon(t) = \mathbb{C}_\alpha \, \varepsilon^i(t) + \xi_\alpha \, \dot{\varepsilon}_\alpha^i(t) \quad \Rightarrow \quad \frac{\varepsilon(t)}{r_\alpha} = \frac{\varepsilon_\alpha^i(t)}{r_\alpha} + \dot{\varepsilon}_\alpha^i(t) \tag{6.37}$$

where $r_\alpha = \xi_\alpha / \mathbb{C}_\alpha$ is the relaxation time in the chain α^{th}. This last differential equation solution leads to the following inelastic strain expression,

$$\begin{cases} \varepsilon_\alpha^i(t) = 0 & \forall \quad \tau < \tau_0 \\ \varepsilon_\alpha^i(t) = \int_{-\infty}^{t} \frac{1}{r_\alpha} e^{-(t-s)/r_\alpha} \cdot \varepsilon(s)\,ds & \forall \quad \tau \geq \tau_0 \end{cases}$$ (6.38)

By substituting this equation into equation (6.36), the following stress expression is obtained,

$$\begin{cases} \sigma(t) = 0 & \forall \quad \tau < \tau_0 \\ \sigma(t) = \mathbb{C}_0 \varepsilon(t) - \sum_{\alpha=1}^{n-Maxw} \mathbb{C}_\alpha \varepsilon_\alpha^i(t) = \\ \quad = \mathbb{C}_0 \varepsilon(t) - \sum_{\alpha=1}^{n-Maxw} \frac{\mathbb{C}_\alpha}{r_1} \int_{-\infty}^{t} e^{-(t-s)/r_\alpha} \cdot \varepsilon(s)\,ds & \forall \quad \tau \geq \tau_0 \end{cases}$$ (6.39)

Getting the following *relaxation uniaxial function* $G(t)$,

$$G(t) = [J(t)]^{-1} = \mathbb{C}_\infty + \sum_{\alpha=1}^{n-Maxw} \mathbb{C}_\alpha e^{-t/r_\alpha}$$ (6.40)

From here the stress can be written in the following compact form,

$$\sigma(t) = \int_{-\infty}^{t} G(t-s) \cdot \frac{d\varepsilon(s)}{ds}\,ds$$ (6.41)

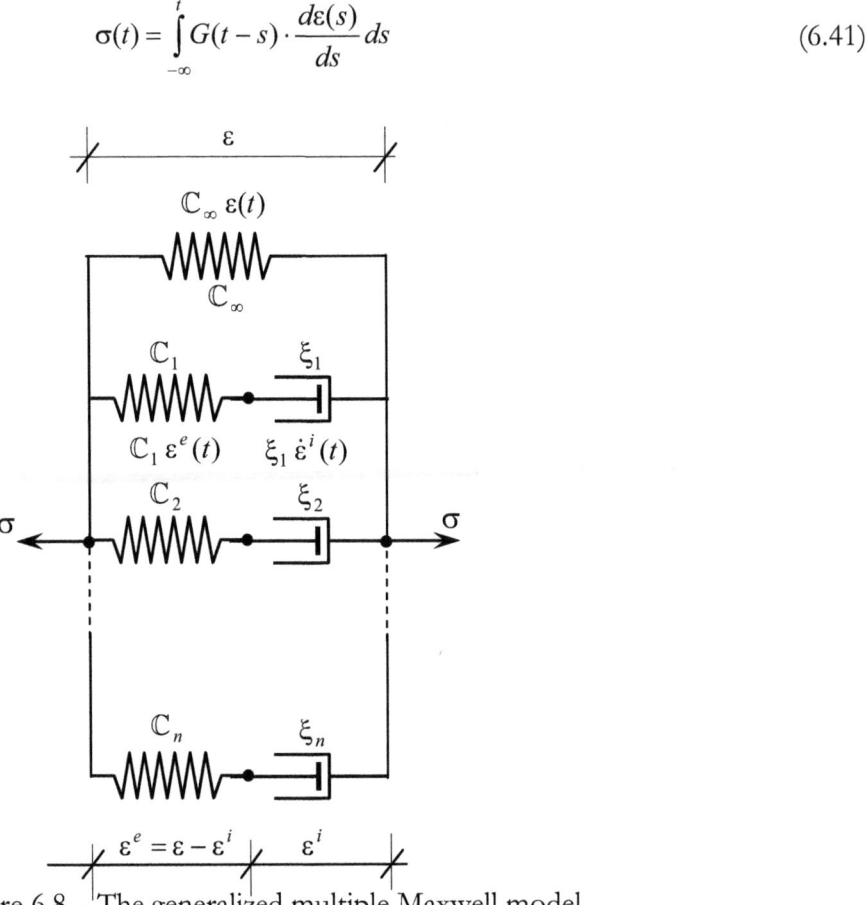

Figure 6.8 – The generalized multiple Maxwell model.

Similarly to the previously presented models, there is a particular response when a strain is applied, $\varepsilon(t)$, that meets the following simplification,

$$\begin{cases} \varepsilon(t) = 0 & \forall \ t < \tau_0 = 0 \\ \varepsilon(t) = \varepsilon_0 & \forall \ t \geq \tau_0 = 0 \end{cases} \tag{6.42}$$

obtaining the following simplified form for the total stress,

$$\begin{cases} \sigma(t) = 0 & \forall \ \tau < \tau_0 \\ \sigma(t) = G(t)\,\varepsilon_0 & \forall \ \tau \geq \tau_0 \end{cases} \tag{6.43}$$

6.2.6.1 Dissipation evaluation

This model is very suitable for the representation of the viscoelastic behavior of various materials. Consequently, a brief review of its thermodynamics will be presented in this section and, more particularly, its dissipation.

The free energy (see Chapter 2) can be defined as,

$$\Psi(\varepsilon, \varepsilon_\alpha^i, t) = \frac{1}{2}\mathbb{C}_\infty \varepsilon(t)^2 + \frac{1}{2}\sum_{\alpha=1}^{n-Maxw}\mathbb{C}_\alpha\left[\varepsilon(t) - \varepsilon_\alpha^i(t)\right]^2 \tag{6.44}$$

And from here, after applying Coleman's deduction[4,5] (equations 2.48 and 2.55), the following secant constitutive equation is obtained,

$$\sigma(t) = \frac{\partial\Psi(\varepsilon,\varepsilon_\alpha^i,t)}{\partial\varepsilon} = \mathbb{C}_\infty\varepsilon(t) + \sum_{\alpha=1}^{n-Maxw}\mathbb{C}_\alpha\left[\varepsilon(t) - \varepsilon_\alpha^i(t)\right] \equiv \sum_{\alpha=1}^{n-Maxw}\xi_\alpha\,\dot{\varepsilon}_\alpha^i(t) \tag{6.45}$$

And also the energy dissipation,

$$\Xi_m(t) = \sum_{\alpha=1}^{n-Maxw}\sigma_\alpha(t)\cdot\varepsilon_\alpha^i(t) \equiv \sum_{\alpha=1}^{n-Maxw}\xi_\alpha\left[\dot{\varepsilon}_\alpha^i(t)\right]^2 \geq 0 \tag{6.46}$$

This same expression can be obtained in a different way and also by its own dissipation definition (Chapter 2, equation 2.51, for constant time variation of the temperature $\dot{\theta} = \text{constant}$). Then,

$$\begin{aligned} \Xi_m &= \sigma(t)\dot{\varepsilon}(t) - \dot{\Psi}(t) = \sigma(t)\dot{\varepsilon}(t) - \frac{\partial\Psi}{\partial\varepsilon}\dot{\varepsilon}(t) - \sum_{\alpha=1}^{n-Maxw}\frac{\partial\Psi}{\partial\varepsilon_\alpha^i}\dot{\varepsilon}_\alpha^i(t) \\ &= \underbrace{\left[\sigma(t) - \frac{\partial\Psi}{\partial\varepsilon}\right]}_{0}\dot{\varepsilon}(t) - \sum_{\alpha=1}^{n-Maxw}\frac{\partial\Psi}{\partial\varepsilon_\alpha^i}\dot{\varepsilon}_\alpha^i(t) \geq 0 \\ \Xi_m &= -\sum_{\alpha=1}^{n-Maxw}\frac{\partial\Psi}{\partial\varepsilon_\alpha^i}\dot{\varepsilon}_\alpha^i(t) = -\sum_{\alpha=1}^{n-Maxw}-\mathbb{C}_\alpha\left(\varepsilon(t) - \varepsilon_\alpha^i(t)\right)\dot{\varepsilon}_\alpha^i(t) = \\ \Xi_m &= \sum_{\alpha=1}^{n-Maxw}\sigma_\alpha(t)\,\dot{\varepsilon}_\alpha^i(t) = \sum_{\alpha=1}^{n-Maxw}\xi_\alpha\left[\dot{\varepsilon}_\alpha^i(t)\right]^2 \geq 0 \end{aligned} \tag{6.47}$$

---

[4] Malvern, L. (1969). *Introduction to the mechanics of a continuous medium.* Prentice Hall, Englewood Cliffs, NJ.
[5] Lubliner, J. (1990). *Plasticity theory.* MacMillan, New York.

As observed, the dissipation expression obtained this way is the same as the one expressed by equation (6.46).

6.3 Multiaxial generalization of the viscoelastic constitutive laws

6.3.1 Multiaxial form of viscoelastic models

There is a rigorous way to write the generalization of any viscoelastic constitutive law obtained through phenomenological concepts based on the "spring-damping" analogy. As a matter of fact, the laws formulated in section 6.2 can be perfectly presented in multiaxial form by using the *creep function* $J_{ijkl}(t) = \mathbf{J}(t)$ and *relaxation function* $G_{ijkl}(t) = \mathbf{G}(t)$, both defined as a fourth-order tensor. This is,

$$\begin{cases} J_{ijkl}(t) = J^V(t)\left(\delta_{ij}\,\delta_{kl}\right) + J^D(t)\left(\delta_{ik}\,\delta_{jl} + \delta_{il}\,\delta_{jk}\right) \\ G_{ijkl}(t) = G^V(t)\left(\delta_{ij}\,\delta_{kl}\right) + G^D(t)\left(\delta_{ik}\,\delta_{jl} + \delta_{il}\,\delta_{jk}\right) \end{cases} \tag{6.48}$$

where δ_{ij} is the "Kronecker delta" and J^V, J^D, G^V, G^D are the scalar functions defining the volumetric viscous (V index) and deviatoric (D index) behaviors of the material. For isotropic materials, the definitions of the creep and relaxation tensors are similar to the elasticity tensor, thus, the definition of the same decomposition can be expressed by a deviatoric and a volumetric part.

Once the functions (6.48) are defined, the constitutive equation for any viscoelastic model formulated either for stresses or strain can be written as,

$$\begin{cases} \sigma_{ij}(t) = \displaystyle\int_{-\infty}^{t} G_{ijkl}(t-s) \cdot \frac{d\varepsilon_{kl}(s)}{ds}\,ds & : \text{Relaxation viscoelastic model} \\[4mm] \varepsilon_{ij}(t) = \displaystyle\int_{-\infty}^{t} J_{ijkl}(t-s) \cdot \frac{d\sigma_{kl}(s)}{ds}\,ds & : \text{Creep viscoelastic model} \end{cases} \tag{6.49}$$

As it can be observed, this is a very general definition, although it is not very useful to define creep and relaxation functions for materials affected by viscoelastic phenomena. Nevertheless, there is another way to define creep and relaxation functions based on experimental work. For example, the form established by the Concrete European Committee (CEB-1978) and the International Federation of Prestressed (FIP-1978) for relaxations functions is introduced as follows,

$$J_{ijkl}(t) = 0.4\,\beta_d\,(t-t_c)\left[\mathbb{C}_{ijkl}^{\,h}\right]^{-1} \tag{6.50}$$

where t_c is the reference initial time, $\mathbb{C}_{ijkl}^{\,h}$ is the constitutive tensor of concrete and β_d is a function considering the rheological effects of the material. This latter form is shown in Figure 6.9.

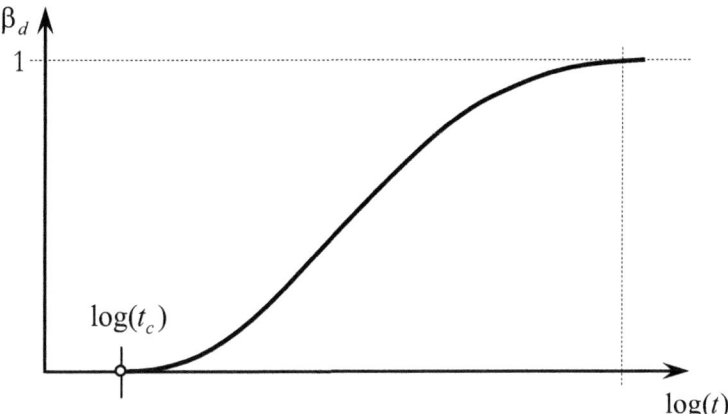

Figure 6.9 – Response function over time in concrete as of (CEB-FIP).

Unfortunately, finding the integral solution (6.49) through this exponential function is not easy, and it is necessary to use a convolution integral.

From equation (6.50) it follows that a creep tensor function or relaxation function depending on a scalar function can be used for multiaxial problem approximations. Accordingly, another approximation can be carried out based on an isotropic material whose Poisson coefficient is $\nu \to 0$. In this case, it is possible to define one single creep strain function for all directions.

This way the strain in multiaxial time can be formulated with the following approximation,

$$\varepsilon_{ij}(t) \cong \left\{ \mathbb{C}_{ijkl}^{-1} \cdot \mathbb{C} \right\} \int_{-\infty}^{t} J(t-s) \cdot \frac{d\sigma_{kl}(s)}{ds} ds \tag{6.51}$$

where $\mathbb{C}$ is Young's elastic modulus. The stress in time is then,

$$\sigma_{ij} = \mathbb{C}_{ijkl}\varepsilon_{kl}(t) = \mathbb{C} \int_{-\infty}^{t} J(t-s) \cdot \frac{d\sigma_{ij}(s)}{ds} ds \tag{6.52}$$

6.3.2 Numerical solution of integrals and algorithms

For the solution of viscoelastic problems, normally convolution integrals such as equations (6.49) must be solved. If functions $J_{ijkl}(t)$ or $G_{ijkl}(t)$ comply with the semi-group[6] properties, in other words, if the functions inside the integral meet the condition,

$$f(a+b) = f(a) \cdot f(b) \tag{6.53}$$

The convolution integrals can be avoided by carrying out the following time integral,

$$I(t) = \int_{-\infty}^{t} f(t-s) \cdot g(s)\, ds = I(t-\Delta t) + \int_{t-\Delta t}^{t} f(t-s) \cdot g(s)\, ds \tag{6.54}$$

The integral of the previous time $I(t-\Delta t)$ is used in each step and the integration is carried out only at the current time interval Δt.

The exponential functions found in viscoelastic models, based on spring-damping analogies, lead to exponential relaxation and creep functions that comply with the semi-

---

[6]. Simo, J. C and Hughes T.J.R, (1998). *Elastoplasticity and Viscoplasticity. Computational Aspects. Springer Verlag.*

group properties. However, for real materials such as the concrete, its function is similar to the one expressed in equation in (6.50), and to use this property to avoid the convolution it is necessary to simplify.

Multiaxial case of the generalized Kelvin model

As an example, an approximation of a multiaxial strain (6.11) obtained for the **generalized Kelvin model** (6.2.3) is presented below. This approximation is written considering that at time $t + \Delta t$, the function to integrate is worth $f(t + \Delta t - s) = \mathbb{C}_{ijkl}^{-1} (\mathbb{C}/\xi) e^{-(t+\Delta t-s)/r_1}$, getting the following strain expression,

$$\varepsilon_{ij}(t + \Delta t) = \underbrace{\mathbb{C}_{ijkl}^{-1} \sigma_{kl}(t + \Delta t)}_{\varepsilon_{ij}^e(t+\Delta t)} + \underbrace{\mathbb{C}_{ijkl}^{-1} \frac{\mathbb{C}}{\xi} \int_{-\infty}^{t+\Delta t} e^{-(t+\Delta t-s)/r_1} \cdot \sigma_{kl}(s)\, ds}_{\varepsilon_{ij}^i(t+\Delta t)} \tag{6.55}$$

Such that the convolution defining the inelastic strain can be solved as,

$$\varepsilon_{ij}^i(t + \Delta t) = \varepsilon_{ij}^i(t) e^{-(\Delta t)/r_1} + \mathbb{C}_{ijkl}^{-1} \frac{\mathbb{C}}{\xi} \int_{t}^{t+\Delta t} e^{-(t+\Delta t-s)/r_1} \cdot \sigma_{kl}(s)\, ds \tag{6.56}$$

With the integral approximation using the Simpson rule, the following expression can be written for very short time increments,

$$\varepsilon_{ij}^i(t + \Delta t) \cong \varepsilon_{ij}^i(t) e^{-(\Delta t)/r_1} + \mathbb{C}_{ijkl}^{-1} \cdot \frac{\mathbb{C}}{\xi} e^{-(\Delta t)/r_1} \cdot \sigma_{kl}(t + \Delta t) \cdot \Delta t \tag{6.57}$$

A possible algorithm to solve this problem is shown in Table 6.1.

---

1. **Initialization**

$$^0\left[\varepsilon_{ij}^i\right]^{t+\Delta t} \equiv \left[\varepsilon_{ij}^i\right]^t$$

$$^0\left[\sigma_{ij}^i\right]^{t+\Delta t} \equiv \left[\sigma_{ij}^i\right]^t$$

2. **Definition of the sub-increment of time**

$$\delta t = \frac{\Delta t}{m} \quad ; \quad \text{Step count}: \ \alpha = 1$$

3. **Start of the process of sub-increment and calculation of the inelastic strains**

$$^\alpha\left[\varepsilon_{ij}^i\right]^{t+\Delta t} \cong {}^{\alpha-1}\left[\varepsilon_{ij}^i\right]^{t+\Delta t} e^{-(\delta t)/r_1} + \mathbb{C}_{ijkl}^{-1} \cdot \frac{\mathbb{C}}{\xi} e^{-(\delta t)/r_1} \cdot \underbrace{\left[\sigma_{kl}\right]^{t+(\alpha-1)\delta t}}_{{}^{\alpha-1}\left[\sigma_{kl}\right]^t}$$

4. **Stress calculation**

$$^\alpha\left[\sigma_{ij}\right]^{t+\Delta t} \cong \mathbb{C}_{ijkl} \cdot \left[\left[\varepsilon_{ij}\right]^{t+\Delta t} - {}^\alpha\left[\varepsilon_{ij}^i\right]^{t+\Delta t}\right]$$

   If $\alpha \leq m$ then $\alpha = \alpha + 1$, return to 3.

5. **End of the integration and proceed with the calculation**

Table 6.1 – Algorithm to obtain the stress of the generalized Kelvin model.

Multiaxial case of the generalized Maxwell model

Another relevant algorithm worth mentioning is the **generalized Maxwell model** algorithm (6.2.5). In this model, the multiaxial expression for stress, equation (6.29) is written at $t + \Delta t$ through the following approximation,

$$\sigma_{ij}(t + \Delta t) = \mathbb{C}_{ijkl}\left[\varepsilon_{ij}(t + \Delta t) - \frac{\mathbb{C}_1}{\mathbb{C}_0 \xi} \int_{-\infty}^{t+\Delta t} e^{-(t+\Delta t - s)/r_1} \cdot \varepsilon_{kl}(s)\, ds \right] \tag{6.58}$$

in which the variables involved are the ones already defined for the uniaxial problem, section 6.2.5. The integral solution can be carried out without a convolution, it is then re-written as,

$$\sigma_{ij}(t + \Delta t) = \mathbb{C}_{ijkl}\, \varepsilon_{ij}(t + \Delta t) - \left[\mathbb{C}_{ijkl}\frac{\mathbb{C}_1}{\mathbb{C}_0 \xi} \int_{-\infty}^{t} e^{-(t-s)/r_1} \cdot \varepsilon_{kl}(s)\, ds \right] -$$
$$- \mathbb{C}_{ijkl}\frac{\mathbb{C}_1}{\mathbb{C}_0 \xi} \int_{t}^{t+\Delta t} e^{-(t+\Delta t - s)/r_1} \cdot \varepsilon_{kl}(s)\, ds \tag{6.59}$$

After integrating the third term using the trapezoidal rule and reordering the expressions, the stress is written as,

$$\sigma_{ij}(t + \Delta t) = \mathbb{C}_{ijkl}\, \varepsilon_{ij}(t + \Delta t) - \mathbb{C}_{ijkl} \cdot \varepsilon_{kl}(t) \cdot e^{-(\Delta t)/r_1} - \sigma_{ij}(t) \cdot e^{-(\Delta t)/r_1} -$$
$$- \mathbb{C}_{ijkl}\frac{\mathbb{C}_1}{\mathbb{C}_0 \xi}\left[e^{-(\Delta t)/r_1} \cdot \varepsilon_{kl}(t) - \varepsilon_{kl}(t + \Delta t) \right] \cdot \frac{\Delta t}{2}$$
$$\sigma_{ij}(t + \Delta t) = -\mathbb{C}_{ijkl}\, \varepsilon_{kl}(t) \cdot e^{-(\Delta t)/r_1}\left[1 + \frac{\mathbb{C}_1}{\mathbb{C}_0 \xi}\frac{\Delta t}{2} \right] + \mathbb{C}_{ijkl}\, \varepsilon_{kl}(t + \Delta t) \cdot \left[1 - \frac{\mathbb{C}_1}{\mathbb{C}_0 \xi}\frac{\Delta t}{2} \right] +$$
$$+ \sigma(t) \cdot e^{-(\Delta t)/r_1} \tag{6.60}$$

A possible algorithm to solve the problem is shown in Table 6.2.

1. Obtaining the strain

$$\left[\varepsilon_{ij} \right]^{t + \Delta t}$$

2. Stress integration,

$$\left[\sigma_{ij} \right]^{t + \Delta t} = \left[\sigma_{ij} \right]^{t} \cdot e^{-(\Delta t)/r_1} - \mathbb{C}_{ijkl}\left[\varepsilon_{kl} \right]^{t} \cdot e^{-(\Delta t)/r_1}\left[1 + \frac{\mathbb{C}_1}{\mathbb{C}_0 \xi}\frac{\Delta t}{2} \right] + \mathbb{C}_{ijkl}\left[\varepsilon_{kl} \right]^{t + \Delta t} \cdot \left[1 - \frac{\mathbb{C}_1}{\mathbb{C}_0 \xi}\frac{\Delta t}{2} \right]$$

Table 6.2 – Algorithm for obtaining the stress of the generalized Maxwell model.

If a force applied leads to a constant stress, the model represents correctly the creep problem in time and if a displacement applied leads to a constant strain, the behavior is recovered by relaxation.

6.4 The Kelvin model in dynamic problems

Viscoelasticity is responsible for the damping phenomenon in the dynamic behavior of the structure. With these models the damping concept can be incorporated in a rational way based on the laws of mechanics. Thus, the damping depends on the properties of the material, and it is not necessary to use artificial formulations such as the formulations introduced by Rayleigh[7].

Then the influence of Kelvin multiaxial model is presented (6.2.2). Its formulation is simple and suitable for structural damping representations and also to compare it with Rayleigh damping concept.

Following the formulation of the model presented in section (6.2.1), its constitutive equation can be synthesized through the following expressions,

$$\sigma(t) = \sigma^e(t) + \sigma^{vis}(t) = \mathbb{C}\,\varepsilon(t) + \xi\,\dot{\varepsilon}(t) = \mathbb{C}\left[\varepsilon(t) + \frac{\xi}{\mathbb{C}}\dot{\varepsilon}(t)\right]$$

$$\varepsilon(t) = \varepsilon^e(t) = \varepsilon^{vis}(t) = \frac{\sigma(t)}{\mathbb{C}} - \frac{\xi}{\mathbb{C}}\dot{\varepsilon}(t) = \int_{-\infty}^{t} \frac{1}{\xi}e^{-(t-s)/r_1} \cdot \sigma(s)\,ds$$

(6.61)

where σ^e and σ^{vis} represent the elastic and viscous part of the stress respectively, $\varepsilon^\infty = \mathbb{C}^{-1}\sigma(t)$ the stationary strain (in infinite time), $\mathbb{C}$ Young's modulus and ξ viscosity. Extending this to the multiaxial case by using the approximation in section 6.3 , the following expression results,

$$\sigma_{ij}(t) = \sigma_{ij}^e(t) + \sigma_{ij}^{vis}(t) = \mathbb{C}_{ijkl}\left[\varepsilon_{kl}(t) + \frac{\xi}{\mathbb{C}}\dot{\varepsilon}_{kl}(t)\right]$$

(6.62)

being $r = \dfrac{\xi}{\mathbb{C}}$ the delayed time defined in the previous sections.

For the solution of the problem in dynamics it is necessary to obtain the displacement field and the velocities from the solution of the motion equation. The Newmark method can be used for this (see chapter B3). In other words, the algorithm is simple and is summarized in the following table,

---

1. Obtaining the strain and strain velocity from the displacement fields and velocities obtained from the integration of the movement equation. In this case the Newmark method is used,

$$\left[\varepsilon_{ij}\right]^{t+\Delta t} = \nabla_i^s\left[u_j\right]^{t+\Delta t}$$

$$\left[\dot{\varepsilon}_{ij}\right]^{t+\Delta t} = \nabla_i^s\left[\dot{u}_j\right]^{t+\Delta t}$$

2. Direct Integration of the stress,

$$\left[\sigma_{ij}\right]^{t+\Delta t} = \mathbb{C}_{ijkl}\left\{\left[\varepsilon_{kl}\right]^{t+\Delta t} + \frac{\xi}{\mathbb{C}}\left[\dot{\varepsilon}_{kl}\right]^{t+\Delta t}\right\}$$

Table 6.3 – Algorithm for obtaining the stress in dynamic problems using the Kelvin model.

---

[7] Clough R. and Penzien J. (1977). Dynamics of Structures. Mc Graw-Hill - New York.

6.4.1 The Kelvin model dissipation

In the generalized Kelvin model for multiaxial cases, the free energy can be written as follows,

$$\Psi = \frac{1}{2}\varepsilon_{ij}\, \mathbb{C}_{ijkl}\, \varepsilon_{kl} \tag{6.63}$$

Obtaining from the second law of thermodynamics (chapter 2) the following expression for the dissipation per unit volume,

$$\Xi_m = \sigma_{ij}\,\dot{\varepsilon}_{ij} - \dot{\Psi} = \left(\sigma_{ij} - \mathbb{C}_{ijkl}\,\varepsilon_{kl}\right)\dot{\varepsilon}_{ij} = \sigma_{ij}^{vis}\,\dot{\varepsilon}_{ij} = \xi\,\dot{\varepsilon}_{ij}\,\mathbb{C}_{ijkl}\,\dot{\varepsilon}_{kl} \tag{6.64}$$

And the dissipation in the whole volume at time t is

$$E_{dis} = \int_0^t \left(\int_V \Xi_m dV \right) dt \tag{6.65}$$

The field of strain velocities is obtained from equation $\dot{\varepsilon}_{ij} = \nabla_i^s \dot{u}_j = \nabla_i^s N_{jk}\dot{U}_k$, such that the velocity is obtained directly from Newmark's solution from the differential equation of dynamics (B2.68) o (B2.72) for the discreet solid.

6.4.2 Equation of the dynamic equilibrium for the Kelvin model

The damping forces term is obtained by writing the dynamic equilibrium equation as a function of the Kelvin stress state. Taking into consideration the dynamic equilibrium equation discretized for a basic domain (2.72) and then applying its formulation to the whole domain Ω, the already known equation (3.1) is written as

$$0 = \Delta f_k = M_{kj}\,\ddot{U}_j(t+\Delta t) + f_k^{\text{int}}(\dot{\mathbf{U}}, \mathbf{U}, t+\Delta t) - f_k^{\text{ext}}(t+\Delta t) \qquad \in \Omega$$

$$0 = \Delta\mathbf{f} = \mathbb{M}\,\ddot{\mathbf{U}}(t+\Delta t) + \mathbf{f}^{\text{int}}(\dot{\mathbf{U}}, \mathbf{U}, t+\Delta t) - \mathbf{f}^{\text{ext}}(t+\Delta t) \qquad \in \Omega$$

$$0 = \underset{\Omega^e}{\overset{i}{\mathbf{A}}}\left[\int_{V^e}\rho\, N_{ki}\,N_{ij}\,dV\right]_{\Omega^e}^{t+\Delta t}\ddot{U}_j\Big|_{\Omega^e}^{t+\Delta t} + \underset{\Omega^e}{\overset{i}{\mathbf{A}}}\left[\int_{V^e}\sigma_{in}\nabla_i^S N_{nk}\,dV\right]_{\Omega^e}^{t+\Delta t} - \underset{\Omega^e}{\overset{i}{\mathbf{A}}}\left[\oint_{S^e}t_i N_{ik}\,dS + \int_{V^e}\rho\, b_i N_{ik}\,dV\right]_{\Omega^e}^{t+\Delta t}$$

Substituting into the latter the viscoelastic constitutive law (6.62),

$$\Delta f_k = M_{kj}\,\ddot{U}_j(t+\Delta t) + \underbrace{\underset{\Omega^e}{\overset{i}{\mathbf{A}}}\left[\int_{V^e}\sigma_{in}\nabla_i^S N_{nk}\,dV\right]_{\Omega^e}^{t+\Delta t}}_{f_k^{\text{int}}} - f_k^{\text{ext}}(t+\Delta t) = 0$$

$$\Delta f_k = M_{kj}\,\ddot{U}_j(t+\Delta t) + \underset{\Omega^e}{\overset{i}{\mathbf{A}}}\left[\int_{V^e}\left[\mathbb{C}_{inrs}\varepsilon_{rs} + \mathbb{C}_{inrs}\frac{\xi}{\mathbb{C}}\dot{\varepsilon}_{rs}\right]\nabla_i^S N_{nk}\,dV\right]_{\Omega^e}^{t+\Delta t} - f_k^{\text{ext}}(t+\Delta t) = 0 \tag{6.66}$$

And then, operating on the internal forces term, the following known expression for the equilibrium equation is obtained,

$$\Delta f_k = M_{kj} \ddot{U}_j (t + \Delta t) + \mathop{\mathbf{A}}_{\Omega^e} {}^{i}\left[\int_{v^e} \left[\mathbb{C}_{inrs} \nabla_r^S N_{sj} U_j + \mathbb{C}_{inrs} \frac{\xi}{\mathbb{C}} \nabla_r^S N_{sj} \dot{U}_j \right] \nabla_i^S N_{nk} dV \right]_{\Omega^e}^{t+\Delta t} - \tag{6.67}$$

$$- f_k^{ext} (t + \Delta t) = 0$$

$$0 = \Delta f_k = M_{kj} \ddot{U}_j (t + \Delta t) + \mathop{\mathbf{A}}_{\Omega^e} {}^{i}\left[\left\{ \int_{v^e} \left[\mathbb{C}_{inrs} \nabla_r^S N_{sj} \right] \nabla_i^S N_{nk} dV \right\} U_j \Big|_{\Omega^e} \right]_{\Omega^e}^{t+\Delta t} +$$

$$+ \mathop{\mathbf{A}}_{\Omega^e} {}^{i}\left[\left\{ \int_{v^e} \left[\mathbb{C}_{inrs} \frac{\xi}{\mathbb{C}} \nabla_r^S N_{sj} \right] \nabla_i^S N_{nk} dV \right\} \dot{U}_j \Big|_{\Omega^e} \right]_{\Omega^e}^{t+\Delta t} - f_k^{ext} (t + \Delta t) = 0 \tag{6.68}$$

Its compact form is written as,

$$0 = \Delta f_k = M_{kj} \ddot{U}_j (t + \Delta t) + K_{kj} U_j (t + \Delta t) + D_{kj} \dot{U}_j (t + \Delta t) - f_k^{ext} (t + \Delta t) = 0$$

$$0 = \Delta \mathbf{f} = \mathbb{M} \ddot{\mathbf{U}}(t + \Delta t) + \mathbb{K} \mathbf{U}(t + \Delta t) + \mathbb{D} \dot{\mathbf{U}}(t + \Delta t) - \mathbf{f}^{ext} (t + \Delta t) \tag{6.69}$$

In short, the elastic and viscous forces are worth, respectively,

$$\mathbf{f}^{elast} (t + \Delta t) = \mathbb{K} \cdot \mathbf{U}(t + \Delta t) = \mathop{\mathbf{A}}_{\Omega^e} \left[\int_{v^e} \mathbf{B} : \mathbb{C} : \mathbf{B} \, dV \right]_{\Omega^e}^{t+\Delta t} \cdot \left[\mathbf{U} \right]_{\Omega^e}^{t+\Delta t}$$

$$\mathbf{f}^{vis} (t + \Delta t) = \mathbb{D} \cdot \dot{\mathbf{U}}(t + \Delta t) = \mathop{\mathbf{A}}_{\Omega^e} \left[\int_{v^e} \mathbf{B} : (r \mathbb{C}) : \mathbf{B} \, dV \right]_{\Omega^e}^{t+\Delta t} \cdot \left[\dot{\mathbf{U}} \right]_{\Omega^e}^{t+\Delta t} \tag{6.70}$$

As observed in the latter equation the damping matrix for the Kelvin model is equal to the initial stiffness matrix multiplied by the delay time

$$\mathbb{D} = r \mathbb{K} \tag{6.71}$$

Thus, Rayleigh's particular case of damping with coefficients $\alpha = 0$ and $\beta = r = \xi/\mathbb{C}$ leads to Kelvin's viscous damping,

$$\mathbb{D} = \alpha \mathbb{M} + \beta \mathbb{K} \xrightarrow{\alpha=0,\beta=r} \mathbb{D} = r \mathbb{K} = 2\nu\omega \tag{6.72}$$

From this equation it follows that using Rayleigh's damping as cited and Kelvin's viscous damping leads to solution in displacements, strains, velocities and identical accelerations,

Kelvin with $r = \dfrac{\xi}{\mathbb{C}}$ en: $\qquad 0 = \Delta \mathbf{f} = \mathbb{M}\,\ddot{\mathbf{U}}(t + \Delta t) + \mathbf{f}^{\text{int}}(\dot{\mathbf{U}}, \mathbf{U}, t + \Delta t) - \mathbf{f}^{\text{ext}}(t + \Delta t)$

$$\equiv$$

Rayleigh with $\alpha = 0$ y $\beta = r$: $\quad 0 = \Delta \mathbf{f} = \mathbb{M}\,\ddot{\mathbf{U}}(t + \Delta t) + \mathbb{K}\,\mathbf{U}(t + \Delta t) + \mathbb{D}\,\dot{\mathbf{U}}(t + \Delta t) - \mathbf{f}^{\text{ext}}(t + \Delta t)$

6.4.3 Stress considerations. The Rayleigh vs. the Kelvin model

As observed in the previous section, it can be concluded that using Rayleigh's damping in the dynamic equilibrium equation is exactly the same as using the Kelvin model. This is true only in the displacements and their derivatives but it is not in the stress field because in the first case a linear elastic constitutive law is carried out and the damping is added as an external force to the system, whereas in the other case a viscoelastic constitutive model is used. As a result the stress obtained is different in both models,

$$\sigma_{ij}(t) = \begin{cases} \mathbb{C}_{ijkl}\,\varepsilon_{kl}(t) & \text{for the elastic model} \\[2mm] \mathbb{C}_{ijkl}\left[\varepsilon_{kl}(t) + \dfrac{\xi}{\mathbb{C}}\dot{\varepsilon}_{kl}(t) \right] & \text{for the viscoelastic model} \end{cases} \tag{6.73}$$

As observed above, the difference between the two stresses is greater as the damping and /or velocity is greater.

Moreover, this difference becomes more significant as the strength of the structure is evaluated or the material is working within a bounded strength limit, which is fracture, damage or plasticity.

6.4.4 Dissipation considerations. The Rayleigh vs. the Kelvin model

Differences can also be found between Rayleigh's damping equilibrium equation and other models incorporating to its internal force one of Kelvin's constitutive models.

The power equilibrium equation B2.32 ($P_d = P_{\text{int}} - \dot{K}$) is met in both cases. However, in the Rayleigh damping case this equation must be enforced to ensure compliance.

Viscoelastic dissipation

By post-multiplying the dynamic equilibrium equation (B2.72) by $\dot{\mathbf{U}}(t + \Delta t)$, expressed in time $t + \Delta t$, the following is obtained

$$0 = \underbrace{\left[\mathbb{M}\,\ddot{\mathbf{U}}\right]\cdot\dot{\mathbf{U}}}_{\dot{K}} + \underbrace{\overbrace{\left[\mathbf{f}^{\text{elast}}(\mathbf{U}) + \mathbf{f}^{\text{vis}}(\dot{\mathbf{U}})\right]}^{\mathbf{f}^{\text{int}}(\dot{\mathbf{U}}, \mathbf{U})}\cdot\dot{\mathbf{U}}}_{P_d} - \underbrace{\left[\mathbf{f}^{\text{ext}}\right]\cdot\dot{\mathbf{U}}}_{P_{\text{int}}} \tag{6.74}$$

where the deformative power is obtained by substituting equation (6.70) into the previous one,

$$P_d = \left\{ \mathbf{f}^{\text{elast}}(\mathbf{U}) \right\} \cdot \dot{\mathbf{U}} = \left\{ \mathbb{K} \cdot \mathbf{U} \right\} \cdot \dot{\mathbf{U}} = \left\{ \underset{\Omega^e}{A} \left[\int_{V^e} \mathbf{B} : \mathbb{C} : \mathbf{B} \, dV \right]_{\Omega^e} \cdot [\mathbf{U}]_{\Omega^e} \right\} \cdot \dot{\mathbf{U}} +$$

$$+ \left\{ \mathbf{f}^{\text{vis}}(\dot{\mathbf{U}}) \right\} \cdot \dot{\mathbf{U}} = \left\{ \mathbb{D} \cdot \dot{\mathbf{U}} \right\} \cdot \dot{\mathbf{U}} = \left\{ \underset{\Omega^e}{A} \left[\int_{V^e} \mathbf{B} : (r\,\mathbb{C}) : \mathbf{B} \, dV \right]_{\Omega^e} \cdot [\dot{\mathbf{U}}]_{\Omega^e} \right\} \cdot \dot{\mathbf{U}}$$

(6.75)

The damping power term arises from this equation.

Rayleigh's damping dissipation

By post-multiplying the dynamic equilibrium equation (B2.72) by $\dot{\mathbf{U}}(t + \Delta t)$, expressed in time $t + \Delta t$, then

$$0 = \underbrace{[\mathbb{M}\,\ddot{\mathbf{U}}] \cdot \dot{\mathbf{U}}}_{\dot{K}} + \underbrace{\overbrace{[\mathbf{f}^{\text{elast}}(\mathbf{U})] \cdot \dot{\mathbf{U}}}^{\mathbf{f}^{\text{int}}(\dot{\mathbf{U}}, \mathbf{U})}}_{P_d} - \underbrace{[\mathbf{f}^{\text{ext}}] \cdot \dot{\mathbf{U}}}_{P_{\text{int}}}$$

(6.76)

where the deformative power is obtained by substituting equation (6.70) into the previous one, with a time delay $r = 0$,

$$P_d = \left\{ \mathbf{f}^{\text{elast}}(\mathbf{U}) \right\} \cdot \dot{\mathbf{U}} = \left\{ \mathbb{K} \cdot \mathbf{U} \right\} \cdot \dot{\mathbf{U}} = \left\{ \underset{\Omega^e}{A} \left[\int_{V^e} \mathbf{B} : \mathbb{C} : \mathbf{B} \, dV \right]_{\Omega^e} \cdot [\mathbf{U}]_{\Omega^e} \right\} \cdot \dot{\mathbf{U}}$$

(6.77)

The damping power term does not exist and then there is no dissipative power into the system. To complete this formulation, a dissipative term must be added to the power balance. However, it is important to point out that it is external to the system and does not come from the material. Thus, equation (6.76) is written as,

$$P_d = \left\{ \mathbf{f}^{\text{elast}}(\mathbf{U}) \right\} \cdot \dot{\mathbf{U}} + P_{disip}^R = \left\{ \mathbb{K} \cdot \mathbf{U} \right\} \cdot \dot{\mathbf{U}} + \left\{ \mathbb{D} \cdot \dot{\mathbf{U}} \right\} \cdot \dot{\mathbf{U}} = \left\{ \underset{\Omega^e}{A} \left[\int_{V^e} \mathbf{B} : \mathbb{C} : \mathbf{B} \, dV \right]_{\Omega^e} \cdot [\mathbf{U}]_{\Omega^e} \right\} \cdot \dot{\mathbf{U}} +$$

$$+ \left\{ [\alpha\mathbb{M} + \beta\mathbb{K}] \cdot [\dot{\mathbf{U}}]_{\Omega^e} \right\} \cdot \dot{\mathbf{U}}$$

(6.78)

The basic difference between the use of a viscoelastic model and an elastic one plus structural damping (see equation (6.73)) is observed when evaluating the dissipation. In the former case, the dissipation comes from the constitutive model. In the second, it comes from an added term. Nevertheless, the dissipated energy for all the structure must be the same in both cases,

$$E_{dis}(t) = \int_0^t \left[\underset{\Omega^e}{A} \left\{ \left[\int_{V^e} \left(\mathbb{C}_{inrs} \frac{\xi}{\mathbb{C}} \nabla_r^S N_{sj} \right) \nabla_i^S N_{nk} \, dV \right] \cdot \dot{U}_j \right\}_{\Omega^e} \cdot \underset{\Omega^e}{A} \{\dot{U}_k\}_{\Omega^e} \right] dt \equiv \int_0^t (\mathbb{D} : \dot{\mathbf{U}}) \cdot \dot{\mathbf{U}} \, dt$$

(6.79)

The considerations highlighted in the Kelvin and Rayleigh models can be observed in the examples shown below.

6.4.5 Cantilever beam

This shows a cantilever beam subjected to a displacement at its end and then it is stopped to let the beam oscillate freely (Figure 6.10). The variation of the displacements in time for the elastic case, the viscoelastic case with $r = \xi/\mathbb{C} = 0.01$ sec and for the structural damping case with $\beta = 0.01$ sec have been represented in the same figure. The longitudinal maximum stress variation in time at the fixed end for the same three cases abovementioned has been represented in figure 6.11. It can be noted that the displacement's response is the same but not for the stresses.

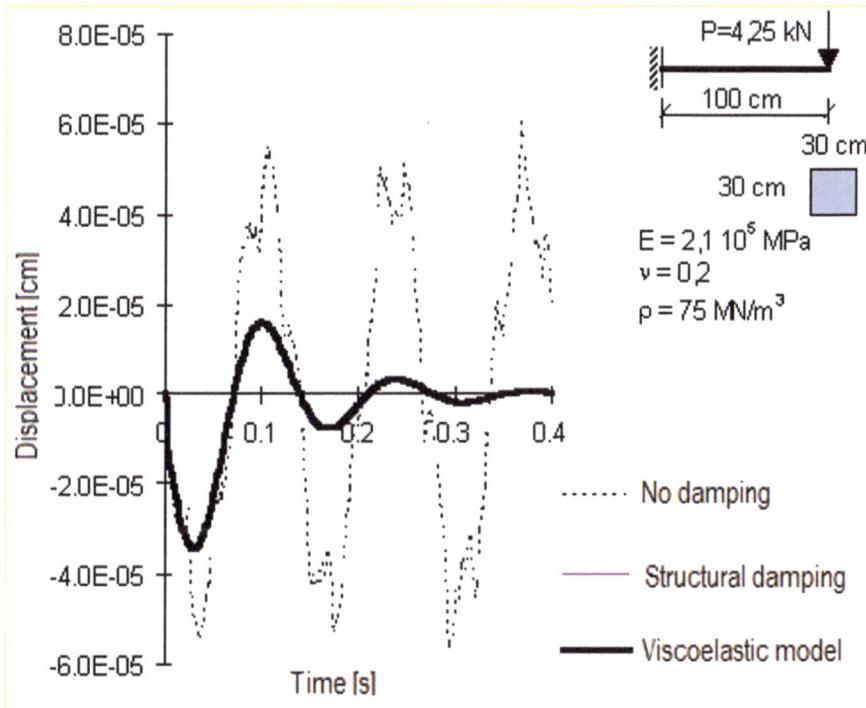

Figure 6.101 – Cantilever beam. Displacements vs. Time for structural damping with $\beta = 0.01$ and viscoelastic model with $r = 0.01$.

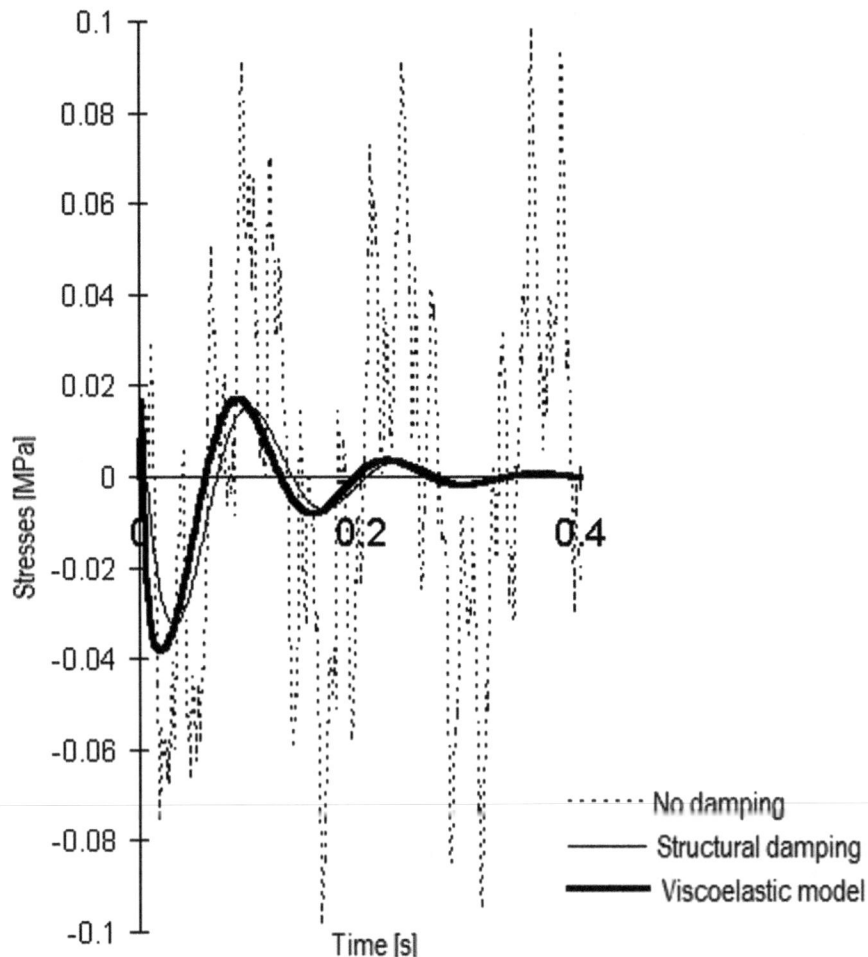

Figure 6.11 – Cantilever beam. Stresses vs. Time for structural damping with $\beta = 0.01$ and viscoelastic model with $r = 0.01$.

6.4.6 Frame with rigid beam and lumped mass

The behavior of the frame shown in figure 6.12, subjected to a load varying in time, is analyzed in this example. The variation of the displacement at the load application point is represented for the non-damping case, the viscoelastic model case with $\tau = \eta/E_o = 0{,}14$s and $\tau = 0{,}25s$ and the structural damping case with $\beta = 0{,}14s$ and $\beta = 0{,}25s$. Again, it can be observed that displacements are the same in both ways to consider the damping.

The variation of the normal maximum stress at the fixed end for the non-damping case (see Figure 6.13), the viscoelastic model case with $\tau = \eta/E_o = 0{,}14$ s and $\tau = 0{,}25$ s and the structural damping case with $\beta = 0{,}14$ s and $\beta = 0{,}25$ s is represented in figure 5.13. It can be observed that the stress results in both cases are different. For larger damping, the difference in stresses is greater.

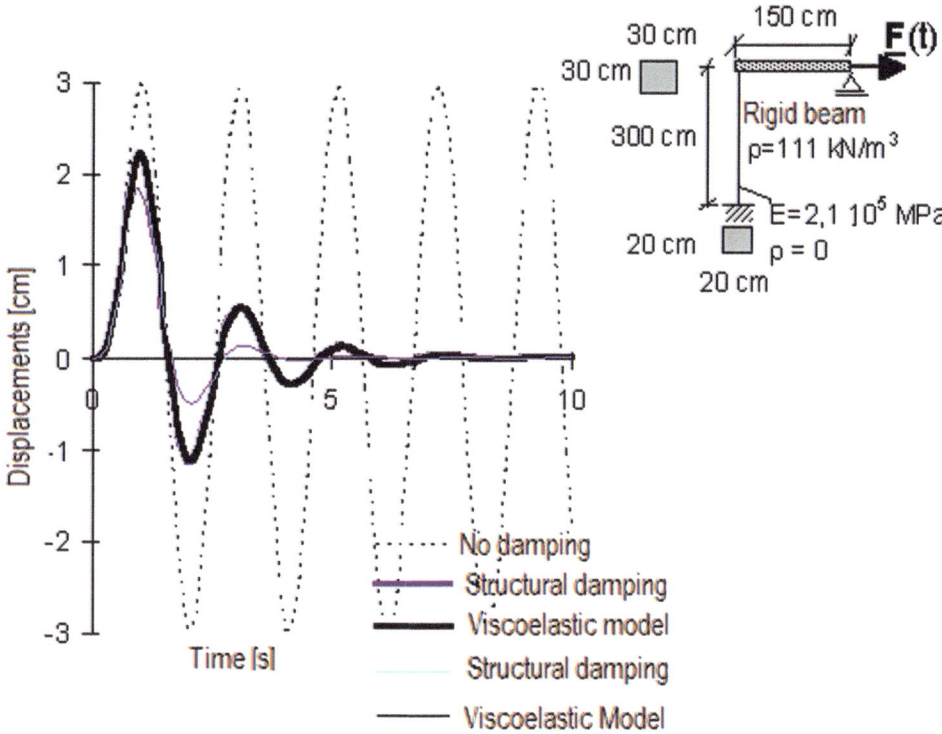

Figure 6.12 – Frame. Displacements vs. Time for: 1) The second and third response with structural damping with $\beta = 0.14$ and the viscoelastic model with $r = 0.14$, respectively. 2) The fourth and fifth response with structural damping with $\beta = 0.25$ and the viscoelastic model with $r = 0.25$, respectively.

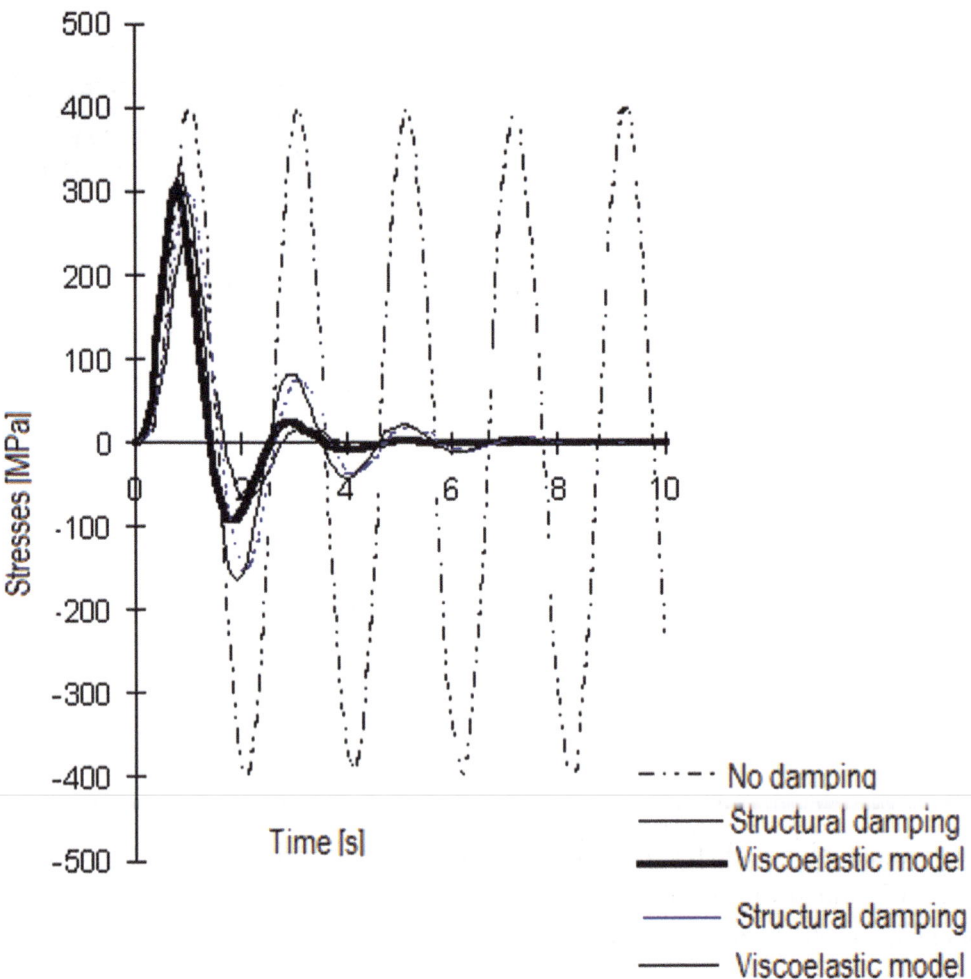

Figure 6.13 – Frame. Stresses vs. Time for: 1) The second and third response with structural damping with $\beta = 0.14$ and the viscoelastic model with $r = 0.14$, respectively. 2) The fourth and fifth response with structural damping with $\beta = 0.25$ and the viscoelastic model with $r = 0.25$, respectively.

6.5 Viscoplasticity

The theory of viscoplasticity is a general case of the theory of elasticity and the theory of plasticity (see Figure 6.14).

Viscoplasticity, unlike elasticity and plasticity, considers the viscosity parameter ξ as a model variable. This makes the plastic model sensitive to time. Like in previous cases, the study herein described is concerned with the viscoplastic solution for small strains.

The basic characteristics of the viscoplastic model for small strains are:

- The total strain is defined in additive form

$$\boldsymbol{\varepsilon} = \boldsymbol{\varepsilon}^e + \boldsymbol{\varepsilon}^{vp} \tag{6.80}$$

where $\boldsymbol{\varepsilon}$ represents the total deformation at the point, and $\boldsymbol{\varepsilon}^e$ and $\boldsymbol{\varepsilon}^{vp}$ are the elastic and viscoplastic components of the total strain.

- The evolution of the viscoplastic internal variable is defined by using a normality rule or viscoplastic flow similar to that used in plasticity (equation 5.45),

$$\dot{\boldsymbol{\varepsilon}}^{vp} = \lambda^{vp} \frac{\partial \mathbb{G}(\boldsymbol{\sigma},\mathbf{q})}{\partial \boldsymbol{\sigma}} = \lambda^{vp} \, \mathbf{g} \;\Rightarrow\; d\dot{\boldsymbol{\varepsilon}}^{vp} = \lambda^{vp} \, \mathbf{g} \, dt \qquad (6.81)$$

where $\mathbb{G}$ is the function of viscoplastic potential, analogous to the function of plastic potential, which when dealing with associated viscoplastic flow is the same as the threshold function of creep $\mathbb{G}(\boldsymbol{\sigma},\mathbf{q}) \equiv \mathbb{F}(\boldsymbol{\sigma},\mathbf{q})$. The viscoplastic parameter λ^{vp} plays an analogous role to the plastic consistency parameter used in plasticity, but in this case its definition is more general,

$$\lambda^{vp} = \frac{\left\langle \Phi\left[\mathbb{F}(\boldsymbol{\sigma},\mathbf{q})\right]\right\rangle}{\xi} \qquad (6.82)$$

where $\Phi\left[\mathbb{F}(\boldsymbol{\sigma},\mathbf{q})\right]$ is the overstress function introduced by Perzyna[8] (1963) and $\mathbb{F}(\boldsymbol{\sigma},\mathbf{q}) = \left[f(\boldsymbol{\sigma})/K\right] - 1$ is the viscoplastic creep function, analogous to the plasticity threshold function (equation 5.59). The Macaulay brackets are defined as $\langle x \rangle = 0{,}5\left[x + |x|\right]$, and K is the hardening internal variable which, like in plasticity, can be related to an uniaxial equivalent strength $\overline{\sigma} = K$. The evolution of this variable depends on a rule defined from its variation in time and can be linear or quadratic or a shape from experiments. The over stress function is defined as

$$\left\langle \Phi\left[\mathbb{F}(\boldsymbol{\sigma},\mathbf{q})\right]\right\rangle = \begin{cases} 0 & \forall \; \mathbb{F} \le 0 \\ \Phi\left[\mathbb{F}(\boldsymbol{\sigma},\mathbf{q})\right] & \forall \; \mathbb{F} > 0 \end{cases} \qquad (6.83)$$

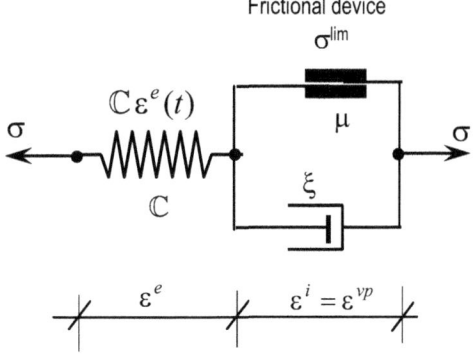

Figure 6.14 – Viscoplastic model

Unlike the elastoplastic problem, the instantaneous compliance of the equality between the creep function and the hardening variable is not required. As a matter of fact, and unlike plasticity, the viscoplastic creep function can be positive $\mathbb{F}(\boldsymbol{\sigma},\mathbf{q}) \ge 0$ during the viscoplastic process. However, this condition should be coincident with that of plasticity at infinite

[8] Perzyna P. (1963). The constitutive equation for rate sensitive plastic materials. *Quart. Appl. Math.*, 20, 321-332.

$t \to \infty$ (Figure 6.15). Thus, the equivalent condition to the plastic creep condition in viscoplasticity can be written from the following equation,

$$\lambda^{vp} = \frac{\langle \Phi[\mathbb{F}(\boldsymbol{\sigma},\mathbf{q})]\rangle}{\xi} = \dot{\mu} \qquad (6.84)$$

by imposing that $\Phi > 0$ be always positive during the viscoplastic behavior and by letting the condition below be met at every time (its compliance ensures the viscoplastic equilibrium):

$$\Phi[\mathbb{F}(\boldsymbol{\sigma},\mathbf{q})] - \xi\,\dot{\mu} = 0 \qquad (6.85)$$

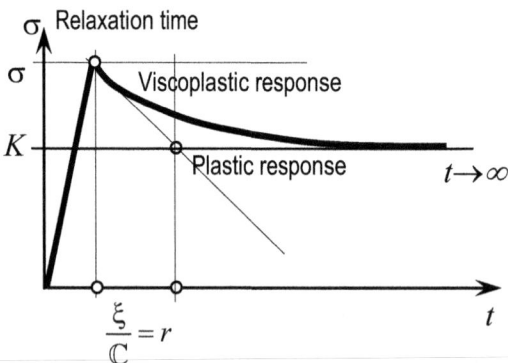

Figure 6.15 – Viscoplastic behavior at a point

In the remaining definitions viscoplasticity can be regarded as a generalization or regularization of the plasticity because the transition from elastic field to the inelastic field *relaxed in time* is allowed by the viscosity.

6.5.1 Limit states of viscoplasticity

Viscoplasticity formulation allows for the transition towards two theories that set up the viscoplastic limits. Depending on the viscosity magnitude the following limits can be defined as follows,

- **Infinite viscosity case** $(\xi \to \infty)$, when viscosity tends to an "elastic behavior" (see Figure 6.14). That is,

$$\lambda^{vp} = \lim_{\xi \to \infty}\frac{\langle \Phi[\mathbb{F}(\boldsymbol{\sigma},K)]\rangle}{\xi} = 0 \;\Rightarrow\; \dot{\boldsymbol{\varepsilon}}^{vp} = 0 \;\Rightarrow\; \dot{\boldsymbol{\varepsilon}} \equiv \dot{\boldsymbol{\varepsilon}}^{e} \qquad (6.86)$$

   It can be observed that the model coincides with the linear elastic one, like the Hooke model.

- **Null viscosity case** $(\xi \to 0)$, when the model tends to an "elastoplastic behavior" (Figure 6.14). That is,

$$\lambda^{vp} = \lim_{\xi \to 0}\frac{\langle \Phi[\mathbb{F}(\boldsymbol{\sigma},K)]\rangle}{\xi} = \lim_{\xi \to 0}\frac{\langle \Phi[\mathbb{F}(\boldsymbol{\sigma},K)=0]\rangle}{\xi} = \dot{\lambda}^{p} \;\Rightarrow\; \dot{\boldsymbol{\varepsilon}}^{vp} \equiv \dot{\boldsymbol{\varepsilon}}^{p} \qquad (6.87)$$

   It can be observed that it is necessary to impose the plastic creep condition $\mathbb{F}(\boldsymbol{\sigma},K)=0$ for the existence of the sought limit. This is the only way to ensure the existence of the viscoplastic parameter in the null viscosity situation, now having the form of the plastic consistency factor $\lambda^{\text{vp}}\, dt = d\lambda^{\text{p}}$. From the optimization point of view, viscoplasticity can be regarded as a regularized elastoplastic problem. It means that the elastoplastic problem is relaxed $\mathbb{F}(\boldsymbol{\sigma}, K) \geq 0$, a solution is left out of the convex elastic space but it is penalized with the viscosity parameter ξ.

These limit situations make the viscoplastic model a versatile formulation capable of dealing with elastic, plastic and viscoplastic problems.

6.5.2 Over stress function

The creep threshold function enables the transition between two different behaviors of the material,

$$\mathbb{F}(\boldsymbol{\sigma}, K) = \left[\frac{f(\boldsymbol{\sigma})}{K}\right] - 1 \qquad \begin{cases} \leq 0 & \text{Elastic state,} \\ > 0 & \text{Viscoplastic state,} \end{cases} \tag{6.88}$$

This over stress function can be defined in two ways,

$$\Phi\big[\mathbb{F}(\boldsymbol{\sigma}, K)\big] = \begin{cases} \mathbb{F}^n(\boldsymbol{\sigma}, K) & \text{Perzyna model,} \\ \mathbb{F}(\boldsymbol{\sigma}, K) & \text{Duvaut-Lyons model,} \end{cases} \tag{6.89}$$

where "*n*" is an exponent to be defined accordingly to the material type to be represented in the numerical simulation. The easy way is to choose "*n*=1", then implicitly assume the over stress function of Duvaut-Lyons[9]. Thus, the viscoplastic temporal strain is expressed as,

$$\dot{\boldsymbol{\varepsilon}}^{\text{vp}} = \lambda^{\text{vp}} \cdot \frac{\partial \mathbb{G}(\boldsymbol{\sigma}, \mathbf{q})}{\partial \boldsymbol{\sigma}} = \left[\frac{\langle \mathbb{F}(\boldsymbol{\sigma}, K)\rangle}{\xi}\right] \cdot \mathbf{g} \tag{6.90}$$

Therefore, the stress is defined as,

$$\dot{\boldsymbol{\sigma}} = \mathbb{C} : \dot{\boldsymbol{\varepsilon}}^e = \mathbb{C} : \big(\dot{\boldsymbol{\varepsilon}} - \dot{\boldsymbol{\varepsilon}}^{\text{vp}}\big) = \mathbb{C} : \left(\dot{\boldsymbol{\varepsilon}}^e - \left[\frac{\langle \mathbb{F}(\boldsymbol{\sigma}, K)\rangle}{\xi}\right] \cdot \mathbf{g}\right) \tag{6.91}$$

6.5.3 Integration algorithm for the viscoplastic constitutive equation

The numerical integration of the differential equation (6.38) can be carried out in different ways, but Euler's implicit time integration scheme is commonly used, whose characteristics are detailed in Table 6.4. In this algorithm, the current time is defined as "$t + \Delta t$" and the "i" variable is the iteration counter of the linearization process,

---

[9] Maugin, G. A. (1992). *The thermodynamics of plasticity and fracture*. Cambridge University Press.

1. **Stress state prediction**
   a. Initialization of the internal variable of viscoplastic strain and calculation of the predictor stress,

   $$^{i-1}\left(\boldsymbol{\varepsilon}^{\mathrm{vp}}\right)^{t+\Delta t} = \left(\boldsymbol{\varepsilon}^{\mathrm{vp}}\right)^{t}$$

   $$^{i-1}\left(\boldsymbol{\sigma}\right)^{t+\Delta t} = \mathbb{C}:\left(\boldsymbol{\varepsilon}^{t+\Delta t} - {}^{i-1}\left(\boldsymbol{\varepsilon}^{\mathrm{vp}}\right)^{t+\Delta t}\right)$$

   b. Viscoplastic creep condition

   $$^{i}\left[F(\boldsymbol{\sigma},K)\right]^{t+\Delta t} = \left[\frac{f\left[{}^{i-1}(\boldsymbol{\sigma})^{t+\Delta t}\right]}{K\left[{}^{i-1}(\overline{\varepsilon}^{\mathrm{vp}})^{t+\Delta t}\right]}\right] - 1 \quad ; \quad \text{where } \overline{\varepsilon}^{\mathrm{vp}} = \sqrt{\gamma(\boldsymbol{\varepsilon}^{\mathrm{vp}}:\boldsymbol{\varepsilon}^{\mathrm{vp}})}$$

   If: $^{i}\left[\mathbb{F}(\boldsymbol{\sigma},K)\right]^{t+\Delta t} < 0$, $\Rightarrow$ elastic behavior, $^{i}(\boldsymbol{\sigma})^{t+\Delta t} \equiv {}^{i}(\boldsymbol{\sigma}^{*})^{t+\Delta t}$, then go to f.

   If: $^{i}\left[\mathbb{F}(\boldsymbol{\sigma},K)\right]^{t+\Delta t} \geq 0$ then viscoplastic behavior.

   $$\text{Do } {}^{i}\left[r\right]^{t+\Delta t} = {}^{i}\left[\langle\mathbb{F}(\boldsymbol{\sigma},K)\rangle\right]^{t+\Delta t}$$

2. **Viscoplastic equation integration**
   c. Plastic flow calculation

   $$^{i}\left[\lambda^{\mathrm{vp}}\right]^{t+\Delta t} = {}^{i}\left[\frac{r}{\xi}\right]^{t+\Delta t} \quad \Rightarrow \quad {}^{i}\left[\Delta\mu^{\mathrm{vp}}\right]^{t+\Delta t} = {}^{i}\left[\lambda^{\mathrm{vp}}\right]^{t+\Delta t} \cdot \Delta t$$

   $$^{i}\left[\boldsymbol{\varepsilon}^{\mathrm{vp}}\right]^{t+\Delta t} = {}^{i-1}\left[\boldsymbol{\varepsilon}^{\mathrm{vp}}\right]^{t+\Delta t} + {}^{i}\left[\lambda^{\mathrm{vp}}\right]^{t+\Delta t} \cdot {}^{i}\left[\mathbf{g}\right]^{t+\Delta t} \Delta t = {}^{i}\left[\Delta\mu^{\mathrm{vp}}\right]^{t+\Delta t} \cdot {}^{i}\left[\mathbf{g}\right]^{t+\Delta t}$$

   d. Stress correction

   $$^{i}\left(\boldsymbol{\sigma}\right)^{t+\Delta t} = {}^{i-1}\left(\boldsymbol{\sigma}\right)^{t+\Delta t} - \mathbb{C}:\underbrace{\left({}^{i}\left[\lambda^{\mathrm{vp}}\right]^{t+\Delta t} \cdot {}^{i}\left[\mathbf{g}\right]^{t+\Delta t} \Delta t\right)}_{{}^{i}\left[\Delta\mu^{\mathrm{vp}}\right]^{t+\Delta t} \cdot {}^{i}\left[\mathbf{g}\right]^{t+\Delta t}}$$

   e. Convergence verification (eq. (6.85))

   If $^{i}\left[\mathrm{r}\right]^{t+\Delta t} = {}^{i}\left[-\Phi[F(\boldsymbol{\sigma},\mathbf{q})] + \xi\frac{\Delta\mu}{\Delta t}\right]^{t+\Delta t} \geq$ Tolerance, $\ i = i+1$ then go to 2

   f. END of integration and continue computation

Table 6.4 – Algorithm to obtain stress in viscoplastic problems.

6.5.4 The particular case of the Duvaut-Lyon model for a Von Mises viscoplastic material

Let a Duvaut-Lion viscoplastic model for small strain be assumed, whose viscoplastic creep function has the following Von Mises form,

$$\Phi[\mathbb{F}(\boldsymbol{\sigma},\mathbf{q})] \equiv \mathbb{F}(\boldsymbol{\sigma},\mathbf{q}) = \left[\frac{f(\boldsymbol{\sigma})}{K(\overline{\varepsilon}^{\mathrm{vp}})}\right] - 1 \equiv \left[\frac{\sqrt{J_2}}{K(\overline{\varepsilon}^{\mathrm{vp}})}\right] - 1 \tag{6.92}$$

where $\overline{\varepsilon}^{\mathrm{vp}} = \sqrt{\frac{2}{3}(\boldsymbol{\varepsilon}^{\mathrm{vp}}:\boldsymbol{\varepsilon}^{\mathrm{vp}})}$ is the effective plastic strain and $K(\overline{\varepsilon}^{\mathrm{vp}})$ is any hardening function that can generally be a hardening internal variable and thus be defined from its variation in time. Taking into consideration that the second invariant of the deviatoric can be written as[10],

$$J_2 = \frac{1}{2}\mathbf{s}:\mathbf{s} = \frac{3}{2}\tau_{oct}^2 \quad \text{or also} \quad \sqrt{J_2} = \frac{\|\mathbf{s}\|}{\sqrt{2}} = \frac{\left(s_1^2 + s_2^2 + s_3^2\right)^{1/2}}{\sqrt{2}} \tag{6.93}$$

where $\mathbf{s} = \boldsymbol{\sigma} - \frac{1}{3}\mathrm{tr}(\boldsymbol{\sigma})\mathbf{I}$, or $s_{ij} = \sigma_{ij} - \frac{1}{3}\sigma_{kk}\delta_{ij}$ is the deviatoric stress.

Considering equation (6.90) and a viscoplastic potential function associated to the creep function (equation (6.92)), the following tensor of viscoplastic flow is obtained,

$$\mathbb{G}(\boldsymbol{\sigma},\mathbf{q}) = J_2 - \hat{K}(\overline{\varepsilon}^{vp}) \quad \Rightarrow \quad \mathbf{g} = \frac{\partial \mathbb{G}}{\partial \boldsymbol{\sigma}} = \frac{\partial \mathbb{G}}{\partial \mathbf{s}}\frac{\partial \mathbf{s}}{\partial \boldsymbol{\sigma}} = \mathbf{s} \tag{6.94}$$

Substituting this last expression into equation (6.90), the following viscoplastic deformation results,

$$\dot{\boldsymbol{\varepsilon}}^{vp} = \lambda^{vp} \cdot \frac{\partial \mathbb{G}(\boldsymbol{\sigma},\mathbf{q})}{\partial \boldsymbol{\sigma}} = \left[\frac{\langle \mathbb{F}(\boldsymbol{\sigma},K)\rangle}{\xi}\right]\cdot\mathbf{g} = \frac{\left\langle\left[\frac{\sqrt{J_2}}{K(\overline{\varepsilon}^{vp})}\right] - 1\right\rangle}{\xi}\cdot\mathbf{s} \tag{6.95}$$

As we are dealing with a von Mises viscoplastic potential, an appropriate law to represent a behavior with dominant deviation is written. Then, an elastic volumetric behavior will be obtained and only the deviatoric part will undergo the viscoplasticity effects[11]. Thus, the deviatoric constitutive viscoplastic law will only be described and integrated, but it is important to remember that the elastic volumetric behavior σ_{ij}^{vol} must be added to obtain the total stress.

Therefore, the deviatoric stress is written as,

$$\mathbf{s} = 2G(\mathbf{e} - \mathbf{e}^{vp}) \quad \Rightarrow \quad \Delta\mathbf{s} = 2G(\Delta\mathbf{e} - \Delta\mathbf{e}^{vp}) \tag{6.96}$$

Also, the deviatoric stress at time $t + \Delta t$ will be,

$$\mathbf{s}^{t+\Delta t} = \mathbf{s}^t + \Delta\mathbf{s}^{t+\Delta t} = \underbrace{\mathbf{s}^t + 2G\,\Delta\mathbf{e}^{t+\Delta t}}_{(\mathbf{s}^*)^{t+\Delta t}} - 2G(\Delta\mathbf{e}^{vp})^{t+\Delta t} \tag{6.97}$$

And substituting the plastic flow rule we will have,

---

[10] Oller S. (2001). *Fractura mecánica – Un enfoque global.* CIMNE – Ediciones UPC. Barcelona.

[11] The constitutive law decomposed into its volumetric and deviatoric parts is written as,

$$\sigma_{ij} = \lambda^{Lame}\,\varepsilon_{kk}^e\,\delta_{ij} + 2\mu^{Lame}(\varepsilon_{ij}^e - \tfrac{1}{3}\varepsilon_{kk}^e\,\delta_{ij}) = \lambda^{Lame}\,\varepsilon_{kk}^e\,\delta_{ij} + 2\mu^{Lame}e_{ij}^e$$

where Lame's constants are $\lambda^{Lame} = E/3(1-2\nu)$ and $\mu^{Lame} = G = E/2(1+\nu)$, E and ν are the Young modulus and the Poisson coefficient respectively. Since the von Mises flow is dominantly deviatoric $(\dot{\varepsilon}_{kk}^{vp} = 0 \Rightarrow \dot{\varepsilon}_{ij}^{vp} \equiv \dot{e}_{ij}^{vp} = [\langle\Phi\rangle/\xi]\mathbf{s})$, the previous constitutive law is particularized in the following as

$$\sigma_{ij} = \lambda^{Lame}\,\varepsilon_{kk}\,\delta_{ij} + 2G(e_{ij} - e_{ij}^{vp}) = \sigma_{ij}^{vol} - s_{ij}$$

$$\mathbf{s}^{t+\Delta t} = (\mathbf{s}^*)^{t+\Delta t} - 2G\frac{\left\langle \Phi\left[\mathbb{F}^{t+\Delta t}\right]\right\rangle}{\xi}\mathbf{s}^{t+\Delta t}\Delta t, \quad with \quad \Phi\left[\mathbb{F}^{t+\Delta t}\right] \equiv \mathbb{F}^{t+\Delta t} = \mathbb{F}(\mathbf{s}^*)^{t+\Delta t} \qquad (6.98)$$

From the above equation, the tensor of deviatoric stresses at current time $\mathbf{s}^{t+\Delta t}$ can be obtained, without fulfilling the consistency condition, which is a characteristic of plasticity. That is,

$$\mathbf{s}^{t+\Delta t} = \frac{(\mathbf{s}^*)^{t+\Delta t}}{1 + \dfrac{\Delta t}{\xi}2G\left\langle\mathbb{F}(\mathbf{s}^*)^{t+\Delta t}\right\rangle} \qquad (6.99)$$

In this last equation, the following characteristics of the viscoplastic behavior can be observed,

1. If $\Delta t \to \infty$ or $\xi \to 0$, then in order to find a solution we must meet the condition $\Phi\left[\mathbb{F}^{t+\Delta t}\right] = 0$ and for that, the plastic consistency condition must be applied. Therefore, the problem becomes an elastoplastic problem (see equation (6.84)).

2. If $\Delta t \to 0$ or $\xi \to \infty$, then the sought stress coincides with the predictor stress $\mathbf{s}^{t+\Delta t} = (\mathbf{s}^*)^{t+\Delta t}$, which implies that the behavior has been linear elastic during the interval Δt.

Subject Index

Author Index